T0418838

Advanced Nanocatalysts for Biodiesel Production

Advanced Nanocatalysts for Biodiesel Production is a comprehensive and advanced book on practical and theoretical concepts of nanocatalysts dealing with future processing techniques towards environmental sustainability. The book critically discusses the latest emerging advanced nanocatalysts for biodiesel production aimed at reducing complexities and costs in the quest to meet future energy demands. Efforts have been made at clarifying the scope and limitations of biodiesel production in large-scale commercialization. The book discusses the size-dependent catalytic properties of nanomaterials and their working mechanisms in biodiesel production. Life cycle assessment of optimized viable feedstock from domestic as well as industrial waste is also addressed to improve the efficiency of biodiesel production.

This book will be a valuable reference for researchers and industrial professionals focusing on elementary depth analysis of nanocatalyst multifunctional technological applications in seeking key ideas for mimicking biodiesel production towards ecology and the economy.

Key Features

- Provides a comprehensive environmental assessment of advanced nanocatalysts for biodiesel production to meet the world's energy demands
- Discusses the green platform-based nanocatalysts like metal oxides/sulphides, 2D layered material synthesis and their relevance for biodiesel production
- Presents a pathway for cheaper, cleaner and more environmentally friendly processing techniques for biodiesel production

Advanced Nanocatalysts for Biodiesel Production

Edited by
Bhaskar Singh and Ramesh Oraon

CRC Press is an imprint of the
Taylor & Francis Group, an **informa** business

First edition published 2023
by CRC Press
6000 Broken Sound Parkway NW, Suite 300, Boca Raton, FL 33487-2742

and by CRC Press
4 Park Square, Milton Park, Abingdon, Oxon, OX14 4RN

CRC Press is an imprint of Taylor & Francis Group, LLC

ISBN: 9780367638245 (hbk)
ISBN: 9780367638252 (pbk)
ISBN: 9781003120858 (ebk)

DOI: 10.1201/9781003120858

Typeset in Times
by Newgen Publishing UK

Contents

Foreword

The use of renewable energy to mitigate environmental pollution and provide a sustainable energy source has caught worldwide attention. Since the United Nations Conference on Human Environment in 1972, renewable forms of energy have received national and international support. This support was renewed in the most recent major global event, that is, the 26th COP in 2021 in Glasgow (UK). Among the renewable fuels, liquid biofuels including biodiesel, are regarded as promising alternatives in the transportation sector to replace petroleum-derived counterparts.

Despite the unique attributes of biodiesel, such as biodegradability, non-toxicity, and a more environmentally friendly emissions profile, the high cost and scarcity of raw materials are major bottlenecks for the commercialization of biodiesel in many parts of the world. This shortcoming further highlights the importance of an effective conversion platform capable of taking in different oil feedstocks, including low-quality ones. Catalysts play an important role in obtaining high biodiesel yields with high quality complying with the established international standards. Nanomaterials have immense utility in various spheres of science and engineering, including producing efficient catalysts for biodiesel production. The book *Advanced Nanocatalysts for Biodiesel Production* strives to address the multiple aspects of nanocatalysts and their applications in developing biodiesel research and use.

Bhaskar Singh and Ramesh Oraon, the book editors, have done commendable work in bringing together researchers and scholars from different parts of the world to present a timely and comprehensive picture of the application of nanocatalysts in the field of biodiesel. Both editors possess good academic and research credentials and expertise in the diverse nanomaterials for energy and environmental applications. They have put substantial effort into compiling the chapters and preparing the book, ensuring that it will serve the community interested in nanocatalysts and their application in biodiesel synthesis.

I want to extend my best wishes to the publisher, the editors, and the contributors for the book's success. I hope the readers and the policy makers and other stakeholders will find it valuable for further research and commercialization aspects of biodiesel.

Meisam Tabatabaei, PhD
Professor
University Malaysia Terengganu (Malaysia)

Preface

This first edition of *Advanced Nanocatalysts for Biodiesel Production* features an up-to-date catalogue for nanocatalysts. The book highlights challenges with respect to feedstock selection criteria, catalysis and related aspects in biodiesel production. Based on these challenges, nanocatalysts are explored and presented within the chapters. Reaction schemes are illustrated in detail along with the structures of compounds. Nanocatalysts for biodiesel production have gained interest from researchers and academicians owing to their intriguing properties which have reduced environmental stress through decreased raw material demand. This can be attributed to the regio-selective behaviour of nanocatalysts which can be tailored through surface modification. The last decade has seen a rise in the use of nanocatalysts owing to sustainability demands. Nanocatalysts present a better regio-selective behaviour, stability and efficiency. In addition, the performance parameters in terms of energy, electronic, optical and photonic efficiencies are also reported in this book. Biodiesel as a fuel must meet the required specifications to be marketed as fuel. Hence, the role of the catalyst becomes important. Owing to the advances in nanotechnology, nanocatalysts could serve an important role in the development of biodiesel.

The primary intention of this book is to provide information to scientists and researchers who can further contribute to the development in this area. The first chapter deals with the overview of the present and future demands of energy and the relevance of biodiesel as a transportation fuel. Chapters 2 and 3 discuss the first- and second-generation feedstocks, respectively, along with their scope and limitations. The role of catalysts in the synthesis of biodiesel and their mechanisms are discussed in Chapter 4. Chapter 5 deals with metal oxide- and metal sulphide-based nanocatalysts in the synthesis of biodiesel. Magnetic-based nanomaterials and their utility towards a green chemistry approach that could enhance its lifetime and thereby its recovery by an external magnetic field are covered in Chapter 6. The usage of bio-based nanocatalysts (derived from chicken eggshells, snail shells, plant and tree leaves, seeds, stones, etc.) are presented in Chapter 7. Many of these currently studied common biomaterials get into the fuel due to leaching and need to be addressed. Chapter 8 stresses the importance and scope of 2D layered materials as catalysts in electrochemical reactions. The chapter also provides a brief account of various techniques being used for the synthesis of 2D layered materials. This chapter highlights the catalytic activity of graphene-supported nanoparticles in electrochemical reactions. In addition, the catalytic activities of graphitic carbon nitride (g-C_3N_4) and hexagonal boron nitride (h-BN) are summarized. Furthermore, the prospects and research directions are described. The efficiency, cost-effectiveness and sustainability aspects of nanomaterials are elaborated upon in Chapter 9. Chapter 10 examines the industrial applications of the by-products obtained from the production of biodiesel. The life cycle assessment of biodiesel production is described in Chapter 11. A comparison of the nano-based catalysts with traditional catalysts is drawn in Chapter 12.

It is a pleasure to express our gratitude to the contributing authors. The authors of the individual chapters have expertise in their areas. They encompass a group of academicians and scientists belonging to reputed universities and organizations worldwide. The editors express their appreciation to all of the contributors for sharing their experience. The editors are also thankful to the administrative assistance provided in the preparation of the contextual indexing. We also acknowledge the time and effort given to the planning, administration and preparation of this book by the publisher.

Dr. Bhaskar Singh
Dr. Ramesh Oraon

Acknowledgements

I would like to take this opportunity to thank those who had a significant contribution in shaping this book. I thank the entire team at CRC Press for collaborating on this book and providing all the necessary assistance. My special thanks go to Ms. Jyotsna Jangra, the Editorial Assistant at CRC Press/Taylor & Francis Books India Pvt. Ltd., who was there to promptly respond and address all my queries and provide the necessary assistance from the commencement of the book until its completion. I acknowledge and thank all the authors who contributed their works in the book. The authors submitted the work very timely, even during the period of difficulty due to the prevailing pandemic. The authors were very kind to address the comments that enhanced the quality of the chapters.

I am very grateful to my fellow co-editor of the book, Dr. Ramesh Oraon. It has been an enriching and wonderful experience working with Dr. Ramesh on this book. I am grateful to my PhD supervisor Prof. Yogesh Chandra Sharma for imbibing in me the zeal to do work on the contemporary issues related to the environment and energy. I also express my thanks to Prof. Manoj Kumar and my colleagues at the Department of Environmental Sciences, Central University of Jharkhand, Ranchi, for supporting me and always being there to encourage my achievements and accomplishments. I would like to acknowledge the support and love of my parents (Shri Sachchida Nand Singh and Smt. Usha Kiran), my loving and beautiful wife Ragini, and other members of my family. I acknowledge the love of my daughter, Riyansika, who brings joy and happiness in my life that rejuvenates me in work.

(**Bhaskar Singh**)

Editing a book on *Advanced Nanocatalysts for Biodiesel Production* has been an impressive journey with my co-editor Dr. Bhaskar Singh. A very special thanks are due to him as none of this would have been possible without his relentless effort and support. I am eternally grateful to my late father, Gulab Oraon, and family for everything they have done to help me succeed in life. I express my heartfelt thanks to Dr. Ganesh Chandra Nayak for his guidance, motivation and affection during all of my struggles and successes. I thank all the contributing authors who have worked hard to make this book a success. There have been tremendous efforts and dedication to work collectively on this book. I also especially thank the publisher CRC Press/Taylor & Francis for giving us the opportunity to contribute through this book information for the entire scientific community. Finally, thanks are due to a number of friends and colleagues who have been a part of my getting there, for showing up every day and helping the authors turn their ideas into this successful compilation.

(**Ramesh Oraon**)

Editors

Dr. Bhaskar Singh received his PhD from the Indian Institute of Technology (BHU) India, in 2010. Dr. Singh also holds an MPhil from Pondicherry University, India, with a Gold Medal (2006). He is currently serving as an Assistant Professor at the Department of Environmental Sciences, Central University of Jharkhand, Ranchi, India. Dr. Singh is currently engaged in teaching in the thrust areas of environmental sciences. He works on various aspects of biodiesel development for its commercial applications. He was recently featured in the list of the world's top 2% of scientists per the Stanford-Elsevier Report. He has published more than 50 research and review papers in peer-reviewed and high-impact international journals.

Dr. Ramesh Oraon is Assistant Professor in the Department of Nanoscience and Technology, Central University of Jharkhand, Ranchi, India. He obtained his PhD from IIT (ISM) Dhanbad, India, in 2018. He has been awarded with several fellowships (Brain Korea Postdoctoral BK21, DST National Postdoctoral Fellowships NPDF, etc.) and projects (DST EMEQ, BSR start-up grant). His research area includes the development of hybrid electrode materials for energy and environment applications. He has over 25 publications which include edited books, book chapters, research papers and review articles of international repute. He teaches various materials science subjects such as self-assembly and molecular engineering, polymer engineering, nanocomposites, and so on.

Contributors

Inbaoli A.
National Institute of Technology
Calicut, India

Km Abida
School of Chemistry & Biochemistry
TIET-VT Center of Excellence in Emerging Materials
Thapar Institute of Engineering & Technology
Patiala, India

Madhu Agarwal
Department of Chemical Engineering
Malaviya National Institute of Technology Jaipur
Jaipur, India

Amjad Ali
School of Chemistry & Biochemistry
TIET-VT Center of Excellence in Emerging Materials
Thapar Institute of Engineering & Technology
Patiala, India

Indra Bahadur
Department of Chemistry
Faculty of Natural and Agricultural Sciences
North-West University
South Africa

Silma de S. Barros
Departamento de Engenharia de Materiais–DEMAR
Escola de Engenharia de Lorena–EEL/USP
Brazil

Jitamanyu Chakrabarty
National Institute of Technology
Durgapur, India

Ingrity Suelen Costa Sá
Programa de Pós-graduação em Química
Universidade Federal do Amazonas
UFAM, Brazil

Patrícia Arrojo da Silva
Bioprocess Intensification Group
Federal University of Santa Maria, UFSM
Santa Maria, RS, Brazil

Michael Daramola
Department of Chemical Engineering
University of Pretoria
Pretoria, South Africa

Ana E.M. de Freitas
Programa de Pós-Graduação em Design
Universidade Federal do Amazonas
UFAM, Brazil

Flávio A. de Freitas
Centro de Biotecnologia da Amazônia–CBA/SUFRAMA
Programa de Pós-Graduação em Química
Universidade Federal do Amazonas
UFAM, Brazil

Mariany Costa Deprá
Bioprocess Intensification Group
Federal University of Santa Maria, UFSM
Santa Maria, RS, Brazil

Rituraj Dubey
Department of Chemistry
B.N.M. College, Barhiya
Munger University, Munger
Bihar, India

Deepak Gusain
Department of Environmental Studies
Ramanujan college
University of Delhi
Nehru Nagar, Delhi, India

Shuit Siew Hoong
Department of Chemical Engineering
Lee Kong Chian Faculty of Engineering and Science
Universiti Tunku Abdul Rahman
Malaysia

Sim Jia Huey
Department of Chemical Engineering
Lee Kong Chian Faculty of Engineering and Science
Universiti Tunku Abdul Rahman
Malaysia

Eduardo Jacob-Lopes
Bioprocess Intensification Group
Federal University of Santa Maria, UFSM
Santa Maria, RS, Brazil

Durgesh Kumar
Department of Chemistry
Atma Ram Sanatan Dharma College
University of Delhi
New Delhi, India

Vinod Kumar
SCNS
Jawaharlal Nehru University
New Delhi, India

Sujith Kumar C.S.
National Institute of Technology
Calicut, India

Kamlesh Kumari
Department of Zoology
Deen Dayal Upadhyaya College
University of Delhi
New Delhi, India

Pushpendra Kushwaha
Department of Chemical Engineering
Malaviya National Institute of Technology Jaipur
Jaipur, India

Paola Lasta
Bioprocess Intensification Group
Federal University of Santa Maria, UFSM
Santa Maria, RS, Brazil

Steven Lim
Department of Chemical Engineering
Lee Kong Chian Faculty of Engineering and Science
Universiti Tunku Abdul Rahman
Malaysia

Pang Yean Ling
Department of Chemical Engineering
Lee Kong Chian Faculty of Engineering and Science
Universiti Tunku Abdul Rahman
Malaysia

Marcia S. F. Lira
Centro de Biotecnologia da Amazônia–CBA/SUFRAMA
Brazil

Karishma Maheshwari
Department of Chemical Engineering
Malaviya National Institute of Technology Jaipur
Jaipur, India

António B. Mapossa
Institute of Applied Materials
Department of Chemical Engineering
University of Pretoria
Pretoria, South Africa

José R. Mora
Universidad San Francisco de Quito
Ecuador

Mayank Pandey
Department of Environmental Studies
P.G.D.A.V. College (Evening)
University of Delhi
Nehru Nagar, Delhi, India

Wanison A. G. Pessoa Jr.
Instituto Federal de Educação
Ciência e Tecnologia do Amazonas–IFAM/CMDI
Brazil

Jayaraj S.
National Institute of Technology
Calicut, India

Edson Pablo Silva
Centro de Biotecnologia da Amazônia–CBA/SUFRAMA
Brazil

Himmat Singh
School of Chemistry & Biochemistry, Thapar Institute of Engineering & Technology
Patiala, India

Laxman Singh
Department of Chemistry
R.R.S. College, Mokama
Patliputra University, Patna
Bihar, India

Prashant Singh
Department of Chemistry
Atma Ram Sanatan Dharma College
University of Delhi
New Delhi, India

Sunita Singh
Department of Chemistry
National Institute of Technology
Durgapur, India

Mitsuo L. Takeno
Instituto Federal de Educação
Ciência e Tecnologia do Amazonas–IFAM/CMDI
Brazil

Juan S. Villarreal
Universidad San Francisco de Quito
Ecuador

Vijay Kumar Vishvakarma
Department of Chemistry
Atma Ram Sanatan Dharma College
University of Delhi
New Delhi, India

Ritu Yadav
Department of Chemistry
Atma Ram Sanatan Dharma College
University of Delhi
New Delhi, India

Leila Queiroz Zepka
Bioprocess Intensification Group
Federal University of Santa Maria, UFSM
Santa Maria, RS, Brazil

1

Energy: Present and Future Demands

Mayank Pandey*[*] and Deepak Gusain

CONTENTS

1.1 Introduction

Since the discovery of fire, the human race has realized the importance of energy and tried to extract or generate energy from natural resources. Globally, many civilizations and faiths have worshipped various natural sources of energy (like wind, water, solar radiation, fire, etc.), since time immemorial. The availability and abundance of sources of energy have been the backbone for the development of state, regional and global economies. Energy in the form of electricity has been and will remain pivotal for sustaining and augmenting the standard of living of the global population, and economic growth is entirely based on the uninterrupted supply of the desired amount energy. With the exponential rise in the global population, rapid industrialization

[*] *Corresponding Author* mayank@pgdave.du.ac.in

DOI: 10.1201/9781003120858-1

and urbanization, advancements in science and technology, the energy demand from each and every sector (such as education, agriculture, industry, health and well-being, urbanization, communication, rapid transportation, storage, etc.) has increased tremendously.

The perception and image of energy generation and consumption have changed from the inception of the first industrial revolution that took place in Europe in the mid-18th century, where machines running on fossil fuels replaced and reduced the required manpower significantly. The exploration and exploitation of fuel reserves within the planet obtained a global thrust, and the excavation of fossil fuels, along with mineral reserves, has been growing at an unprecedented rate until today. However, the gap between energy demand and supply is widening consistently and putting huge pressure on the conventional resources. In 2018, the global energy consumption reached 13864.9 Mtoe (million tonnes oil equivalent), which was 2.9 percent higher as compared to the year 2017 (BP Statistical Review of World Energy, 2019). However, until 2016, almost 13% of the global population had no access to energy. A significant fraction of the global population uses conventional fuel woods for cooking and other domestic purposes. The global population will cross to the nine billion mark by 2050, which means that ensuring food security, water and hygiene, and energy accessibility will be huge challenges for the global economies and states. The energy sector and environment are interlinked by a two-way path where energy generation causes an adverse impact on the environment mainly by greenhouse gas (GHG) emission. On the other hand, extreme climatic conditions (global warming causing climate change and related events such as extreme weather, natural disasters, etc.) can have a serious impact on the energy infrastructure. Also, the energy sector is facing new threats in the forms of of cybersecurity and pandemic outbreaks like that of COVID-19. Therefore, it is crucial for the world to explore new methods of energy generation, transmission, storage and consumption, keeping in mind environmental conservation and sustainable growth while restricting the rise in average global temperature to below 2 °C compared to the pre-industrial era.

1.2 Types of Energy Sources

All the available energy resources can be broadly categorized into:

1.2.1 Conventional or Non-Renewable Sources of Energy

Energy sources which, after their consumption, cannot be renewed or replenished within a short span of time are called non-renewable sources of energy. As these sources have been used conventionally for an extremely long time, these are also called conventional sources of energy. These sources have been formed within the Earth by the gradual decay of biotic components over a geological time scale and hence are known as fossil fuels. The most common examples of fossil fuels are coal, coke, petroleum oil, natural gases, etc. The extraction, processing and consumption of fossil fuels give rise to various environmental (pollution, biodiversity loss, global warming/climate change, etc.) and socio-economic and cultural issues (land acquisition, relocation and rehabilitation etc.).

1.2.2 Non-Conventional or Renewable and Clean Sources of Energy

These energy sources are continuously available in nature or are replenished before the previous energy packet is exhausted, and hence they can be used time and again. Therefore, these sources are believed to be infinite and inexhaustible. Examples of renewable energy sources are solar energy, wind energy, hydropower, geothermal energy, tidal and wave energy, biofuels, hydrogen fuel cells, etc. As these sources of energy cause minimal adverse impact on the environment, they are also called 'clean sources of energy'. However, since ancient times, these sources have been used in different civilizations in a sporadic manner, with the systematic and scientific harvesting of these energy sources at a large scale being only a few decades old. Hence, these sources are also called non-conventional sources of energy. The global energy sector is witnessing a transitional phase where the world is focussing increasingly on renewable and clean fuels to generate energy.

1.3 Energy Demand: Present and Future Scenario

The demands for energy and energy sources have continued to increase with the inception of the first industrial revolution. Although biomaterials were also being used to generate fuels, the discovery of coal fields and crude oil/natural gas reserves changed the consumption pattern of the energy sources. Believing that coal, crude oil and natural gases are inexhaustible, the over-exploitation of fossil fuels started and, with advancements in excavation technologies, the extraction of fossil fuels saw a new spike between 1900 and 1950, and specially in the latter half of the twentieth century (Figure 1.1 and Figure 1.2). This continued until 1973 when, for the first time, the

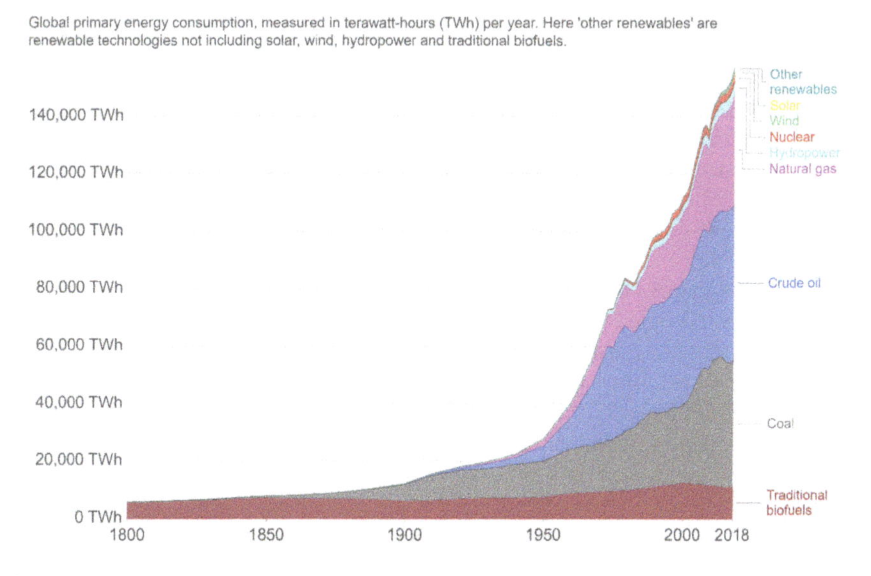

FIGURE 1.1 Global primary energy consumption (1800–2018). (https://ourworldindata.org/energy.)

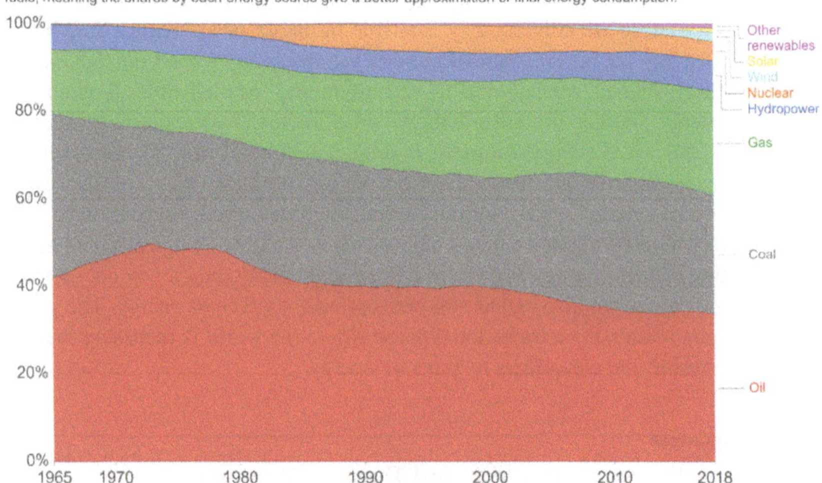

FIGURE 1.2 Global energy consumption by source (1965–2018). (https://ourworldindata.org/energy.)

global socio-economic and political structures witnessed an *oil crisis* or *oil shock*, when the Organization of the Petroleum Exporting Countries (OPEC) increased oil prices by up to 400 percent. Later, the world saw a large number of oil shocks for various geo-political and economic reasons. Also, epidemic outbreaks can have an uneven and adverse impact on the energy sector. For instance, during the current COVID-19 pandemic, the international community has observed that, on one hand, the global energy demand declined by 3.8% in the first quarter of 2020, while on the other hand, the disease containment measures demanded up to 50 percent more energy in some parts of the world. Similarly, India, the fastest growing global economy, witnessed a significant reduction in energy demand by approximately 30 percent in the first quarter of 2020 (IEA Global Energy Review, 2020). In spite of the tremendous efforts made worldwide to make electricity accessible to all, sub-Saharan and South Asian countries still have the largest portion of their populations without electricity and so using biomass for heating and cooking purposes, which in turn deteriorates the indoor air quality and causes adverse health impacts. The State of Electricity Access report (2017) claims that '1.06 billion people still do not have access to electricity, and 3.04 billion people still rely on solid fuels and kerosene for cooking and heating' and predicts that 'universal electricity access will not be met by 2030, unless urgent measures are taken'. Energy-related emissions reached a recorded high in 2018 (IEA World Energy Outlook, 2019). There is also consistent, continued and vast diversity in per capita energy consumption. Per capita primary energy consumption was recorded to be highest in the United States (294.8 Gigajoule), Trinidad & Tobago (465.5 Gigajoule) and China (175.1 Gigajoule), while it was significantly lower in India (25 Gigajoule), the fastest growing global economy (BP Statistical Review of World Energy, 2019). Among the non-OECD region, Asia is the highest consumer of energy

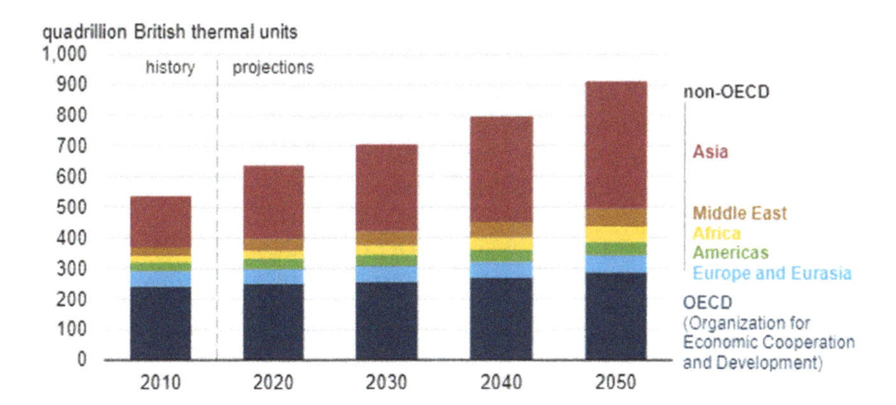

FIGURE 1.3 Global primary energy consumption by region (2010–2050). (U.S. Energy Information Administration, International Energy Outlook 2019.)

as this region has some of the fastest growing economies (Figure 1.3). Therefore, it has become crucial to increase the share of renewable and clean energy sources in the global energy market, coupled with energy efficiency and its judicial use to attain universal access to sustainable energy.

For the next thirty years (2020–2050), fossil fuels will remain the backbone of the global energy sector and its consumption rate will be more than 4 percent. Globally, oil and gas remain the largest sources of energy. A report also has suggested that, in the coming decades, power plants in Africa and Asia will run on natural gas (Enerdata Global Energy Statistical Yearbook 2020). As compared to 2016, the demand for gas and oil will increase significantly, by 53 percent and 2 percent, respectively, whereas the demand for coal will see a sharp decline of 28 percent by 2050 (IRENA, 2020a). The year 2018 saw an unprecedented rise in natural gas consumption (195 billion cubic metres) at the rate of 5.3 percent, which is one of the fastest rates of growth since 1984. At present, in terms of total energy consumption, China consistently tops the list, with 3284 Mtoe, followed by the United States (2213 Mtoe), India (913 Mtoe), Russia (779 Mtoe) and Japan (421 Moe) (Enerdata Global Energy Statistical Yearbook 2020). By 2050, electricity will become the central energy carrier and its share of final consumption will reach around 50 percent, as compared to the present 20 percent (IRENA, 2019b). In 2017, Brazil, Russia, India and China were the largest non-OECD consumers of electricity as they together consumed 67.2 percent of total non-OECD electricity final consumption (or 37.2% of global consumption) (IEA Electricity Information Overview, 2019). Although energy consumption growth has slowed recently (2018–19) due to the economic downturn, the world witnessed a constant rise in energy consumption pattern by two percent per year from 2000 to 2018 (Enerdata Global Energy Statistical Yearbook, 2020).

World Energy Outlook report (2019) by IEA categorizes the future energy demand under three scenarios: current policies scenario, stated policies scenario and sustainable policies scenario. The current policies scenario depicts the conditions where the globe follows its present path without additional changes to policies, under which the energy demand may increase by 1.3 percent per year up to 2040. The stated policies scenario represents the specific policy initiatives which have been announced so far, and it is predicted under this scenario that energy demand will see a growth of 1 percent per year until 2040. The sustainable development scenario, however, represents an ideal situation which targets meeting the Sustainable Development Goals (SDGs) and universal electricity access to comply with the Paris Agreement objectives by incorporating vibrant changes within the energy sector. Under the stated policy scenarios, by 2040, the United States will be the principal global player in oil and gas trading and marketing, and will overtake the OPEC countries and Russia. However, the Middle-East region will remain the main supplier of oil to the world, and Asia, with India in particular, will emerge as the main consumer of oil by 2040. Although Africa may very cheaply harvest the solar energy, it will overtake China in oil consumption by 2040 under the stated policies scenario. The report also strongly suggests that the globe should opt for and adapt to electrified vehicles instead of large personal vehicles (sports utility vehicles – SUVs) which consume extra conventional fuels. By 2040, the global gross energy demand under the current, stated and sustainable policies scenarios will be 19,177 Mtoe, 17,723 Mtoe and 13,279 Mtoe, respectively (IEA World Energy Outlook, 2019).

The gross domestic product (GDP) per person clearly indicates the living standard of the population of a state and their consumption of energy. International Energy Outlook (2019) has projected that the GDP per person will rise in non-OECD countries (by 150 percent) at a higher proportion as compared with the OECD countries (less than 50 percent) between 2018–2050. Asia and Africa will remain the home for largest proportion of natural resources and the global population. ASEAN region's energy requirement may reach up to 1070 Mtoe by 2040 (Erdiwansyaha et al., 2019). India and China are, and will remain, the fastest growing economies worldwide (particularly in non-OECD countries) and hence, the greatest demands for energy will arise from these two nations. Similarly, it has been projected that the energy consumption in non-OECD countries will rise by as much as 70 percent as compared with OECD countries (15 percent) from 2018 to 2050. The industrial sector has emerged as one of the largest consumers of energy and will reach the figure of 315 quadrillion British thermal units (Btu) by 2050. Similarly, electricity consumption by the residential and commercial sectors will increase tremendously in non-OECD countries in the coming years. China is the largest consumer of energy while India has the highest demand of energy with the growth of economy. India and China will jointly consume more than two-thirds of global industrial coal by 2050 (IEO, 2019).

Globally, India has emerged as one of the largest consumers of energy and hence, it is relevant to throw some light on India's energy sector. India is the seventh largest country in land size, has the second largest population (approximately 1.4 billion), with rapid urbanization (2.4 percent per year) and is the third largest and one of the fastest growing major economies of the world. India has framed a set of goals, including 175 GW of installed capacity of renewable energy by 2022, uninterrupted and reliable power access to all by 2022, housing for all by 2022, 100 Smart Cities Mission, cleaning of rivers, providing clean cooking fuels, highway construction and expansion, AMRUT (Atal Mission for Rejuvenation & Urban Transformation), National Smart Grid Project, Green Energy Corridor and meeting the Intended Nationally Determined Contribution (in the Paris Agreement) and is committed to achieving the same in the near future (NITI and IEEJ, 2017). At the time of independence (1947), India's total installed capacity was 1362 megawatt (MW), which has been increased to 371,054.12 MW at present (Ministry of Power, Govt. of India). The Ministry of Power, Govt. of India has revealed that the overall growth rate in the energy sector was recorded at 0.95 percent with the generation of 1390.467 billion units (BU) during 2019–20 and it has observed a rise in the share of renewable energy (9.12 percent), hydroelectricity (15.48 percent) and nuclear energy (22.90 percent), while thermal energy has been reduced by 2.75 percent over the same period. From being an importer of electricity for many decades, India began exporting electricity in 2016 (IEA Electricity Information Overview, 2019). In almost two decades (2000–2019), approximately 750 million of the Indian population have gained access to electricity (IEA India 2020, 2020). Hence, India is not far from achieving universal household electrification. Being the fastest growing economy, India's share of global activity is forecast to reach up to 25 percent by 2050. The World Energy Outlook report (2019) by the IEA predicts that, by 2040, the primary energy demand of the country will reach 2055 Mtoe, 1841 Mtoe and 1294 Mtoe under the current, stated and sustainable policies scenarios, respectively (IEA World Energy Outlook, 2019).

Initiatives and incentives to the schemes and policies such as 'Make in India', 'Zero Defect and Zero Effect' and 'Vocal for Local', that promote the Indian manufacturing and handicraft sectors, will demand more energy in the near and more distant future. Since 2008, the industrial sector in India has emerged as the largest consumer of energy, followed by the residential, commercial and other sectors. Industries like iron and steel, chemical and petrochemical, textiles and leather are the largest electricity consumers. International Energy Outlook (2019) predicted that in the residential and commercial sectors, India will observe the fastest relative growth between 2018 and 2050. It clearly indicates that India's energy consumption will increase threefold, with an annual average rate of 3.4 percent and will reach 47 quadrillion Btu by 2050. India, of all the non-OECD Asian countries, will experience the fastest relative growth in commercial energy consumption between 2018 and 2050. India's natural gas production will observe a 1.5 percent growth rate per year between 2018 and 2050 as India is determined to increase the share of natural gas in the country's energy mix from the current 6 percent to 15 percent by 2030 (International Energy Outlook, 2019; IEA: India 2020, 2020). In 2018, India was the second-largest coal consumer globally. International Energy Outlook (2019) prediction forecasts that India's coal consumption will rise at a rate of 3.1 percent per year and will reach 2.9 billion short tons (46 quadrillion Btu) by 2050, as compared to the present 1.1 billion short tons (17 quadrillion Btu). India's coal imports will grow by 4.1 percent while Australia and Indonesia will remain the largest exporters. Coal consumption by India's steel and cement sectors are making India one of the largest consumers of coal worldwide. An International Index of Energy Security Risk report (2020) revealed that India's rank has consistently improved since 1995 while the United States, New Zealand and Canada are the top performers.

1.4 Transformation of the Energy Sector: Conventional to Clean and Alternate Sources of Energy

The exponential rise in energy demand at the global level has adversely impacted on biotic and abiotic components of nature and environment, in totality. Although the share of clean and renewable energy sources (solar, wind, geothermal, tidal, biofuels, hydrogen fuel cells, etc.) in energy generation has increased significantly, conventional fossil fuels (coal, petroleum oil, natural gas, etc.) remain the largest source of energy generation worldwide. The use of fossil fuels as energy sources has resulted in global threats such as pollution, global warming/climate change, biodiversity loss, etc., which in turn give rise to a variety of social issues and adversely impact the economy at the macro and micro levels. The world has witnessed an average rise of 1.3 percent annually in energy-related carbon dioxide emissions over the last five years (IRENA, 2019b). Evidently, the globe is becoming warmer as the average global temperature was recorded at 0.86 °C and 0.93 °C above the pre-industrial baseline for the decades 2006–2015 and 2009–2018, respectively. In five years (2014–2018), it increased to 1.04 °C above the baseline (IPCC, 2018). Therefore, it has become critical to provide alternate and clean sources of energy to meet the Sustainable Development Goals (SDGs; 17 goals and 196 associated targets) which declare the importance of ensuring access to affordable, reliable, sustainable and modern energy for all (SDG 7). As the rest of the SDGs are directly/indirectly interlinked with SDG 7, it has becomes crucial for urgent arrangements to be made so that the global population may have access to safe and clean energy by 2030.

SUSTAINABLE DEVELOPMENT GOAL 7

'Ensure access to affordable, reliable, sustainable and modern energy for all'

- *7.1 By 2030, ensure universal access to affordable, reliable and modern energy services*
- *7.2 By 2030, increase substantially the share of renewable energy in the global energy mix*
- *7.3 By 2030, double the global rate of improvement in energy efficiency*
  - *7.a By 2030, enhance international cooperation to facilitate access to clean energy research and technology, including renewable energy, energy efficiency and advanced and cleaner fossil-fuel technology, and promote investment in energy infrastructure and clean energy technology*
  - *7.b By 2030, expand infrastructure and upgrade technology for supplying modern and sustainable energy services for all in developing countries, in particular least developed countries, small island developing States, and land-locked developing countries, in accordance with their respective programmes of support*

Emissions from the energy sector and related processes need to be curtailed urgently. The Global Energy Statistical Yearbook 2020 by Enerdata revealed that, in 2019, the top five CO_2 emitters were China (9729 $MtCO_2$), the United States (4920 $MtCO_2$), India (2222 $MtCO_2$), Russia (1754 $MtCO_2$) and Japan (1045 $MtCO_2$), while New

Zealand (33 $MtCO_2$), Sweden (39 $MtCO_2$) and Norway (40 $MtCO_2$) emerged as the lowest emitters. Carbon emissions by China increased steadily by 2.8 percent in 2019. To achieve the 1.5 °C target set by the Paris Agreement, the G20 Nations must reduce the global total CO_2 emissions, to below 2010 levels, by 45 percent by 2030 and reach net zero by 2050. Similarly, CO_2 emissions from the energy sector and related processes from G20 Nations have to be reduced by 40 percent to below 2010 levels by 2030, and reach net zero by 2060 (Brown to Green, 2019). The reduction of emissions from the transportation sector is equally indispensable. The Brown to Green report (2019) states that, by 2050, the share of low-carbon fuel in the transport sector should be increased to at least 60 percent and the international community should aim to ensure that no internal-combustion-engine vehicles are sold after 2035. Also, there is an urgent need to fully decarbonize freight trucks by 2050. In 2018, emissions from the transportation sector saw a growth of 1.2 percent in G20 Nations (Brown to Green, 2019). These goals can only be achieved by increasing the share of renewable and clean sources for industrial, domestic and transportation sectors.

The IRENA (2019b) report claims that if policies and programs are made and implemented in accordance with the Paris Agreement targets, the share of renewable energy in the primary energy supply would be around two-thirds in 2050 as compared to the present contribution of one-sixth. The contributions of renewable and clean sources is increasing significantly in the global energy market as the world has seen a tremendous increase in solar (30 Mtoe) and wind (32 Mtoe) energy until the year 2018 (BP Statistical Review of World Energy, 2019). By 2050, wind (6.7 trillion-kilowatt hour) and solar (8.3 trillion-kilowatt hour) energy will contribute more than 70 percent of the energy generation at the global level (IEO 2019). By 2050, regions like the EU, Latin America and Caribbean, Oceania, South-East Asia and sub-Saharan Africa will increase their share of renewable energy in the total primary energy supply (TPES) by more than 70 percent, while the share of renewable energy will be more than 80 percent of power generation in East Asia, the EU, Latin America and the Caribbean, North America, Oceania, the rest of Asia and Europe, South-East Asia and sub-Saharan Africa (IRENA, 2020a). Considering the exponential electricity demand at the global level, the consumption of renewable energy will increase by 3 percent annually between 2018–2050, while nuclear energy consumption will grow by only 1 percent annually during the same period (International Energy Outlook 2019). Renewable energy and energy efficiency jointly provide more than 90 percent of the measures needed to mitigate carbon emissions (IRENA, 2020a).

India gave its Nationally Determined Contributions (NDC) commitments at the Paris Agreement (2015) which, in addition to other targets, focus on a reduction of the emissions intensity of GDP by 33–35% from 2005 levels by the year 2030; generation of 40 percent electrical energy from non-fossil fuel-based energy resources by 2030; an increase in the forest cover to create an additional carbon sink of 2.5–3 billion tonnes of CO_2 equivalent by 2030; and to increase the investment in climate change vulnerable sectors to make them more adaptable and manageable, etc. Renewable energy installed capacity in India has been expanded by approximately 226 percent in the last five years. Globally, India has now reached fifth position in overall installed renewable energy. Until 2017, India's renewable energy share in the total final energy consumption had reached 32.2 percent (https://trackingsdg7.esmap.org//). Also, India's initiatives on energy conservation and efficiency have fruitfully resulted in significant decreases in annual energy demand (approximately 15 percent) and carbon emissions (approximately 300 million tonnes) (IEA India 2020, 2020). Based on the indicators of the Regulatory Indicators for Sustainable Energy (RISE), India has recently performed fairly well in electricity access (71), renewable energy (87) and clean cooking (72) with an overall score of 75. However, India is working hard to improve its score in energy efficiency (67) (https://rise.esmap.org/country/india). Compared to the present contribution (10 percent), solar and wind energy jointly will contribute more than 50 percent of energy generation by 2050 due to which the share of coal-fired generated electricity will fall from 73% in 2018 to 38% in 2050 (IEO, 2019). The renewable energy sector will create a large number of job opportunities in the future as the IRENA (2018) report reveals that 10.3 million jobs were created by this sector in 2017 with a growth of 5.3% compared to the previous year, with the leading countries creating employment in this sector being China, Brazil, the United States, India, Germany and Japan (IRENA, 2018).

1.5 Biofuel

Fuels which are chemically derived from biotic materials (plant, seeds, algae, etc.) or waste from biotic materials (agricultural waste such as sugarcane bagasse) are called biofuels. These fuels are being used as a partial substitute for conventional fuels like petrol, diesel, gasoline, etc. On August 10, 1893, Rudolf Diesel ran an automobile engine using biofuel derived from peanut oil. Later, the conventional fuel diesel was named after Rudolf Diesel, with August 10 being observed as International Biofuel Day every year. The United States observes National Biodiesel Day every March 18, the birthday of Rudolf Diesel. Later, Henry Ford designed the Model T car, which ran on hemp-derived biofuel. With the discovery of crude oil reserves, biofuel synthesis did not continue to grow. However, realizing the exhaustible nature of crude oil, biofuel projects have been started by different countries, especially the United States, Brazil and the European Union (EU). In the 1950s and 1960s, the food versus fuel debate was ignited globally as the world was facing a serious food grain crisis but many countries were emphasizing the production of food crops (maize, sugarcane, etc.) to produce biofuels, with the practice facing serious criticism worldwide. However, at present, many workers suggest that the use of biofuels will in turn significantly augment the quality of the environment and food production (Subramaniam et al., 2020).

Biofuels are commonly used in a blended form with conventional transportation fuels. At present, even the aviation sector is using a bio jet fuel blended aviation fuel. Biofuels are renewable (as the feedstocks are renewable), eco-friendly and cost-effective. Therefore, globally, biofuels have emerged as one of the best alternative sources of energy, at least for the transportation and industry sectors. Also, the biofuel sector, among other renewable sources, has great potential to increase employment opportunities. The IRENA (2018) report shows that 1.93 million jobs have been created by the biofuel sector. Latin America alone created almost half these jobs, while Asia (known for its cheap labour and feedstock) created 21 percent of the jobs, followed by North America (16 percent) and Europe (10 percent). Brazil, the

United States, the EU and Southeast Asian countries have thus emerged as the largest employers in this sector (IRENA, 2018).

The most common types of biofuels are biodiesel, ethanol and biogas, which are produced by various chemical reactions (microbial conversion of lignocellulosic biomass, transesterification, hydrotreatment-alkane isomerization and thermochemical) and refinement. Based on the type of feedstock used, biofuels have been categorized as the first-generation biofuels (using edible biomass such as corn, sugarcane, sugar beet, etc.), Second-generation biofuels use non-edible biomass, food and agro-waste, third-generation biofuels use microorganisms and algae and fourth-generation biofuels use genetically modified microbes (Alalwan et al., 2019; Chowdhury and Loganathan, 2019; Ganguly et al., 2021). The spectrum of feedstocks continues to evolve, mainly due to the food versus fuel debate and to achieve higher energy efficiency.

Bioenergy is an important source of energy for the power, industry and transport sectors. Hence, it has become critical for the global and regional energy sectors to increase the share of biofuel in gross demand in these sectors. Under the transforming energy scenario, primary modern bioenergy demand would rise from around 30 exajoules (EJ) (2016) to 125 EJ by 2050 (IRENA, 2020a). Liquid biofuels consumption would reach 652 billion litres, up from 129 billion litres in 2016. The United States is the top producer of biofuels among the developed countries, while Brazil, Indonesia, China, Argentina and Thailand are developing economies that have boosted biofuel production (Subramaniam et al., 2020). The share of biofuels is over 60 percent of the total renewable energy sources in the European Union (EU) (Dafnomilis et al., 2017). The share of bioenergy in the total global energy demand is forecast to be 1736 Mtoe, 1828 Mtoe and 1628 Mtoe under current, stated and sustainable policies scenarios by 2040 (IEA World Energy Outlook, 2019). Between 2010 and 2019, with the rise in bioenergy projects, the global weighted-average levelized cost of electricity (LCOE) of bioenergy for power projects declined from USD 0.076/kWh to USD 0.066/kWh (IRENA, 2020b). The share of total bioenergy in the total primary energy supply (TPES) is set to rise by 9 and 10 percent by the years 2030 and 2050, respectively. Similarly, liquid biofuel production will reach 285 billion litres (2030) and 393 billion litres (2050) in the planned energy scenario (IRENA, 2020a). A recent study was conducted by IRENA on seven countries in southern Africa (Eswatini, Malawi, Mozambique, South Africa, Tanzania, Zambia and Zimbabwe) to assess the potential of bioethanol production using sugarcane as the base feedstock. This report revealed that about 4.8 Mha agriculture land is suitable for sugarcane production and, if the plantation was to be scientifically managed, 1.4 billion litres of ethanol per year could generated from sugarcane, in addition to satisfying the domestic and export sugar requirements (IRENA, 2019a).

In India, the road transport sector shares 6.7 percent of thecountry's gross domestic product (GDP) and conventional fuels remain the most highly consumed, such as diesel (almost 72 percent) followed by petrol (around 23 percent) and other fuels (CNG, LPG, biofuels, etc.) (5 percent). At the same time, India has decided to reduce the imports of oil and natural gas by adopting five principal strategies, namely increasing domestic production, adopting biofuels and renewables, energy efficiency norms, improvements in refinery processes and demand substitution (National Policy on Biofuel, 2018). India has started various programmes, and strategies have been adopted such as the Ethanol Blended Petrol Programme, National Biodiesel Mission,

Biodiesel Blending Programme, pricing, incentives, opening alternative routes for ethanol production, sale of biodiesel to bulk and retail customers, focus on R&D, etc. to promote biofuel production and consumption for more than a decade. In order to incentivize the production and consumption of biofuels in India, the Government of India has framed the National Policy on Biofuel (2018) with the objective being 'to enable availability of biofuels in the market thereby increasing its blending percentage'. The goal of this policy is to achieve 20 percent blending of ethanol in petrol and 5 percent blending of biodiesel in diesel by 2030, as compared to the present 2 percent ethanol blending in petrol and less than 0.1 percent biodiesel blending in diesel. By 2040, the share of bioenergy may grow from 10 percent to 13 percent under the current, stated and sustainable policies scenarios (IEA World Energy Outlook, 2019). On August 27, 2018, the first biofuel-based Indian civilian aircraft (Spicejet Q400) flew successfully between Dehradun and Delhi. India also has initiated the 'Repurpose of Used Cooking Oil' program where the used oil (from the domestic and hospitality sectors) will be collected and sent to biodiesel plants for blending.

1.6 Conclusion

The global energy sector has been continuously evolving and developing with diverse generation, storage, transmission and application means. It is necessary to assess the present generation and demand for energy with future projections as it would be helpful in policy formulation, demand side management, mitigation of unforeseen adverse conditions and selection of energy sources. Energy demand is exponentially increasing with the high population growth rate and economic developments, and it is indispensable to raising the living standards of the global population. The largest part of the energy market is utilized by industry, and the urban ecosystem (domestic cooling and/or heating) and transportation sectors. Fossil fuels (oil, natural gas and coal) have conventionally been the backbone of the energy sector, but these are exhaustible in nature and create environmental hazards. Therefore, energy sources should be used judicially in the coming decades and centuries. The world is observing a great paradigm shift, where the share of renewable and clean sources of energy is rising at an unprecedented rate and will surpass the contribution percentage of conventional sources of energy in the next few decades. Transformation of the energy sector is crucial because over-exploitation of non-renewable resources will lead to extreme environmental issues such as pollution, biodiversity loss, global warming/climate change, etc. and the related aftereffects. This will ultimately leading to an existential crisis for the human race. Also, transformation of the consumer sectors (industries, residential and agriculture) is equally important. Initiatives such as the design and construction of 'green' buildings, linked with renewable energy systems and energy efficiency features, for industrial, commercial and residential purposes should be promoted and implemented on a large scale.

1.6.1 Challenges

There are some serious challenges before the energy sector, which demand urgent attention and need to be settled and satisfied in the near future. Some of the notable hurdles include the following: widening gap of demand and supply of energy; achievement of maximum energy security and efficiency; environmental degradation and emissions from energy and related processes; necessity of safe, sustainable and durable techniques for the generation, storage, transmission and consumption of energy; energy–water nexus; constraints of land and water resources in setting up large projects; consumer-friendly market and price mechanism; effective, efficient and eco-friendly management of waste generated from the energy and related processes, etc.

1.6.1.1 Demand and Supply of Energy

There is a huge gap between the global demand and supply of electricity. A significant proportion of the global population still does not have access to energy, electricity in particular. Also, there is a noteworthy discrepancy in the consumption pattern of energy at the global level. The per capita energy consumption of developed nations has always been exceptionally high. However, in the recent past, developing nations and fast-growing economies (India, China, etc.) have emerged as the largest producers and consumers of energy. Also, less developed sections of the globe, especially in the tropical and sub-tropical region (e.g., Africa), have shown great potential to fill the energy demand and supply gap at regional and global levels in the future. Effective and efficient energy generation, storage, transmission and demand side management are key to energy security.

1.6.1.2 Technology Advancements and Energy Efficiency

Universal access to electricity, energy security and energy conservation cannot be achieved without energy efficiency. Therefore, old technologies, machineries and appliances should be phased out and replaced with eco-friendly and energy-efficient techniques. This is not an easy task as environment-friendly and energy-efficient technologies are generally costlier than conventional ones and emerging economies are unable to renovate the entire energy sector with up-to-date technologies in a short period of time.

1.6.1.3 Limited Resources and Environmental Degradation

Economic growth cannot be accepted at the cost of environmental degradation. The Paris Agreement goals and SDGs should be targeted in parallel. Population pressure, coupled with rapid economic growth, adversely affects nature and natural resources, to a lesser or greater extent. Also, the reduction of the use of energy-related greenhouse gas emissions is another great challenge which the world is attempting to tackle. Land resources are limited and the growing energy demand is negatively affecting land use and land-use change at the global level. This eventually leads to land degradation and loss of forest cover. The debate about food and fuel remains relevant as the demand for food grain is increasing, while the size and productivity of agricultural lands are decreasing at a fast pace. Land acquisition for mega projects is also a challenging area for states and authorities as it requires the relocation and rehabilitation of people on a large scale. Similarly, on one hand, the energy sector demands a huge volume of water for various purposes, while, on the other hand, the global population is facing an unprecedented crisis due to a shortage of safe and potable drinking water. Therefore, a balancing act is required for a sustainable energy future.

1.6.1.4 Favourable Markets, Prices and Policy Mechanisms

Sustainable Development Goals (SDG 7, in particular) could not be imagined or achieved without the availability of low-cost energy. International cooperation and coordination, along with favourable state policies, are a must to make the market and price mechanism consumer friendly. Investments and subsidies on renewable and clean sources of energy and the imposition of penalties and fees for fossil fuels use should be promoted. National commitment and social acceptance are also required for a cleaner future.

1.6.1.5 Extreme Weather and Climatic Conditions

The historic rise in the GHG concentration in the atmosphere, frequent episodes of extreme weather and climatic conditions (cloud bursts, floods, droughts, landslides, cyclones, extreme temperatures, etc.) adversely and severely affect the infrastructure and availability of energy. Global warming and extreme weather conditions are increasing the cooling and heating energy demand worldwide. Recent experiences suggest that the energy sector is prone to sudden pandemic outbreaks and cyber threats.

1.6.1.6 Waste Management

The high demand for and generation of energy may also give rise to the generation of a huge amount of solid, liquid or gaseous wastes (e.g., used PV panels, spent nuclear fuel rods, used batteries, discharge of wastewater and the emission of noxious gases, etc.). Local and global authorities should have a blueprint to manage waste effectively, efficiently and in an eco-friendly manner using various mechanisms such as reduce, reuse, recycle, refurbish and refuse.

1.7 The Road Ahead

Sustainable Development Goals (SDGs) cannot be imagined and achieved without transforming and decarbonizing the energy sector. The transformation and decarbonization of the energy sector can be attained by the decentralized (generation of energy at the community or individual level) energy generation using renewable sources (solar PV, small hydro and wind turbines, biomass, geothermal, etc.). Biofuel production from biomass, agricultural and other solid wastes shows a promising future. Decarbonization of the transportation sector is equally important, and could be achieved by promoting mass public transport means (such as metro, rapid metro), and using electric motor vehicles and low-carbon fuels or biofuel. As the world is observing an ever-increasing demand for electricity, this sector (renewable and non-renewable energy) has huge potential to create a high number of job opportunities in the near and more distant future. Hence, the availability of clean and cheap energy will augment the living standards of the global population.

ANNEXURE 1.1

Suffixes and Units of Energy

Kilo	**$\times 10^3$**
Mega and Million	**$\times 10^6$**
Giga and Billion	**$\times 10^9$**
Tera and Trillion	**$\times 10^{12}$**
Peta and Quadrillion	**$\times 10^{15}$**
Exa	**$\times 10^{18}$**
Zetta	**$\times 10^{21}$**
Yotta	**$\times 10^{24}$**
1 Tonne	**1000 kg**
1 Watt	**1 Joule per second**
1 Unit of electricity	**1 kilowatt hour**
1 Calorie	**4.184 Joule**
1 Barrel of Oil	**42 US gallons or 158.9873 litres**
Barrel of oil equivalent (BOE)	**6.12×10^9 J or 5.8×10^6 $BTU_{59\,°F}$**
Tonne oil equivalent (toe)	**41.868×10^9 J**
Tonne of coal equivalent (tce)	**29.288×10^9 J**
British Thermal Unit (BTU)	**1.0545×10^3 J**

REFERENCES

Alalwan, H.A., Alminshid, A.H., Aljaafari, H.A.S. 2019. Promising evolution of biofuel generations. Subject review. *Renewable Energy Focus* 28: 127–139.

BP 2019. *Statistical Review of World Energy*.

Brown to Green: The G20 Transition towards a Net-Zero Emissions Economy. 2019. Climate Transparency, Berlin, Germany.

Chowdhury, H., Loganathan, B. 2019. 3rd generation biofuels from microalgae: A review. *Current Opinion in Green and Sustainable Chemistry*, https://doi.org/10.1016/j.cogsc.2019.09.003.

Dafnomilis, G., Hoefnagels, R., Pratama, Y.W., Schott, D.L., Lodewijks, G., Junginger, M. 2017. Review of solid and liquid biofuel demand and supply in Northwest Europe towards 2030 – A comparison of national and regional projections. *Renewable and Sustainable Energy Reviews* 78: 31–45.

Enerdata Global Energy Statistical Yearbook. 2020. https://yearbook.enerdata.net/total-energy/world-consumption-statistics.html visited on July 13, 2020

Erdiwansyaha, M., Mamat, R., Sani, M.S.M., Khoerunnisa, F., Kadarohman, A. 2019. Target and demand for renewable energy across 10 ASEAN countries by 2040. *The Electricity Journal* 32: 106670.

Ganguly, P., Sarkhel, R., Das, P. 2021. The second- and third-generation biofuel technologies: comparative perspectives. In Dutta, S., Mustansar Hussain, C. (eds.), *Sustainable Fuel Technologies Handbook*, 29–50. Elsevier and Academic Press. doi:10.1016/b978-0-12-822989-7.00002-0

IEA (International Energy Agency). 2019. Electricity Information Overview 2019.

IEA (International Energy Agency). 2019. *World Energy Outlook 2019*. IEA, Paris. www.iea.org/reports/world-energy-outlook-2019. ISBN 978-92-64-97300-8.

IEA (International Energy Agency). 2020. IEA: Global Energy Review 2020; The impacts of the Covid-19 crisis on global energy demand and CO_2 emissions.

IEA (International Energy Agency). 2020. *India 2020; Energy Policy Review*. IEA, Paris. www.iea.org/reports/india-2020

IEO (International Energy Outlook). 2019 with projections to 2050, US Energy Information Administration (US EIA).

International Index of Energy Security Risk. 2020 Global Energy Institute, US Chamber of Commerce.

IPCC (Inter-governmental Panel for Climate Change). 2018. *Global Warming of 1.5°C*. An IPCC Special Report on the impacts of global warming of 1.5°C above pre-industrial levels and related global greenhouse gas emission pathways, in the context of strengthening the global response to the threat of climate change, sustainable development, and efforts to eradicate poverty, IPCC, Geneva.

IRENA (International Renewable Energy Agency). 2018. *Renewable Energy and Jobs – Annual Review 2018*. International Renewable Energy Agency, Abu Dhabi.

IRENA (International Renewable Energy Agency). 2019a. *Sugarcane bioenergy in southern Africa: Economic potential for sustainable scale-up*. International Renewable Energy Agency, Abu Dhabi.

IRENA (International Renewable Energy Agency). 2019b. *Global Energy Transformation: A roadmap to 2050* (2019 edition). International Renewable Energy Agency, Abu Dhabi.

IRENA (International Renewable Energy Agency). 2020a. *Global Renewables Outlook: Energy transformation 2050 (Edition: 2020)*. International Renewable Energy Agency, Abu Dhabi.

IRENA (International Renewable Energy Agency). 2020b. *Renewable Power Generation Costs in 2019*. International Renewable Energy Agency, Abu Dhabi.

Ministry of Power, Govt. of India www.powermin.ac.in visited on 26.07.2020

National Policy on Biofuel, India. 2018. http://petroleum.nic.in/national-policy-biofuel-2018- 0, visited on 30.07.2020

NITI and IEEJ. 2017. Energizing india; A Joint Project Report of NITI Aayog and IEEJ.

State of Electricity Access Report 2017 (Vol. 2): Full Report (English). World Bank Group, Washington, DC. http://documents.worldbank.org/curated/en/364571494517675149/full-report

Subramaniam, Y., Masron, T.A., Azman, N.H.N. 2020. Biofuels, environmental sustainability, and food security: A review of 51 countries. *Energy Research & Social Science* 68: 101549. DOI: 10.1016/j.erss.2020.101549

2

Biodiesel from First-Generation Feedstock: Scope and Limitations

Madhu Agarwal,[*] Pushpendra Kushwaha, and Karishma Maheshwari

CONTENTS

2.1 Introduction

Global environmental issues and a dwindling supply of non-renewable sources of energy, like fossil fuel, have increased public attention around sources of ‘green’ goods. Investigators have analysed many ways to upgrade current technologies with safe and green alternatives, such as petro-diesel (Rorrer et al., 2019). Petro-diesel produces a moderate amount of pollutant emissions, which have detrimental effects on humans and the climate. It should be noted that petro-diesel can emit harmful

[*] *Corresponding Author* magarwal.chem@mnit.ac.in

DOI: 10.1201/9781003120858-2

pollutants including carbon monoxide, hydrocarbons, particulate matter, sulphur oxide, and nitrogen oxides. The toxicity of these variable and risky contaminants to humans has been identified. Energy and a clean environment are two of the most necessary human needs, both of which are required for socio-economic sustainability. Increased energy and environmental demands have derived from rapid population growth, industrial development, and modernization (Mahlia et al., 2020).

In 2019, global biodiesel production increased by 13%, i.e. up to 47.40 billion litres. Biodiesel manufacturing is more difficult geologically than ethanol production, with the top five countries accounting for 57% of its global manufacture in 2019. While there has been a drop in demand in the United States, biofuel production rose by 5%, with Indonesia showing rapid improvements in biodiesel production. Indonesia has surpassed the United States (14%) and Brazil (17%) as the largest producer in the world (12%), with the next biggest producers being Germany (8%), France (6.3%), and Argentina (5.3%). The generation of ethanol by molasses and other sugar industry by-products has increased in India, where the government has prioritized biofuels as a way to minimize oil imports. In 2019, India's ethanol production increased by 70% to 2 billion litres, and Canada and Thailand have been replaced as the world's fourth-largest manufacturers (Henner and REN21, 2017).

2.1.1 Generation of Biodiesel

There are four generations of biodiesel, as elaborated in Table 2.1, in consideration of the feedstocks used for processing and the various biodiesel production technologies. First-generation biodiesel can be made from edible feedstocks including palm (*Arecaceae*), rapeseed (*Brassica napus*), soybean (*Glycine max*), and corn (*Zea mays*) oils (Gupta et al., 2018). Second-generation biodiesel can also be made from non-edible feedstocks such as jatropha (*Jatropha curcas*), neem (*Azadirachta indica*), rubber crop (*Ficus elastica*), and karanja (*Millettia pinnata*) oils. Since microalgae and waste cooking/frying oil do not compete with crops for space, they can be used to make third-generation biodiesel. In fourth-generation biodiesel, photo-biological solar fuels and electro-fuels are also used, and can be classified as a new field of study that needs to be investigated (Syafiuddin et al., 2020).

Almost all biodiesels are produced from the same renewable source and contain the same basic ingredients. Biodiesel is identified as mono-alkyl esters of long-chain fatty acids formed by transesterification with methanol and catalyst from edible, non-edible, and waste cooking/frying oil, according to the American Society for Testing and Materials (ASTM) (Gupta et al., 2019). During the transesterification process, glycerol (glycerine) is formed as a by-product. B100 stands for 100 percent Fatty Acid Alkyl Ester (FAAE), while B20 stands for 'biodiesel blends' amounting to 20% FAAE and 80% mineral diesel. Solar energy is used for the manufacturing of biodiesel, which is the backbone of a sustainable bio-economy. Biodiesel remains very important today, especially in the transportation industry. The demand for agricultural land and the development of renewable energy is a major concern. Biodiesel development includes future research such as feedstock development, optimum production process, quantity and quality enhancement, and a carbon-neutral economy (Singh et al., 2020).

TABLE 2.1

First-, Second-, Third-, and Fourth-Generation Biodiesel Feedstocks, Benefits, and Challenges

Generation of biodiesel	Feedstock	Technology for processing	Benefits	Challenges	References
First generation	Edible vegetable oils	Esterification and transesterification of oils	1. Renewable energy source 2. It is eco-friendly 3. Biofuel conversion is easy	1. It is a competitor to food crops 2. Food prices rise as a result of increased competition 3. The shortage of land	(Goh et al., 2019)
Second generation	Non-edible vegetable oils, excess cooking oil, feedstock products, animal fats	Oil/seed esterification and transesterification (using organic catalysts and additives)	1. Renewable energy source 2. Ecologically friendly 3. Does not present a hazard to food crops 4. Effective land use (non-arable lands)	1. Rivalry for land and water 2. Specific downstream processing technologies are required 3. Expensive to manufacture 4. Oil yield availability is uncertain in the long term	(Mofijur et al., 2020)
Third generation	Aquatic feedstock that has been harvested (microalgae)	Algae cultivation, harvesting, oil extraction, and all steps in the process are known as transesterification	1. Source of renewable energy 2. It is environmentally friendly 3. There are no issues with food or land use 4. Tendencies to have a faster rate of growth 5. Lipid concentration in the cells is high	1. Biofuel production is inadequate for commercialization 2. For economic viability, high initial production and setup costs are required (large scale)	(Datta et al., 2019)
Fourth generation	Aquatic feedstock that has been harvested (microalgae)	Microalgae genetic modification	1. Source of sustainable energy 2. It is environmentally conscious 3. It is capable of capturing more CO_2 4. Generated at a greater rate than other biofuels	1. One of the most difficult aspects of algae cultivation is the high rate of the initial investment 2. Algae production progress is already in its early phases	(Tulashie and Kotoka, 2020)

2.1.2 Feedstocks for Biodiesel Production

Algae, plant, microbial oil, and animal fats can all be used to produce biodiesel. The purity and structure of biodiesel made from various feedstocks vary. The selection of feedstock is one of the most significant aspects in the production of biodiesel, as it affects several factors such as biodiesel purity, cost, composition, and yield. The abundance and type of feedstock supply are the most important factors in classifying biodiesel into edible, non-edible, and waste-based origins (Gupta et al., 2017). The feedstocks used in biodiesel production are also affected by geography. Before choosing a feedstock, the availability and economic factors of the country are taken into consideration. Karanja and jatropha oils are being studied as possible feedstocks for biodiesel production in India (Singh et al., 2020; Kegl, 2008). Soybean (*Glycine max*), rapeseed (*Brassica napus*), sunflower (*Helianthus annuus*), and mustard (*Brassica*) oils were earlier utilized as biodiesel feedstocks, but their use as biodiesel feedstocks has slowed due to their suitability for human consumption. The use of edible oils as biodiesel feedstocks thus presented a major problem in the form of the 'food versus fuel' debate (Gupta et al., 2016; Kegl, 2008). Biodegradability, low sulphur content, low aromatic content, and wide abundance are among the benefits of using non-edible oils as a biodiesel feedstock, according to numerous reports. Tallow oil, fish oil, animal fats, and microalgae are some of the emerging feedstocks that can be used for the synthesis of biodiesel (Singh et al., 2020).

2.2 First-Generation Biodiesel

Edible oils and food crops are commonly used to manufacture first-generation biodiesel. Soybean (*Glycine max*), palm (*Arecaceae*), sunflower (*Helianthus annuus*), rapeseed (*Brassica napus*), and cottonseed (*Gossypium hirsutum*) oils are common feedstocks (Gupta et al., 2019). Growing food crops, on the other hand, has been argued to reduce the amount of food available for human consumption, resulting in rising global food prices. As a result, while first-generation biodiesel helps meet rising alternative fuel demand, it also clashes with rising food demand, sparking a 'food versus fuel' debate (Mahlia et al., 2020).

2.2.1 The Feedstock of First-Generation Biodiesel

There are a variety of sources of feedstock among the edible oils for production of biodiesel. These are grouped under first-generation biodiesel proliferation and widely consumed due to ease of accessibility, non-toxicity, renewability, and eco-friendly aspects (Ferrero et al., 2021). They are used in biodiesel generation due to their flow attributes, oxidative stability, flash point, cloud point, saponification value, cetane number, calorific value, etc. They comply with the characteristics of available commercial fuel (Masripan et al., 2020). Research has been carried out to explore the properties of oils, namely, sunflower (*Helianthus annuus*), palm (*Arecaceae*), soybean (*Glycine max*), rapeseed (*Brassica napus*), jatropha (*Jatropha curcas*), mahua (*Madhuca longifolia*), waste oil, etc. (Gupta and Agarwal, 2017). Of these collections, soybean (*Glycine max*), sunflower (*Helianthus annuus*), coconut (*Cocos nucifera*), and palm (*Arecaceae*) oils gained greater attention and their properties, modifications, and yields are described in the following subsections of this chapter.

2.2.1.1 *Soybean* (Glycine max)

Soybean (*Glycine max*) is rapidly establishing itself as a significant crop in the oilseed and grain sector, with annual production growing in parallel with the global market supply. The United States, Brazil, and Argentina together provide about 81% of world soybean production, and are the major drivers of rising crop yields (Aransiola et al., 2019). Soybean has a high nutritional quality, particularly protein, which contributes up to 34% of the total, and is in significant need as a supplier of reasonably affordable high-quality protein compared to wild protein (Adie and Krisnawati, 2017). Soybean oil has a higher iodine number of about 128–143, with this higher iodine number ensuring that biodiesel has acceptable oxidative stability (Karmakar et al., 2010). Soybean plants range in height from 0.5 to 1.2 metres, and have a lower oil yield per hectare than other crops (Singh et al., 2020). Depending on the variety, soybean crop yields range from 1800 to 3500 kg per hectare. The average soybean oil yield is 445 kg/hectare (Aransiola et al., 2019). Soybean also helps to replace nitrogen in soil. Soybean needs less fertilizer because it develops a positive fossil energy balance and enables the fixation of nitrogen (Singh et al., 2020). Biodiesel derived from soybean is lacking in real applications due to its poor flow attributes and lower oxidative stability. Therefore, modifications have been attempted by many researchers, for instance, low-cost catalyst (marble slurry derived) has been introduced in a work by Gupta et al. (2018), revealing a strong biodiesel yield of 94±1% (Gupta et al., 2018). Another recent investigation by Adu-Mensah et al. (2021) resulted in a 127% enhancement of the methyl oleic acid composition on partial hydrogenation of soybean biodiesel, which paved the pathway for its implementation in engines. The oxidative stability and flow attributes were massively increased, namely cetane number, kinematic viscosity, etc., with negligible variations in the amount of oxygen. However, the hydrogenation led to a decreased iodine value (Adu-Mensah et al., 2021). Therefore, looking at the properties of soybean feedstock, it has immense potential to be utilized for biodiesel production through modifications to the manufacturing process.

2.2.1.2 *Sunflower* (Helianthus annuus)

Sunflower (*Helianthus annuus*) is used to make biodiesel, which has a high oil content. Sunflower is a main oilseed used to produce edible oil, with an annual output of 25.1 million tonnes. The oxidation stability of oil is affected by the linoleic, oleic, and linolenic acid contents. Sunflower oil has about 70% linoleic acid and is highly sensitive to lipid oxidation (Saydut et al., 2016). Sunflower is the world's fifth most widely grown oilseed harvest (Gupta et al., 2019). The cloud point of the synthesized sunflower oil biodiesel was reported to be –15 °C. As per the specification, there is no restriction for the cloud point of biodiesel in ASTM D 6751, however its value must be provided. For example, an investigation proposes the cloud point of –15 °C for sunflower oil and reports it to be safe for use as fuel in

Turkey, a tropical country, on most days (Saydut et al., 2016). Researchers have examined the properties of sunflower oil and assessed the quantity of fatty acids in sunflower oil, illustrating that it has tremendous potential to be used as a feedstock for biodiesel oil. Research incorporating sunflower oil discharged from deodorizer distillate for biodiesel production yields with 94.32% efficacy within 9 h of reaction time at 65 min, suggesting that this is another sustainable method for good-quality biodiesel production (Diger Kulaksiz and Paluzar, 2021). Another report concluded that transesterification using kaolin-based impregnated catalyst revealed a higher biodiesel yield of 96% (Jalalmanesh et al., 2021). From the literature reports, one can see that sunflower-derived biodiesel has tremendous scope for producing the desirable quality of biodiesel.

2.2.1.3 *Coconut* (Cocos nucifera)

Coconuts are a significant global crop, grown in moist and semi-desert tropics and subtropics, and they are also salt and wind resistant. After a disaster, immature coconuts can provide clean fresh water and can fulfil the demand for drinking water (Aransiola et al., 2019). Coconut is one of the feedstocks used in the manufacture of biodiesel, and in the Philippines, it is allowed for biodiesel manufacturing. Coconut oil is a triglyceride with a high proportion of saturated fatty acids (86%) (Singh et al., 2020). Coconut husks are a rich biomass resource used for composting, horticulture fertilizer, fuel, and activated carbon filters, and they are the world's foremost continuous and rich source of fibre. Coconut yield ranges between 2–6 tonnes of copra. Coconut oil production average 2.9 tonnes per hectare globally (Aransiola et al., 2019). It primarily contains three acids: palmitic (8%), lauric (45%), and myristic (2%), as reported by Singh et al. (2020). A study by Rajesh et al. (2021) revealed coconut fatty acid distillate as a source for biodiesel, reporting an excellent yield of 92.6% with a reaction time of 90 min at a temperature of 60 °C (Rajesh et al., 2021). Therefore, coconut oil has raw attributes suitable for biodiesel production.

2.2.1.4 *Palm Oil* (Arecaceae)

Malaysia's palm (*Arecaceae*) oil sector is a vital part of the nation's economy. Through the exportation of palm oil and its important products to the international market, the palm oil sector plays a critical role in ensuring a steady flow of foreign investment and profits (Mekhilef et al., 2011). Over the last decade, Indonesia and Malaysia have been the two leading producers of palm oil. The market for palm biodiesel oil is rapidly increasing in Europe (Singh et al., 2020). According to Wahid et al., (2008), Brazil, the European Union (EU), and the United States are among the parts of the world that have started large-scale biofuel strategies to minimize their dependency on imported oil. According to statistics from Oil World, Malaysia contributed 46.7% of palm oil imported to the EU in 2007, followed by Indonesia with 37.6% and other nations with 5.8% (Wahid, 2008). Recent investigations have resulted in biodiesel yields of 88% and 96% derived from a calcium-modified novel catalyst and banana weevil ash, respectively (Meriatna et al., 2021; Qu et al., 2021). These results illustrate the acceptability of palm oil in producing biodiesel with a high yield.

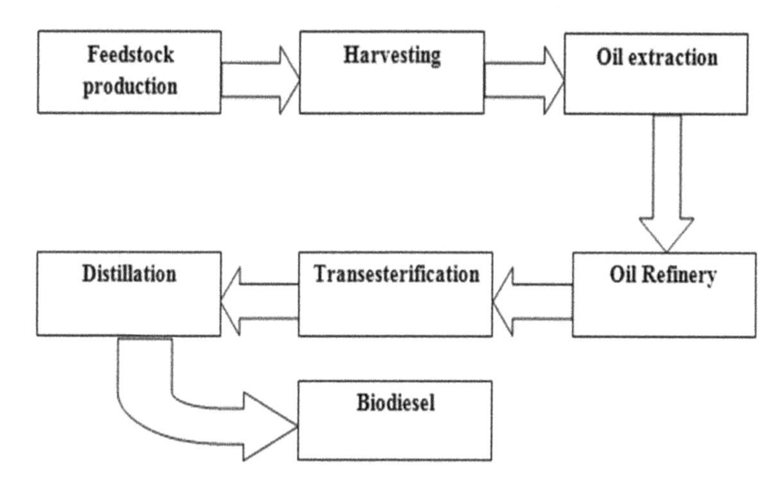

FIGURE 2.1 Steps in the production of biodiesel. (From Ferrero et al., 2021; Gupta et al., 2018.)

2.2.2 Biodiesel Production Method Steps

Biodiesel can be generated from vegetable oil, animal fat, food crops, lignocellulose material, micro- and macroalgae, and other sources (Tulashie and Kotoka, 2020). Biodiesel is usually made via the transesterification process, which includes reacting vegetable oils with alcohol (ethanol or methanol) to make alkyl esters and glycerol with the aid of a catalyst, as illustrated in Figure 2.1, in a similar manner to that represented in literature (Ferrero et al., 2021; Gupta et al., 2018). A number of reversible reactions constitute transesterification. Each step removes one mole of alkyl esters from the triglyceride, resulting in diglyceride, monoglyceride, and finally glycerol formation. The reaction mechanism for alkali-catalysed transesterification has been proposed as three steps (Demirbas and Karslioglu, 2007), whereas esterification is a single-step reaction:

$$\text{Fatty acid } (R_1COOH) + \text{Alcohol } (CH_3OH) \leftrightarrow \text{Ester } (R_1COOCH_3) + \text{Water } (H_2O) \text{ (Esterification Reaction)}$$

$$\text{Triglyceride} + \text{Alcohol } (CH_3OH) \leftrightarrow \text{Diglyceride} + CH_3COOR_1$$

$$\text{Diglyceride} + \text{Alcohol } (CH_3OH) \leftrightarrow \text{Monoglyceride} + CH_3COOR_2$$

$$\text{Monoglyceride} + \text{Alcohol } (CH_3OH) \leftrightarrow \text{Glycerol} + CH_3COOR_3$$

Transesterification is a three-step reversible reaction that takes place in a sequential order. In the first step, triglycerides in the oil are converted to diglycerides, in the second step to monoglycerides, and then to glycerol in the third step. In the transesterification process, an appropriate supply of alcohol to oil molar ratio (greater than 3:1) is typically controlled to favour the forward reaction. A large additional amount of alcohol is normally used to tip the balance on the product side. At the end of the transesterification step, the reaction stream is divided into two layers: the glycerol-rich phase (bottom layer) and the biodiesel phase (top layer). The excess alcohol is dispersed in both the biodiesel and glycerol layers (Chozhavendhan et al., 2020).

2.2.3 Biodiesel Yield

Biodiesel yield is the quantity of biodiesel made from raw oil represented as the percentage of FAME. Biodiesel yield and quality are determined using chromatographic or spectroscopic analysis. Thin-layer chromatography is used to examine the quality of biodiesel. This approach was used as the primary tool to evaluate glycerides (mono-, di-, and tri-) and fatty acids (FA). However, it has drawbacks including humidity susceptibility and a lack of accuracy (Gupta and Agarwal, 2017).

The ester layer (250 mg) has to be dissolved in n-hexane (5 mL) with an internal natural methyl heptadeconate solution (10 gm/L C17 ester in hexane) for gas chromatograph (GC) analysis using the European controlled protocol EN-14103. GC is a type of chromatography that uses gas to separate molecules using a capillary column with a typical length of 30 m, an inner diameter of 0.32 mm, and a film thickness of 0.25 m. The yield of biodiesel using GC analysis can be calculated using equation (2.1):

$$\text{Yield\%} = \frac{\sum \alpha - \alpha\text{mh}}{\alpha\text{mh}} \times \frac{\text{cmh} - \text{vmh}}{\text{M}} \times 100\% \tag{2.1}$$

where $\sum\alpha$ = the total peak area of methyl ester, αmh = area of internal standard (methyl heptadeconate), cmh = methyl heptadeconate concentration in (10 mg/mL), vmh = volume of the internal standard of solution (5 mL), and M = the sample weight of biodiesel in mg (250 mg) (Chen et al., 2016).

2.2.4 Biodiesel Properties and Standards

2.2.4.1 Biodiesel Yield and Characteristics

Determination of the physical and chemical properties of raw materials and biodiesel fuel derived from various feedstocks is essential in order to use the product as a biofuel or blend with petroleum diesel. For instance, the fatty acid composition of feed results in variation of the FAME composition in the produced biodiesel, resulting in varied fuel characteristics (Singh et al., 2019). Moreover, investigations have indicated that various antioxidant additives have been incorporated as per requirements in order to obtain a natural quality of FAME content and to enhance the attributes such as thermal efficacy, calorific value, and fuel consumption (Agarwal et al., 2012). Significantly, the composition profile of FAME is directly related to several parameters, including viscosity, flash point, cetane number, cloud point, calorific weight, and also the cold filter plugging point, pour point, iodine number, and basic gravity (Santos et al., 2020). Furthermore, aside from inherent characteristics, handling and packaging techniques also play a prominent role in terms of the safe transportation of biodiesel. Likewise, cold soak filter strength, acid amount, ash content, methanol content, water content, sediment, glycerine content, and metals are some of the characteristics of biodiesel produced by different feedstocks (Singh et al., 2019). Detailed descriptions of the key physicochemical characteristics of biodiesels made from various feedstocks are given in the following subsections.

2.2.4.1.1 Density

Density is the key attribute for evaluating the performance of an engine. It determines the estimated amount of fuel supplied for precise combustion by injection systems (Sakthivel et al., 2018). Since fuel supplied by fuel injection pumps is found by volume, it may deliver a minimum or maximum amount of fuel. Compared to low-density fuel, high-density fuel has more bulk, for instance, petroleum diesel has lower densities than biodiesel (Gupta et al., 2017). Accordingly, the EN 14214 and ASTM D6751-02 test procedures offer a wide range of options for assessing biodiesel density. In terms of pure oils, for example, groundnut (*Arachis hypogaea*) oil has a maximum density of 970 kg/m^3, while jojoba (*Simmondsia chinensis*) oil has the lowest density of 868 kg/m^3. In a report by Taylor et al. (2014), jatropha (*Jatropha curcas*) biodiesel has a higher density than commercial petro-diesel (Taylor et al., 2014).

2.2.4.1.2 Calorific Value (CV)

The quantity of heat released during the burning of a fuel is its heat value. For an IC engine, fuel with a higher calorific value is more efficient. When the unsaturation level remains stable, the carbon chain length of fatty acids increases, while the mass fraction of oxygen decreases, and the calorific value increases. The unsaturation level of biodiesel has a significant impact on heating values (Singh et al., 2019). The higher value of calorific value in the instance of raw oil is 46.47 MJ/kg for jojoba oil, and the lowest is 35.992 MJ/kg for karanja oil (Canoira et al., 2006). The calorific value of biodiesel is lower than that of diesel because it includes around 11% more oxygen, however this improves the combustion characteristics (Taylor et al., 2014).

2.2.4.1.3 Kinematic Viscosity

The viscosity of fuel is a critical parameter in determining its flow capability. Biodiesel has a viscosity that is about an order of magnitude lower than that of the parent oil and is regulated by the content of alkyl esters (Ramos et al., 2009). The viscosity of the fuel spray has a most significant role in its atomization and penetration. The fuel spray atomizes, vaporizes, and blends with the air. A higher viscosity causes insufficient atomization of the fuel, resulting in dirt deposition and a reduction in the thermal efficiency (Panwar et al., 2010). According to the European standard EN ISO 3104, kinematic measurement was performed with a Canon–Fenske capillary viscosimeter immersed in a specific temperature (40 °C) bath (Ramos et al., 2009). According to Indian standards, the range of kinematic viscosity varies from 2.5 mm^2/s to 6 mm^2/s. The issues caused by higher viscosity are evident through cold start conditions and lower ambient temperatures (Kegl, 2008). The degree of unsaturation has a stronger correlation with viscosity: higher unsaturation equals lower viscosity. The position of the double bond in the fatty acid chain has very little effect on the viscosity. Viscosity is also affected by the double bond arrangement; *trans* configuration has higher viscosity than *cis* configuration (Singh et al., 2019).

2.2.4.1.4 Lubricity

Lubricity refers to the reduction in friction force between the rigid parts of a machine in relative motion. Low-sulphur diesel can benefit from fatty acid methyl esters (FAMEs), that are the major elements of biodiesel (Hong et al., 2020). Hydrodynamic and boundary lubrication are two common types of lubrication that offer accurate knowledge about the entire lubricity. With the aid of a liquid film between the solid sections of the system, such as fuel in the hydrodynamic injector, lubrication is used to minimize wear between them. At the boundary lubrication becomes successful as the hydrodynamic lubricant emerges or separate from machine components (Gupta and Agarwal, 2017). To minimize wear in the fuel dosing part of a diesel engine, a higher quality of lubricant is needed (Knothe, 2005). Proper lubrication is critical in current fuel injection systems. A big challenge in the field of lubricants has been faced in terms of multiple injections, injection rates, high pressure, etc. Lubricity efficacy decreases in the following order: O > N > S > C (Singh et al., 2019). ASTM D6751 is a U.S. standard that defines the specification for B6–B20, wherein the sulphur grades range from B6 to B20, with less than 15 ppm sulphur content (Chandran, 2020). The low sulphur content fuel lubricity was evaluated first by measuring the ball wear scar diameters according to ASTM D6079 and EN ISO 12156-1 (Hong et al., 2020).

2.2.4.1.5 Acid Number (AN)

The acid number is a measurement of the amount of acidic compounds in biodiesel (Kumar, 2017). The acid value, also known as the neutralizing number, is measured in mg of potassium hydroxide (mg KOH) used to neutralize 1 gram of fatty acid methyl esters, with a maximum value of 60.5 mg KOH/g in the standard specifications (EN 14214) (Ramos et al., 2009). The ASTM D664 reference standards method to determine the acid number of biodiesel and petro-diesel designates procedures for the estimation of acidic compounds in biodiesel and petro-diesel. The reliability of ASTM D664 for B20 in the minimum acid number range of 0.123–0.332 mg KOH/g was evaluated to be within 4.13% (Wang et al., 2008), while rubber (*Ficus elastica*) oil has a maximum value of 25.67 mg KOH/g (Dhawane et al., 2015). For all biodiesels, the acid number value rises as the storing time increases. As a result of increased hydroperoxides, which will be further oxidized into acids, the acid number increases. Acids can be produced when water causes the esters to hydrolyse into alcohols and acids (Wang et al., 2008).

2.2.4.1.6 Boiling Point (BP)

The temperature at which a substance's vapour pressure equals the surrounding pressure is defined as the boiling point of that substance. The standard boiling point can also be used to calculate an element's total volatility (Ramos et al., 2009). The type of bond which occurs between the molecules of a substance affects the boiling point of biodiesel (Santos et al., 2020). GC is used to determine the boiling point spectrum using the ASTM-D7398 standard. According to the ASTM-D7398 standard, the boiling point range is 100–615 °C (Singh et al., 2019).

2.2.4.1.7 Cetane Number (CN)

The cetane number (CN) is the criterion describing the ignitable characteristics of diesel fuel. When the cetane number is low, it signifies there are more emissions from the engine, more deposits from incomplete combustion, and more knocking (Acharya et al., 2017). Meanwhile, a higher cetane number results in an increased degree of saturation and the length of the fatty acid chain also increases. ASTM D6751 and EN 14214 both provide for the use of a cetane engine, that is a specific engine developed for analysing the cetane number (Ramos et al., 2009). Since biodiesel contains more oxygen and therefore burns more effectively, it has a higher cetane number, examples include 54.6 for hazelnut (*Corylus*) (Jamail et al., 2017) and a lower cetane number value of 21 for jatropha (Taylor et al., 2014).

2.2.4.1.8 Cloud Point (CP)

The cloud point is the lowest temperature at which wax in fuel crystallizes and takes on a cloudy appearance. Suspended substances that have been fused to the fuel molecules dissolve into the fuel, causing it to have a cloudy appearance. Biodiesel is made from a variety of feedstocks, each of which has a different fatty acid composition. As a result, the cloud point of the biodiesel produced varies with the feedstock used (Akhihiero and Ebhodaghe, 2020). The ASTM D2500 standard technique for cloud point (for biodiesel) specifies a temperature range of 3–12 °C. In the case of pure oils, such as palm (*Arecaceae*) and camelina (*Camelina sativa*) oils, the highest cloud point value is 19.8 °C for palm oil and the lowest is –10 °C for camelina oil (Singh et al., 2019).

2.2.4.1.9 Oxidation Stability

Oxidation stability is a significant factor in determining the degree of biodiesel reaction with air and the extent of oxidation. Biodiesel's oxidative stability is intimately link to the chemical composition of its species composition, and as a result, the correct modification of its composition could improve its stability (Kumar, 2017). The oxidative stability of biodiesel fuel is a significant parameter in determining its efficiency (Gupta et al., 2017). Mixtures containing 80–90% fatty acid methyl esters of animal fat and 10–20% fatty acid methyl esters of vegetable oil with antioxidant compounds added have been used to gain the maximum oxidation stability (Sendzikiene et al., 2005). According to EN 14112, the shortest acceptable induction period for biodiesel fuel is 3 h (Singh et al., 2019).

2.2.4.1.10 Cold Filter Plugging Point (CFPP)

Due to the obvious increasing amount of biodiesel in diesel fuel, low-temperature operability of diesel vehicles is an issue. Cold flow properties are a test of a fuel's ability to operate at low temperatures and illustrate how it works in cold weather (Barba et al., 2020). The cold filter plugging point value of a sample fuel is lower than the cloud point value (Acharya et al., 2017). For the biodiesel cold filter plugging point, the EN 14214 standards are used (Barba et al., 2020). Tallow and palm oil biodiesel have a high maximum temperature point (poor performance), while rapeseed biodiesel has a low maximum temperature point (Gupta et al., 2017). CFPP tests below the cloud point by more than 10 ºC should be interpreted with caution, as they do not accurately represent the low-temperature operability limit (US Department of Energy, 2008).

2.2.4.1.11 Flash Point (FP)

When vapours of fuel come into contact with a fire source, they reach the lowest temperature at which they can ignite. The temperature at which the vapour over the liquid ignites when exposed to an ignition source is known as the oil fuel flash point (Bhuiya et al., 2016). Due to the high risk of ignition, the flash point is a critical feature for safe handling of flammable goods, especially in high-temperature environments (Santos et al., 2020). Biodiesel contains a flash point of about 130—170 °C, whereas traditional diesel has a flash point of 55–65 °C (Gupta et al., 2019). It can be proposed that biodiesel is safer to transport than petroleum-based fuel due to its less combustible nature. EN ISO 3679 and ASTM D93 specify a framework for evaluating flash point (Santos et al., 2020). Biodiesel produced by neem seed pyrolysis oil has the lowest flash point of 55 °C, whereas biodiesel made from linseed (*Linum usitatissimum*) oil has a maximum FP value of 241 °C (Alagu and Ganapathy Sundaram, 2018).

2.2.4.1.12 Iodine Number (IN)

This is a determination of how much iodine is consumed by the binary bonds of FAME molecules in a 100 g biodiesel fuel oil sample. The value of iodine is determined using a Metrohm 702 SM Titrino and the European standard EN 14111 (Ramos et al., 2009). The iodine number has been also reported to be related to the cetane number, cold filter plugging point, and biodiesel viscosity (Atabani et al., 2013). The iodine number has been found to have a high of 156.74 for linseed oil and a minimum of 10 for coconut oil while researching raw oil (Hotti and Hebbal, 2015).

2.2.4.1.13 Peroxide Value (PV)

The peroxide value is a measurement of the concentration of peroxide in biodiesel that can be used to track oxidation over time. Although PV is not included in the biodiesel fuel specification, it is worth exploring as it has an indirect effect on other qualities such as cetane number and so on (Kumar, 2017). The standard technique for determining peroxide values for traditional diesel fuel is ASTM D3703-13 (Syafiuddin et al., 2020).

2.2.4.1.14 Pour Point (PP)

The minimum temperature at which liquid fuel loses its flow properties is known as the pour point (Sakthivel et al., 2018). For the cold flow process, the pour point is a crucial property. The pour point and cloud point of diesel fuel are usually lower than those of biodiesel fuels. The ASTM D97 approach is used to assess the value of pour point for biodiesel fuel. The European Union and Indian specifications have not defined any pour point requirements (Armendáriz et al., 2015). Biodiesel made from castor (*Ricinus communis*) oil contains the lowest pour point of 20 °C, while biodiesel made from dromedary camel (*Camelus dromedarius*) fat has a maximum pour point of 15.5 °C (Sbihi et al., 2014; Singh et al., 2019).

2.2.4.1.15 Saponification Value (SV)

The total number of saponifiable units of oil per unit weight of an oil is the saponification value: the amount of potassium hydroxide required to saponify 1 g of an ester in mg. The existence of a large amount of low-molecular-weight fatty acid chains is demonstrated by a high saponification value (Sakthivel et al., 2018). A superior saponification value of raw oil suggests a high fatty acid proportion, which could contribute to soap formation through the transesterification method. The saponification value varies from 0 to 370 mg KOH/g, according to the ASTM D5558-95 standard (Singh et al., 2019).

2.2.4.1.16 Sulphur Content

Biodiesel has a lower sulphur content than fossil diesel. However, with the exception of petroleum refineries, the current biodiesel sector does not have the capability of refining fuel to remove sulphur to a level of 15 ppm as required by the EPA (Ducoste, 2009). The EN 14214 and ASTM D6751 requirements specify the methodology for determining the amount of water and deposits in biodiesel (Taylor et al., 2014). Sour plum (*Prunus domestica*) oil has the lowest sulphur content (0%), while coconut (*Cocos nucifera*) oil has the highest (1.3%) (Singh et al., 2019).

2.2.4.1.17 Water and Sediment Content

The quantity of water and sediment in biodiesel fuel indicates the purity of the fuel (Jhalani, 2019). Water can be present in biodiesel as dissolved or suspended droplets. The performance of biodiesel reduces as water is present in the gasoline, and its presence often corrodes engine components (Singh et al., 2019). Dirt particles may be found in the sediment in biodiesel fuel, creating difficulties in fuel pipelines. An increased quantity of water promotes the hydrolysis reaction of FA in biodiesel (Jhalani, 2019). The EN 14214 and ASTM D6751-02 requirements specify the methodology for determining the amount of water and deposits in biodiesel (Taylor et al., 2014).

2.2.4.2 Standards of Biodiesel

The most important aspect to evaluate biodiesel efficiency is to check whether the physical and chemical properties are in compliance with the acceptable limit set as per standard norms. Quality norms for biodiesel are rapidly changing due to advancements in ignition engines, compression, re-evaluation, and ever-stricter pollution criteria of the suitability of feedstocks used in biodiesel production (Knothe, 2006). The specifications of existing diesel fuel requirements, the predominance of the most typical types of diesel engines in the region, the pollution laws controlling such engines, the manufacturing process, and the climatic conditions of the region are all variables (such as sulphur content, acid number, iodine number, water and sediment content, etc.) that differ by region, as depicted in Table 2.2 (Montero and Stoytcheva, 2015) and the international standards with attributes as depicted in Figure 2.2.

TABLE 2.2

International Standards for Biodiesel Fuel

S. No.	Country/Area, Specification and Title	Feedstock	Properties of biodiesel produced	References
1	U.S., ASTM D6751 & Biodiesel Fuel Blend with middle distillate fuel	Soybean (*Glycine max*) oil	Heating value = 40.6 kJ/kg Sulphur content = <15 ppm Flash point = 130 °C Viscosity at 40 °C = 6 mm^2/s Cetane number = 55 Density at 15 °C = 0.88 kg/m^3	(Chandran, 2020)
		Almond (*Prunus dulcis*) oil (B20)	Density = 0.9 g/cm^3 Cloud point = 6 °C Pour point = 1 °C	(Akhihiero and Ebhodaghe, 2020)
		Jatropha (*Jatropha curcas*) oil	Flash point = >130 °C Water content = <0.03% Ash content = <0.02% Viscosity = 1.9–6.0 mm^2/s Acid value = <0.8 mgKOH/g	(Taylor et al., 2014)
2	European countries, EN14214 & Referring to fatty acid methyl esters (FAME) as fuel	Mahua (*Madhuca longifolia*) oil	Density at 15 °C = 860–900 kg/m^3 Flash point = >101 °C Sulphur content = 10 ppm (max) Kinematic viscosity at 40 °C = 3.5–5 mm^2/s	(Acharya et al., 2017)
		Jatropha (*Jatropha curcas*) oil	Flash point = >120 °C Density at 15 °C = 860–900 kg/m^3 Viscosity = 3.5–5 mm^2/s Acid value = <0.5 mgKOH/g Water content = <0.05% Ash content = <0.02% Carbon residue = <0.3%	(Taylor et al., 2014)

		Sunflower (*Helianthus annuus*) oil	Flash point = >177 °C Cetane number = 50 Kinematic viscosity at 40 °C = 4.2 mm^2/s Free glycerol = 0.00% Acid number = 0.15 mgKOH/g	(Ramos et al., 2009)
		Peanuts (*Arachis hypogaea*) oil	Flash point = >176°C Cetane number = 53 Kinematic viscosity at 40 °C = 4.6 mm^2/s Free glycerol = 0.01% Acid number = 0.10 mgKOH/g	(Ramos et al., 2009)
3	India, IS15607 and biodiesel blend stock for fuel	Mahua (*Madhuca longifolia*) oil	Density at 15 °C = 860–900 kg/m^3 Flash point = >120°C Sulphur content = 50 ppm (max) Kinematic viscosity at 40 °C = 2.5–6 mm^2/s Cetane number = 51	(Acharya et al., 2017)
		Jatropha (*Jatropha curcas*) oil	Density at 15 °C = 850.67 kg/m^3 Flash point = >88°C Fire point = >98 °C Kinematic viscosity at 40 °C = 3.13 mm^2/s	(Sharma Dugala et al., 2021)
		Palm (*Arecaceae*) oil	Flash point = 175 °C Water and sediment = <0.05% Cetane number = 49 Density at 40 °C = 880 kg/m^3 Pour point = 19.7°C Acid value = 0.2 mgKOH/g	(Verma et al., 2016)
4	Brazil, ANP-42 and biodiesel quality is regulated by Brazilian specification	Beef tallow oil (B2)	Density at 20 °C = 0.844 kg/L Flash point = 43 °C Kinematic viscosity at 40 °C = 2.7 mm^2/s Cetane number = 47.15	(da Cunha et al., 2009)
		Beef tallow oil (B100)	Density at 20 °C = 0.872 kg/L Flash point = 156 °C Kinematic viscosity at 40 °C = 5.3 mm^2/s Cetane number = 60.35	(da Cunha et al., 2009)

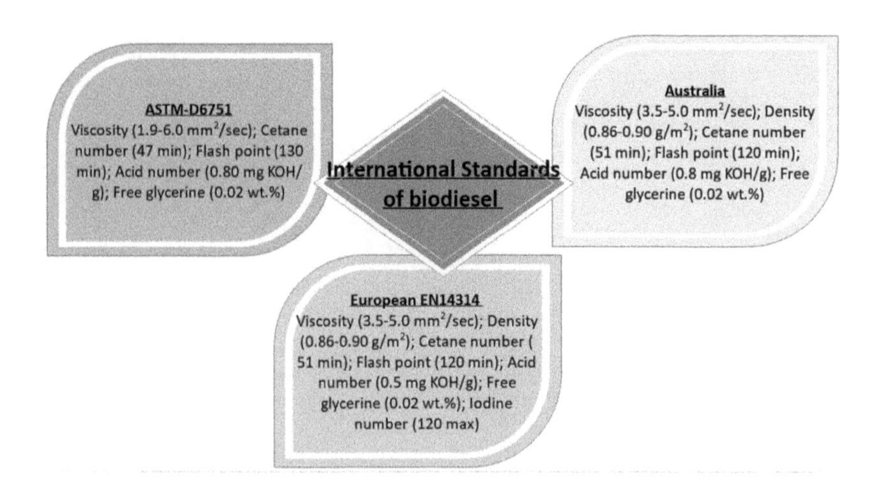

FIGURE 2.2 International standards for biodiesel with their attributes. (From Azad et al., 2016; Mahlia et al., 2020.)

2.3 Scope and Limitations of Biodiesel

Biodiesel has many benefits over diesel fuel, including high flash point, renewability, non-toxicity, environmental friendliness, and biodegradability. Biodiesel has properties equivalent to petroleum diesel and emits less pollutants, making it a viable alternative to diesel fuel in the transportation sector. Increased use of biodiesel could help to reduce pollution and movable carcinogens (Gupta et al., 2017). The key downside to using these feedstocks in that they can raise the cost of food goods might lead to the possibility of shortages in food supply. Biodiesel production from edible feedstock faces challenges such as restricted cultivation land area, high cost, and adaptability to environmental conditions. Due to these disadvantages, consumers (end users of biodiesel fuel) often look for an alternative biodiesel feedstock option (Singh et al., 2020).

2.3.1 Challenges for Sustainable Biodiesel Production

The main drawback of first-generation biodiesel is that it is made from edible oils harvested from food crops, which could lead to an increase in global food prices. Although second-generation biodiesel is not produced from food crops, it still necessitates plant production, which can compete with food crops for arable land. Furthermore, there are a number of flaws in the biodiesel manufacturing process (Agarwal et al., 2012). The two main problems for esterification and transesterification processes are the presence of water in the feedstock and the absence of acid and alcohol from the final product. Alkaline-catalysed processing processes can be hampered by the presence of water, free fatty acids, and glycerine (Mahlia et al., 2020). Furthermore, enzyme-catalysed manufacturing processes are more costly and take longer for completion (Azad et al., 2016). However, above all, the best approach for biodiesel development is transesterification, as this process results in reduced viscosity of the oil resulting in similar characteristics to traditional diesel. A higher reactant/oil ratio, as well as the need for high pressures and temperatures, are some of the challenges of the transesterification approach (Gupta et al., 2017). There are several difficulties faced while producing biodiesel (Mahlia et al., 2020), with some major challenges including the following:

1. They use additional power or energy;
2. They have a high initial investment; and
3. Their optimization processes can be costly, making them inefficient and time-consuming.

2.3.2 Current Perspective and Future Trends

A major challenge to the long-term implementation of developed technology includes technology transformation from developed to developing countries (Gupta et al. 2018). To solve this problem, countries must invest in technological development and implementation systems, as well as national innovation systems. To achieve a shared goal, the implementation of an innovation mechanism generally requires the participation of individuals and organizations in a variety of activities (Gupta et al., 2020). International and national agencies play significant roles in an innovation system for collaborations all over the world. Organizations, for example, have functions in establishing critical information and adapting information from the broader society. Industries' roles are to introduce improvements and expertise into common technologies, while national governments' roles are to recognize and support biodiesel-related research and technology. Incentives from government in developing countries play an important role in supporting domestic biodiesel industries. Subsidies from government, product promotion, and national biodiesel mandates will all help in the development of first-generation biodiesel. Direct financial funding, such as technology development grants and biodiesel purchase subsidies, may be included in such provisions (Mahlia et al., 2020).

2.4 Conclusion

As the emerging scenario of increasing energy demand, biodiesel is a renewable and green energy alternative at present. The chemical composition and physicochemical properties of first-generation biodiesel are briefly addressed. The effects of chemical and physical properties of biodiesel on the compositional factor of FAME are also included. Changes in compositional elements including chain branching, unsaturation and chain length, effect both desirable and undesirable changes in FAME properties. Compositional characteristics of FAME, which improve fuel stability and sustainability, can lead to producing biodiesel at low temperatures. Fatty acid compositions have an opposite effect on biodiesel characteristics, making it difficult to determine the desirable fatty acid composition. Characteristics such as flow, lubricity, viscosity, cetane number, and oxidative stability are used to indicate the suitability of FAME as a biodiesel feedstock. Two of these characteristics, oxidation stability and cold flow, are particularly important because changes occur with the feedstock used. FAME of good quality should have a lower polyunsaturated fatty acid content, which results in less oxidation instability in biodiesel production. The availability of feedstocks, climate conditions, type of activity, and other factors play a key role in biodiesel selection. As a result, the availability of feedstocks, oxidation tolerance, low-temperature operability, and adequate performance must all be considered when choosing a biodiesel feedstock. Metabolic engineering can be used to boost the yield of biodiesel manufacturing units and improve biodiesel production. Because the cost of generating biodiesel remains high, there is a need to reduce the price of biodiesel, which should be supported by a techno-economic analysis.

REFERENCES

Acharya, N., Nanda, P., Panda, S., Acharya, S., 2017. Analysis of properties and estimation of optimum blending ratio of blended mahua biodiesel. *Eng. Sci. Technol. an Int. J.* 20, 511–517. https://doi.org/10.1016/j.jestch.2016.12.005

Adie, M.M., Krisnawati, A., 2017. Soybean opportunity as source of new energy in Indonesia. *Int. J. Renew. Energy Develop.* 3(1), 37–43. https://doi.org/10.14710/ijred.3.1.37-43

Adu-Mensah, D., Mei, D., Zuo, L., Dai, S., Gao, Y., 2021. Improving biodiesel made from cottonseed and soyabean oil with partial hydrogenation. *Proc. Inst. Civ. Eng. Energy* 174, 131–136. https://doi.org/10.1680/jener.20.00002

Agarwal, M., Singh, K., Chaurasia, S.P., 2012. Kinetic modeling for biodiesel production by heterogeneous catalysis kinetic modeling for biodiesel production by heterogeneous catalysis. 4, 013117. https://doi.org/10.1063/1.3687941

Akhihiero, E.T., Ebhodaghe, S.O., 2020. Effect of blending ratio on the fuel properties of almond biodiesel. *Eur. J. Sustain. Dev. Res.* 4, em0119. https://doi.org/10.29333/ejosdr/7804

Alagu, R.M., Ganapathy Sundaram, E., 2018. Preparation and characterization of pyrolytic oil through pyrolysis of neem seed and study of performance, combustion and emission characteristics in CI engine. *J. Energy Inst.* 91, 100–109. https://doi.org/10.1016/j.joei.2016.10.003

Aransiola, E.F., Ehinmitola, E.O., Adebimpe, A.I., Shittu, T.D., Solomon, B.O., 2019. *Prospects of biodiesel feedstock as an effective ecofuel source and their challenges, Advances in Eco-Fuels for a Sustainable Environment*. Elsevier Ltd. https://doi.org/10.1016/b978-0-08-102728-8.00003-6

Armendáriz, J., Lapuerta, M., Zavala, F., García-Zambrano, E., del Carmen Ojeda, M., 2015. Evaluation of eleven genotypes of castor oil plant (*Ricinus communis* L.) for the production of biodiesel. *Ind. Crops Prod.* 77, 484–490. https://doi.org/10.1016/j.indcrop.2015.09.023

Atabani, A.E., Silitonga, A.S., Ong, H.C., Mahlia, T.M.I., Masjuki, H.H., Badruddin, I.A., Fayaz, H., 2013. Non-edible vegetable oils: a critical evaluation of oil extraction, fatty acid compositions, biodiesel production, characteristics, engine performance and emissions production. *Renew. Sustain. Energy Rev.* 18, 211–245. https://doi.org/10.1016/j.rser.2012.10.013

Azad, A.K., Rasul, M.G., Khan, M.M.K., Sharma, S.C., Mofijur, M., Bhuiya, M.M.K., 2016. Prospects, feedstocks and challenges of biodiesel production from beauty leaf oil and castor oil: a nonedible oil sources in Australia. *Renew. Sustain. Energy Rev.* 61, 302–318. https://doi.org/10.1016/j.rser.2016.04.013

Barba, J., Lapuerta, M., Cardeño, F., Hernández, J.J., 2020. Are cold filter plugging point and cloud point reliable enough to prevent cold-start operability problems in vehicles using biodiesel blends? *Proc. Inst. Mech. Eng. Part D J. Automob. Eng.* 234, 2305–2311. https://doi.org/10.1177/0954407020915101

Bhuiya, M.M.K., Rasul, M.G., Khan, M.M.K., Ashwath, N., Azad, A.K., Hazrat, M.A., 2016. Prospects of 2nd generation biodiesel as a sustainable fuel—Part 2: properties, performance and emission characteristics. *Renew. Sustain. Energy Rev.* 55, 1129–1146. https://doi.org/10.1016/j.rser.2015.09.086

Canoira, L., Alcántara, R., Jesús García-Martínez, M., Carrasco, J., 2006. Biodiesel from Jojoba oil-wax: transesterification with methanol and properties as a fuel. *Biomass Bioenergy* 30, 76–81. https://doi.org/10.1016/j.biombioe.2005.07.002

Chandran, D., 2020. Compatibility of diesel engine materials with biodiesel fuel. *Renew. Energy* 147, 89–99. https://doi.org/10.1016/j.renene.2019.08.040

Chen, G., Shan, R., Yan, B., Shi, J., Li, S., Liu, C., 2016. Remarkably enhancing the biodiesel yield from palm oil upon abalone shell-derived CaO catalysts treated by ethanol. *Fuel Process. Technol.* 143, 110–117. https://doi.org/10.1016/j.fuproc.2015.11.017

Chozhavendhan, S., Vijay Pradhap Singh, M., Fransila, B., Praveen Kumar, R., Karthiga Devi, G., 2020. A review on influencing parameters of biodiesel production and purification processes. *Curr. Res. Green Sustain. Chem.* 1–2, 1–6. https://doi.org/10.1016/j.crgsc.2020.04.002

da Cunha, M.E., Krause, L.C., Moraes, M.S.A., Faccini, C.S., Jacques, R.A., Almeida, S.R., Rodrigues, M.R.A., Caramão, E.B., 2009. Beef tallow biodiesel produced in a pilot scale. *Fuel Process. Technol.* 90, 570–575. https://doi.org/10.1016/j.fuproc.2009.01.001

Datta, A., Hossain, A., Roy, S., 2019. An overview on biofuels and their advantages and disadvantages. *Asian J. Chem.* 31, 1851–1858. https://doi.org/10.14233/ajchem.2019.22098

Demirbas, A., Karslioglu, S., 2007. Biodiesel production facilities from vegetable oils and animal fats. Energy sources, Part A Recover. *Util. Environ. Eff.* 29, 133–141. https://doi.org/10.1080/009083190951320

Dhawane, S.H., Kumar, T., Halder, G., 2015. Central composite design approach towards optimization of flamboyant pods derived steam activated carbon for its use as heterogeneous catalyst in transesterification of *Hevea brasiliensis* oil. *Energy Convers. Manag.* 100, 277–287. https://doi.org/10.1016/j.enconman.2015.04.083

Diger Kulaksiz, B., Paluzar, H., 2021. Sunflower oil deodorizer distillate as novel feedstock for biodiesel production and its characterization as a fuel. *Biomass Convers. Biorefinery.* https://doi.org/10.1007/s13399-021-01596-6

Ducoste, J., 2009. Fats, roots, oils, and grease (FROG) in Centralized and decentralized systems. *Water Intelligence* 25, 2006–2009.

Ferrero, G.O., Faba, E.M.S., Eimer, G.A., 2021. Biodiesel production from alternative raw materials using a heterogeneous low ordered biosilicified enzyme as biocatalyst. *Biotechnol. Biofuels* 14, 1–11. https://doi.org/10.1186/s13068-021-01917-x

Goh, B.H.H., Ong, H.C., Cheah, M.Y., Chen, W.H., Yu, K.L., Mahlia, T.M.I., 2019. Sustainability of direct biodiesel synthesis from microalgae biomass: a critical review. *Renew. Sustain. Energy Rev.* 107, 59–74. https://doi.org/10.1016/j.rser.2019.02.012

Gupta, J., Agarwal, M., 2017. Biodiesel production from a mixture of vegetable oils using marble slurry derived heterogeneous catalyst. *Curr Trends Biomedical Eng & Biosci.* 5, 1–5. https://doi.org/10.19080/CTBEB.2017.05.555651

Gupta, J., Agarwal, M., Chaurasia, S.P., Dalai, A.K., 2018. Preparation and characterisation of CaO nanoparticle for biodiesel production from mixture of edible and non-edible oils. *Int. J. Renew. Energy Technol.* 9, 50. https://doi.org/10.1504/ijret.2018.10011067

Gupta, J., Agarwal, M., Dalai, A.K., 2020. An overview on the recent advancements of sustainable heterogeneous catalysts and prominent continuous reactor for biodiesel production. *J. Ind. Eng. Chem.* 8, 58–77. https://doi.org/10.1016/j.jiec.2020.05.012

Gupta, J., Agarwal, M., Dalai, A.K., 2019. Intensified transesterification of mixture of edible and nonedible oils in reverse flow helical coil reactor for biodiesel production. *Renew. Energy* 134, 509–525. https://doi.org/10.1016/j.renene.2018.11.057

Gupta, J., Agarwal, M., Dalai, A.K., 2018. Marble slurry derived hydroxyapatite as heterogeneous catalyst for biodiesel production from soybean oil. *Can. J. Chem. Eng.* 96, 1873–1880. https://doi.org/10.1002/cjce.23167

Gupta, J., Agarwal, M., Dalai, A.K., 2016. Optimization of biodiesel production from mixture of edible and nonedible vegetable oils. *Biocatal. Agric. Biotechnol.* 8, 112–120. https://doi.org/10.1016/j.bcab.2016.08.014

Gupta, J., Agarwal, M., Dohare, R.K., Upadhyaya, S., 2017. A review on process system engineering for biodiesel refineries. *International Journal of Advanced Technology and Engineering Exploration* 4, 42–47.

Henner, D., REN21, 2019. *Global Trends in Renewable Energy Investment* 1–76.

Hong, F.T., Alghamdi, N.M., Bailey, A.S., Khawajah, A., Sarathy, S.M., 2020. Chemical and kinetic insights into fuel lubricity loss of low-sulfur diesel upon the addition of multiple oxygenated compounds. *Tribol. Int.* 152, 106559. https://doi.org/10.1016/j.triboint.2020.106559

Hotti, S.R., Hebbal, O.D., 2015. Biodiesel production and fuel properties from non-edible champaca (*Michelia champaca*) seed oil for use in diesel engine. *J. Therm. Eng.* 1, 330–336. https://doi.org/10.18186/jte.67160

Jalalmanesh, S., Kazemeini, M., Rahmani, M.H., Zehtab Salmasi, M., 2021. Biodiesel production from sunflower oil using K2CO3 impregnated kaolin novel solid base catalyst. *JAOCS, J. Am. Oil Chem. Soc.* 98, 633–642. https://doi.org/10.1002/aocs.12486

Jamail, N., Ishak, M.H., Muhamad, N.A., 2017. Insulation characteristic analysis of coconut oil and palm oil as liquid insulating material. *Int. J. Simul. Syst. Sci. Technol.* 17, 49.1–49.6. https://doi.org/10.5013/IJSSST.a.17.41.49

Jhalani, A., 2019. A comprehensive review on water-emulsified diesel fuel: chemistry, engine performance and exhaust emissions. *Environmental Science and Pollution Research*, 26, 4570–4587.

Karmakar, A., Karmakar, S., Mukherjee, S., 2010. Bioresource technology properties of various plants and animals feedstocks for biodiesel production. *Bioresour. Technol.* 101, 7201–7210. https://doi.org/10.1016/j.biortech.2010.04.079

Kegl, B., 2008. Biodiesel usage at low temperature. *Fuel* 87, 1306–1317. https://doi.org/10.1016/j.fuel.2007.06.023

Knothe, G., 2006. Analyzing biodiesel: standards and other methods. *JAOCS* 83, 823–833.

Knothe, G., 2005. The lubricity of biodiesel. *SAE Tech. Pap.* https://doi.org/10.4271/2005-01-3672

Kumar, N., 2017. Oxidative stability of biodiesel: causes, effects and prevention. *Fuel* 190, 328–350. https://doi.org/10.1016/j.fuel.2016.11.001

Mahlia, T.M.I., Syazmi, Z.A.H.S., Mofijur, M., Abas, A.E.P., Bilad, M.R., Ong, H.C., Silitonga, A.S., 2020. Patent landscape review on biodiesel production: technology updates. *Renew. Sustain. Energy Rev.* 118, 109526. https://doi.org/10.1016/j.rser.2019.109526

Masripan, N.A., Salim, M.A., Omar, G., Mansor, M.R., Hamid, N.A., Syakir, M.I., Dai, F., Mekanikal, K., Teknikal, U., Jaya, H.T., Centre, A.M., Teknikal, U., Jaya, H.T., 2020. Vegetable oil as bio-lubricant and natural additive in lubrication: a review. *Int. J. Nanoelectron. Mater.* 13, 161–176.

Mekhilef, S., Siga, S., Saidur, R., 2011. A review on palm oil biodiesel as a source of renewable fuel. *Renew. Sustain. Energy Rev.* 15, 1937–1949. https://doi.org/10.1016/j.rser.2010.12.012

Meriatna, M., Husin, H., Riza, M., Faisal, M., Jakfar, J., Khairunnisa, K., Syafitri, R., 2021. Biodiesel production from palm oil using banana weevil ash as a solid catalyst. *IOP Conf. Ser. Mater. Sci. Eng.* 1098, 022008. https://doi.org/10.1088/1757-899x/1098/2/022008

Mofijur, M., Siddiki, S.Y.A., Shuvho, M.B.A., Djavanroodi, F., Fattah, I.M.R., Ong, H.C., Chowdhury, M.A., Mahlia, T.M.I., 2020. Effect of nanocatalysts on the transesterification reaction of first, second and third generation biodiesel sources-A mini-review. *Chemosphere.* https://doi.org/10.1016/j.chemosphere.2020.128642

Montero, G., Stoytcheva, M., 2015. *Biodiesel – quality, emissions and by-products.* InTechOpen.

Panwar, N.L., Shrirame, H.Y., Rathore, N.S., Jindal, S., Kurchania, A.K., 2010. Performance evaluation of a diesel engine fueled with methyl ester of castor seed oil. *Appl. Therm. Eng.* 30, 245–249. https://doi.org/10.1016/j.applthermaleng.2009.07.007

Qu, T., Niu, S., Zhang, X., Han, K., Lu, C., 2021. Preparation of calcium modified Zn-Ce/Al2O3 heterogeneous catalyst for biodiesel production through transesterification of palm oil with methanol optimized by response surface methodology. *Fuel* 284, 118986. https://doi.org/10.1016/j.fuel.2020.118986

Rajesh, K., Natarajan, M.P., Devan, P.K., Ponnuvel, S., 2021. Coconut fatty acid distillate as novel feedstock for biodiesel production and its characterization as a fuel for diesel engine. *Renew. Energy* 164, 1424–1435. https://doi.org/10.1016/j.renene.2020.10.082

Ramos, M.J., Fernández, C.M., Casas, A., Rodríguez, L., Pérez, Á., 2009. Influence of fatty acid composition of raw materials on biodiesel properties. *Bioresour. Technol.* 100, 261–268. https://doi.org/10.1016/j.biortech.2008.06.039

Rorrer, J.E., Bell, A.T., Toste, F.D., 2019. Synthesis of biomass-derived ethers for use as fuels and lubricants. *ChemSusChem.* https://doi.org/10.1002/cssc.201900535

Sakthivel, R., Ramesh, K., Purnachandran, R., Shameer, P.M., 2018. A review on the properties, performance and emission aspects of the third generation biodiesels. *Renew. Sustain. Energy Rev.* 82, 2970–2992. https://doi.org/10.1016/j.rser.2017.10.037

Santos, S.M., Nascimento, D.C., Costa, M.C., Neto, A.M.B., Fregolente, L. V., 2020. Flash point prediction: reviewing empirical models for hydrocarbons, petroleum fraction, biodiesel, and blends. *Fuel* 263, 116375. https://doi.org/10.1016/j.fuel.2019.116375

Saydut, A., Erdogan, S., Kafadar, A.B., Kaya, C., Aydin, F., Hamamci, C., 2016. Process optimization for production of biodiesel from hazelnut oil, sunflower oil and their hybrid feedstock. *Fuel* 183, 512–517. https://doi.org/10.1016/j.fuel.2016.06.114

Sbihi, H.M., Nehdi, I.A., Tan, C.P., Al-Resayes, S.I., 2014. Production and characterization of biodiesel from *Camelus dromedarius* (Hachi) fat. *Energy Convers. Manag.* 78, 50–57. https://doi.org/10.1016/j.enconman.2013.10.036

Sendzikiene, E., Makareviciene, V., Janulis, P., 2005. Oxidation stability of biodiesel fuel produced from fatty wastes. *Polish J. Environ. Stud.* 14, 335–339.

Sharma Dugala, N., Singh Goindi, G., Sharma, A., 2021. Evaluation of physicochemical characteristics of Mahua (*Madhuca indica*) and Jatropha (*Jatropha curcas*) dual biodiesel blends with diesel. *J. King Saud Univ.—Eng. Sci.* 33, 424–436. https://doi.org/10.1016/j.jksues.2020.05.006

Singh, D., Sharma, D., Soni, S.L., Sharma, S., Kumar Sharma, P., Jhalani, A., 2020. A review on feedstocks, production processes, and yield for different generations of biodiesel. *Fuel* 262. https://doi.org/10.1016/j.fuel.2019.116553

Singh, D., Sharma, D., Soni, S.L., Sharma, S., Kumari, D., 2019. Chemical compositions, properties, and standards for different generation biodiesels: a review. *Fuel* 253, 60–71. https://doi.org/10.1016/j.fuel.2019.04.174

Syafiuddin, A., Chong, J.H., Yuniarto, A., Hadibarata, T., 2020. The current scenario and challenges of biodiesel production in Asian countries: a review. *Bioresour. Technol. Reports* 12, 100608. https://doi.org/10.1016/j.biteb.2020.100608

Taylor, P., Datta, A., Mandal, B.K., 2014. Use of jatropha biodiesel as a future sustainable fuel use of jatropha biodiesel as a future sustainable fuel, 37–41. https://doi.org/10.1080/23317000.2014.930723

Tulashie, S.K., Kotoka, F., 2020. The potential of castor, palm kernel, and coconut oils as biolubricant base oil via chemical modification and formulation. *Therm. Sci. Eng. Prog.* 16, 981–993. https://doi.org/10.1016/j.tsep.2020.100480

US Department of Energy, 2008. Biodiesel handling and use fuide—Fourth edition. *Natl. Renew. Energy Lab.* 1–56.

Verma, P., Sharma, M.P., Dwivedi, G., 2016. Evaluation and enhancement of cold flow properties of palm oil and its biodiesel. *Energy Reports* 2, 8–13. https://doi.org/10.1016/j.egyr.2015.12.001

Wahid, M.B., 2008. EU's renewable energy directive: possible implications on Malaysian palm oil trade. *Oil Palm Ind. Econ. J.* 6, 1–7.

Wang, H., Tang, H., Wilson, J., Salley, S.O., Ng, K.Y.S., 2008. Total acid number determination of biodiesel and biodiesel blends. *JAOCS, J. Am. Oil Chem. Soc.* 85, 1083–1086. https://doi.org/10.1007/s11746-008-1289-8

3

Biodiesel from Second-Generation Feedstock: Role of Fat, Oil and Grease (FOG) as a Viable Feedstock

Amjad Ali,* Km Abida, and Himmat Singh

CONTENTS

* *Corresponding Author* amjadali@thapar.edu

DOI: 10.1201/9781003120858-3

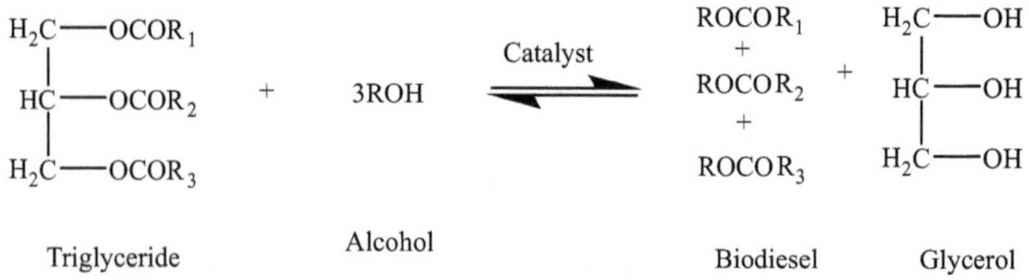

$$\begin{array}{l} H_2C-OCOR_1 \\ HC-OCOR_2 \\ H_2C-OCOR_3 \end{array} + 3ROH \xrightleftharpoons{\text{Catalyst}} \begin{array}{c} ROCOR_1 \\ + \\ ROCOR_2 \\ + \\ ROCOR_3 \end{array} + \begin{array}{l} H_2C-OH \\ HC-OH \\ H_2C-OH \end{array}$$

Triglyceride Alcohol Biodiesel Glycerol

SCHEME 3.1 Reaction showing transesterification of triglycerides with alcohol ($R_{1\text{-}3}$ are various fatty acids, generally having 12–22 C-atoms).

3.1 Introduction

The world is constantly confronting the energy and environmental crisis due to the inevitable and continuing decline in inadequate fossil fuel resources and the increase in environmental pollution due to fuel burning.[1] The combustion of fossil fuel causes the emission of unburnt hydrocarbons, particulate matters, CO, CO_2, SO_3, SO_2, NO_x, etc., which are responsible for environmental pollution as well as global warming.[2] Global warming and the continuous depletion of fossil fuel reserves have encouraged researchers to search for renewable energy resources. In this context, biodiesel (BD) fuel, obtained from renewable sources, has emerged as a promising choice as an eco-friendly, renewable, and sulphurless alternative to fossil-based diesel fuel.[3,4] Moreover, up to 20 vol% blending of BD in fossil fuel-based diesel fuel does not require any modification to conventional diesel engines.[5]

The direct utilization of vegetable oils (VOs) as engine fuels is generally avoided because of their high kinematic viscosity (42 cp), and high atomization temperature as compared to diesel fuel.[6] Thus, VOs need to be transesterified with alcohol (usually ethanol or methanol) to obtain glycerol and BD, a combination of fatty acid methyl or ethyl esters (FAMEs or FAEEs), as shown in Scheme 3.1.

After the completion of the reaction BD, being lower in density than glycerol, forms the upper layer, and hence can be recovered easily.

The BD production processes are usually cleaner as compared to those for conventional fuels. Moreover, CO_2 released during BD combustion is engrossed in the process of photosynthesis, resulting in a carbon-neutral route cycle.[7] Every kilogram of BD burnt can reduce CO_2 production by 3.2 kg.[8] Currently, the major feedstock employed for industrial-scale BD production, across the globe, is edible vegetable oils (VOs). However, in India, the use of waste cooking oil, non-edible oil, and animal fat is encouraged for commercial BD production owing to the scarcity of edible VOs and to avoid the food versus fuel dilemma. Such feedstock is inexpensive and thus, its use for commercial BD production may considerably reduce the cost of production.[9]

3.2 Biodiesel

Rudolph Diesel invented the diesel engine in 1893, and demonstrated the use of peanut oil to fuel the engine. However, the high viscosity (42 cSt) of VOs, in comparison to diesel fuel (< 4 cSt), resulted in poor flow in the engine combustion chamber and inefficient fuel–air mixing.[6] Further, the relatively high molecular weight of VOs causes poor atomization and carbon deposition due to incomplete fuel burning. The discovery of cost-effective and more suitable fossil-based diesel fuel in the early 1900s paused the application of VOs or their derivatives as engine fuel.

The fuel crisis in the 1990s, spiralling crude oil prices, and environmental concerns have reintroduced attention to renewable fuels such as biodiesel. To reduce the viscosity and to improve performance, VOs could be chemically transformed into monoalkyl esters *via* a transesterification reaction. Monoalkyl esters have a lower molecular weight (one-third of VO) as well as viscosity (5–6 cSt) than VOs and possess better flow properties even at low temperatures. E. Duffy and J. Patrick conducted the first transesterification reaction of VOs as early as 1853, several years before the invention of the diesel engine. The term BD was introduced for a fuel consisting of fatty acid alkyl esters in the United States in 1992 by the National Soy Diesel Development Board, currently known as the National Biodiesel Board. The board pioneered the BD commercialization in the United States.[10] The use of biodiesel reduces the engine emissions of unburned hydrocarbons (68%), CO (44%), particulars (40%), SOx (100%), and polycyclic aromatic hydrocarbons (80–90%).[11,12] A BD–diesel fuel blend, having up to 20% (v/v) BD, represented as B20, does not demand any modification in current diesel engines. Some countries nations (e.g., Brazil, the United States, Canada, the UK, Hong Kong, Norway) have made it mandatory to use B5 (5% biodiesel in conventional diesel fuel) in conventional diesel engines. BD characteristics are primarily governed by the fatty acid composition of the triglyceride from which it has been derived. At present, the United States, Europe, Brazil, and Malaysia, are the chief forces in the biodiesel market. Currently, BD production at the industrial level is subjugated by the use of edible Vos, *viz.* palm (6.34 million tons), soybean (7.08 million tons), and rapeseed (6.01 million tons). The prime expense (~75%) of BD production depends upon the feedstock cost employed for transesterification. Furthermore, the BD market is dominated by food-grade edible oils that are costlier than petroleum-based diesel, and consequently, its economic viability is in question in many countries, including India.[13]

3.3 Second-Generation Feedstock for Biodiesel Production

Theoretically, any triglyceride (vegetable oil or animal fat) could be used as feedstock for BD production. The feedstock choice is mainly limited by the cultivation of vegetable oil in a specific geographical area. For instance, soybean oil is the main feedstock in South America and the United States, rapeseed oil in European countries, canola oil in Canada, palm oil in Malaysia and Indonesia, and jatropha oil in India and the African region.[14] Soybean and rapeseed oils are responsible for ~85% of global BD production. Edible VOs constitute the *first-generation* feedstock for the production of biodiesel fuel. However, the application of such VOs is impracticable in many countries, including India, due to their limited accessibility and high price.[15]

Biodiesel usage and production have recently gained significant criticism as it has mainly utilized edible VOs, causing global vegetable oil scarcity and a significant price escalation.[16] Additionally, the cost of BD is not competitive with petroleum diesel fuel due to the high price of feedstock and expensive processing. In this context, the application of *second-generation* feedstock such as non-edible VOs (e.g., castor, jatropha, karanja, microalgae, etc.), waste frying oils, acid oils, and animal fats could address the issues associated with the use of first-generation feedstock.[17–21] Moreover, the former is cheaper than edible oils and could avoid the food versus fuel conflict.[22] Indian BD policy also encourages the use of non-edible VOs such as jatropha and karanja for BD production. Nearly 15 million tons of waste cooking oil are disposed of annually around the globe, which otherwise could also be engaged as a low-cost feedstock for BD generation. According to assessments, the BD production cost could be halved by using waste oils as feedstock. Nonetheless, waste cooking and non-edible oils usually contain a rather higher amount of FFA (2.5–40 wt%) and/or moisture content (0.3–61 wt%).[23,24] The presence of these impurities in feedstock deactivates the homogeneous alkali catalysts *via* saponification.[25] Thus, it is preferred to develop new heterogeneous catalysts to exploit low-quality second-generation feedstock for BD production.

Another triglyceride source, algal oil (also known as third-generation feedstock), has also received noteworthy attention as a BD feedstock.[26,27] Microalgae could be grown at a faster pace on waste land without requiring fresh water and yielding higher oil contents. However, BD produced from algal oil, due to its high saturated fatty acid content, demonstrates poor low-temperature operability.[28]

The fatty acid profile of VOs and animal fats influences the chemical composition as well as physicochemical properties of BD fuel. VOs and fats usually comprise five major fatty acid components: palmitic (16:0), stearic (18:0), oleic (18:1), linoleic (18:2), and linolenic (18:3) acids. Table 3.1 demonstrates the fatty acid composition of VOs/fats frequently employed as feedstock for BD production.[29,30]

3.3.1 Effect of Fatty Acid Composition on BD Fuel Properties

VOs possess a high concentration of unsaturated fatty acids, while animal fats contain a relatively higher concentration of saturated fatty acids. The difference in the feedstock fatty acid composition influences various properties of BD, such as viscosity, oxidation stability, cold flow property, etc.

3.3.1.1 Effect of Unsaturation

Double bonds are more vulnerable towards oxidation and hence, BD with high unsaturated fatty acid contents is degraded/oxidized easily.[31] BD produced from feedstocks that have linoleic and linolenic acids in greater amounts undergo oxidation easily. The oxidation rates of linolenates and linoleates are correspondingly 98 and 41 times higher than those of monounsaturated oleates. However, the cold flow properties improve with unsaturation as BD with a greater number of double bonds solidifies at much lower temperatures. Twists occur in the structure of unsaturated fatty acids and hence, there is less efficient packing of such molecules and they tend to remain in a liquid state. The unsaturation also decreases the viscosity, which in turn increases the lubricity of the resulting product.[32] Another important fuel property, the centane number, in general, increases with saturation.[33]

TABLE 3.1

Properties of Frequently Employed Feedstocks for Biodiesel Production

Feedstock	Country	Percentage fatty acid composition (Cxx:y)										Density	Flash point	Viscosity	Acid value
		C14:0	C16:0	C16:1	C18.0	C18:1	C18:2	C18:3	C20:0	C20:1	C22:0	(g/cm^3)	(°C)	(at 40 °C)	(MJ kg^{-1})
First-generation feedstock															
Cottonseed (*Gossypium* spp.)	Canada, Cambodia	0.8	20.10	0.4	2.60	19.20	55.20	0.60	–	–	–	0.91	234	18.2	NF
Peanut (*Arachis hypogaea*)	China, India	–	8.23	0.3	2.46	58.69	21.77	0.34	1.83	1–2	3.89	0.93	315	–	3.5
Soybean (*Glycine max*)	US, Brazil, China	0.1	11.0	–	4.0	23.40	53.20	7.80	–	–	–	0.91	254	32.9	0.2
Rapeseed (*Brassica napus*)	Germany	–	3.5	–	0.9	64.1	22.3	8.2	–	–	–	0.91	246	35.1	0.92
Sunflower (*Helianthus annuus*)	Brazil	–	–	0.4	4.50	21.10	66.20	–	0.30	2.0	–	0.92	274	32.6	NF
Canola (*Brassica campestris*)	Canada	–	4.0	0.2	2.0	62.0	20.0	9.0	–	–	–	NF	NF	38.2	0.4
Corn (*Zea mays*)	US	0.1	11.67	1.0	1.85	25.16	60.6	0.48	0.24	–	–	0.91	277	34.9	NF
Palm (*Elaeis guineensis*)	Malaysia, Costa Rica	1.1	44.0	0.2	4.50	39.20	10.10	0.41	–	–	–	0.92	267	39.6	0.1
Camelina (*Camelina sativa*)	US	–	5.4	–	2.6	14.3	2.9	38.4	0.25	16.8	1.4	0.91	NF	NF	0.76

(continued)

TABLE 3.1 (Continued)

Properties of the Frequently Employed Feedstocks for Biodiesel Production

		Percentage fatty acid composition (Cxx:y)													
Feedstock	**Country**	**C14:0**	**C16:0**	**C16:1**	**C18.0**	**C18:1**	**C18:2**	**C18:3**	**C20:0**	**C20:1**	**C22:0**	**Density (g/cm³)**	**Flash point (°C)**	**Viscosity (at 40 °C)**	**Acid value (MJ kg⁻¹)**
Second-generation feedstock															
Jatropha (*Jatropha curcas*)	India	0.1	14.20	0.7	7.0	44.70	32.80	0.20	–	–	–	0.92	225	29.4	28
Karanja (*Pongamia pinnata*)	India, Brazil	1.4	3.7–7.9	0.7	2.4–8.9	45.71	11–18.3	0.2	–	–	–	0.92	225	27.82	12
Tallow	US, India, Brazil	3–6	23.30	–	19.40	42.40	2.90	0.90	–	–	–	0.87	150	51.15	–
Poultry	US, India, Russia	–	22.20	–	5.10	42.30	19.30	1.0	–	–	–	0.91	150	59.20	17.9
Used cooking oil	US	Depends upon fresh cooking oil													

Note: Cxx:y = # of carbon atoms: # of double bonds present in the fatty acid, 14:0 – miristic, 16:0 – palmitic acid, 16:1 – palmitoleic, 18:0 – stearic acid, 18:1 – oleic acid, 18:2 – linoleic acid, 18:3 – linolenic acid, 20:0 – arachidic acid, 20:1 – eicosenoic acid, and 22:0 – behenic acid.

3.3.1.2 Effect of Carbon Chain Length

With the increase in fatty acid hydrocarbon chain length, the viscosity of BD increases which, consequently, again influences the fuel lubricity and flow properties of the product.[34] In general, a longer fatty acid chain enhances BD lubricity. High-lubricity BD can be mixed with conventional diesel fuel to improve the overall lubricity of the blend. The degree of unsaturation and chain length together influence the cetane number (CN) of the resulting BD fuel.[33]

3.4 Alcohol Used for Biodiesel Production

Methanol is the most commonly employed alcohol in existing BD production technologies due to its suitable physicochemical properties, low cost, mild reaction conditions requirement, and ease of BD and glycerol separation. However, there are some drawbacks linked to the use of methanol, such as toxicity, explosion risk (due to the low boiling point), and non-renewability, as it is chiefly produced from the refining of crude oil.[35] Over the last few years, researchers have been striving to substitute methanol with non-toxic and renewable alcohol, such as ethanol, for biodiesel production.

Bioethanol (ethanol derived from biomass) is not only a non-toxic and 'green' substrate, but also renewable as it is mainly derived from biomass fermentation. Triglyceride transesterification with ethanol leads to the formation of fatty acid ethyl esters (FAEE) which have better low-temperature operability than the corresponding methyl esters. Nevertheless, presently ethanol is more costly than methanol, less reactive, and FAEEs are difficult to separate from glycerol. Due to its lower reactivity, ethanol is not used in BD production often in the presence of heterogeneous catalysts. Thus it has become essential to develop novel and efficient solid catalysts which can utilize ethanol for biodiesel production.[36] Although ethoxide ions are better nucleophiles than methoxide ions, the longer carbon chain length of the former makes it less mobile as well as less reactive during the reaction.[37] Our group has developed heterogeneous catalysts which were found to be efficient for VO transesterification with ethanol to yield FAEE.[38] The application of bioethanol for FAEE production also increases the production cost by almost 100% as compared to methyl esters.

3.5 Catalysts for Second-Generation Feedstock

VO transesterification mainly exploits two types of catalysts, *viz.*, chemical and enzyme based (Figure 3.1), which further can be categorized as homogeneous and heterogeneous.[39] Due to the exceedingly high price of lipase catalyst, its application in BD production is limited to academic purposes only.[40–42] The Chemical Engineering Division along with the National Collection of Industrial Microorganisms Center, NCL Pune, have developed a bio-catalytic process that utilizes two bacterial strains to convert crude glycerol, a by-product of the BD industry, into commercially valuable compounds – 2,3-butanediol and 1,3-propanediol, along with acetoin and ethanol.[43]

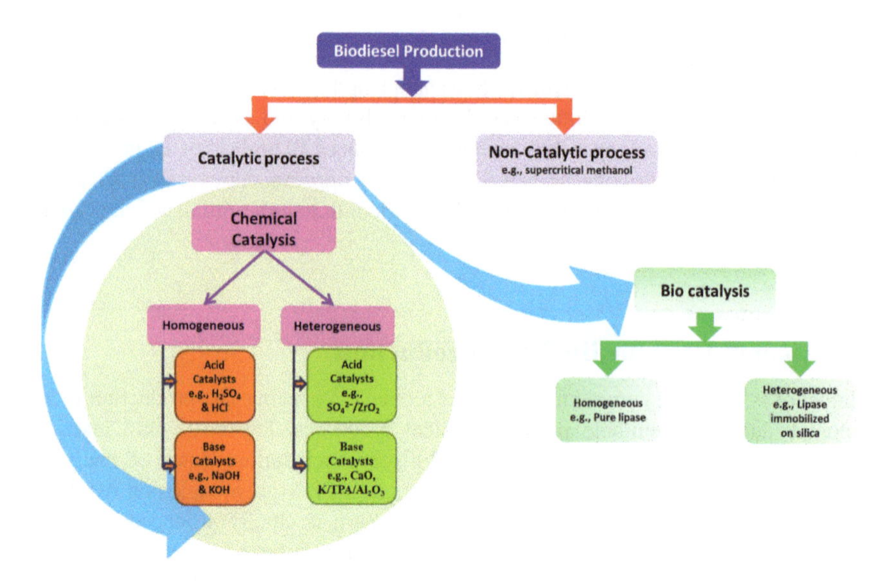

FIGURE 3.1 Different catalysts employed for biodiesel production.

At present, BD manufacturing is dominated by homogeneous acid- or alkali-catalysed transesterification routes due to cost-effectiveness, simple usage, higher activity, and the formation of acceptable conversion levels (> 96.5%) under mild reaction conditions. Heterogeneous catalysts, on the other hand, have the benefits of ease of separation, reusability, and non-contaminated product formation. However, very few industrial plants have utilized them for BD production due to their low activity, which is often supplemented with high temperature and pressure to attain the prescribed FAME yield of > 96.5%.

3.5.1 Homogeneous Catalysts

At an industrial scale, BD is frequently produced in homogeneous alkali (e.g., NaOH, KOH, CH_3Ona, etc.) catalysed reactions as such catalysts are cost effective and generate the product molecules at a high conversion rate under less energy-intensive reaction conditions.[44,45] Nonetheless, homogeneous alkali catalysts demand costly first-generation feedstock for transesterification, consisting of < 0.5 wt% FFA and/or < 0.3 wt% moisture content. At an industrial level, base-catalysed biodiesel production is quite a lengthy procedure that involves multiple steps, as shown in the schematic flow diagram in Figure 3.2[46] the employed alkali catalysts must be neutralized after completion of the reaction and the salt formed during the process needs to be separated from the product by filtration.

The transesterification of feedstock having > 0.5 wt% FFA content (found in second-generation feedstock) is deactivated *via* a saponification reaction, in alkali-catalysed reaction, as indicated in chemical equation (3.1).

$$RCOOH + NaOH \rightarrow RCOONa + H_2O \tag{3.1}$$

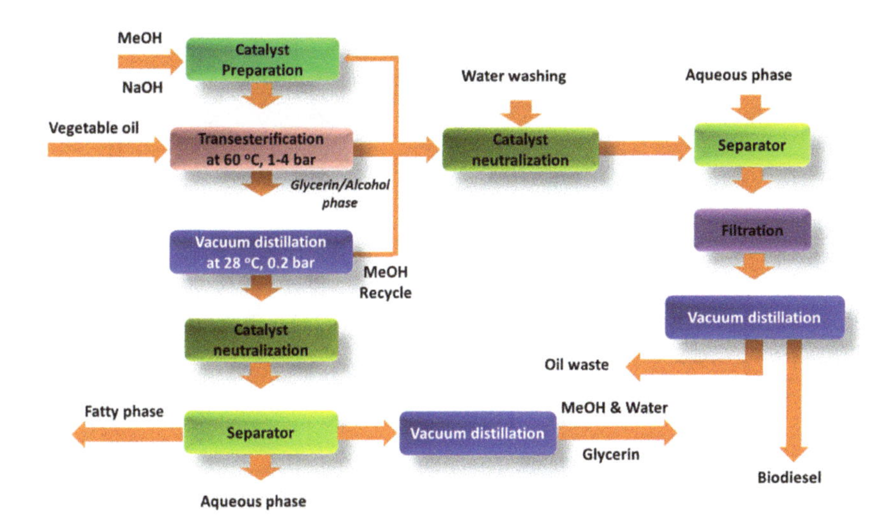

FIGURE 3.2 Block diagram for biodiesel production using alkali catalyst. (Reproduced from reference 46 with permission from Elsevier.)

The presence of water in the reaction mixture could further hydrolyse the fatty esters into fatty acid and methanol as shown in chemical equation (3.2). The fatty acid form in equation (3.2) could again follow equation (3.1) to form more soap.

$$\mathrm{RCOOMe} + \mathrm{H_2O} \rightarrow \mathrm{RCOOH} + \mathrm{MeOH} \tag{3.2}$$

Unwarranted soap formation causes catalyst deactivation, reduced biodiesel yield, and BD emulsification, to make its separation extremely difficult and costly.[47–49]

Homogeneous acid (e.g., H_2SO_4 and HCl) catalysts, on the other hand, could be used for BD production from high-FFA-containing second-generation feedstock as they can catalyse the simultaneous esterification of fatty acid and transesterification of triglycerides, as shown in Figure 3.3.[46] Thus, such catalysts could be useful for the generation of BD from second-generation feedstock with a high FFA content. However, these catalysts are less effective as the rate of acid-catalysed transesterification is ~ 4000 times slower than the alkali-catalysed ones.[39] For this reason, acid catalysts usually required a relatively high temperature (100–120 °C), a high alcohol/oil molar ratio, and a longer reaction duration for completion of the reaction. The presence of water, which is usually found in second-generation feedstock and a by-product of the esterification reaction, was also found to reduce the activity of the acid catalysts to a significant extent. Moreover, acid catalysts are highly corrosive and hence, acid-resistant, complicated, and costly reactors are required to perform the reaction. The acid catalysts, after the reaction, must be neutralized by suitable alkali (e.g., CaO) and the resulting formed salt must be separated from the reaction mixture.

In another approach, second-generation feedstock could be utilized for BD production in a two-step process involving (i) acid-catalysed FFA esterification which follows the (ii) base-catalysed triglyceride transesterification.[50,51] At an industrial

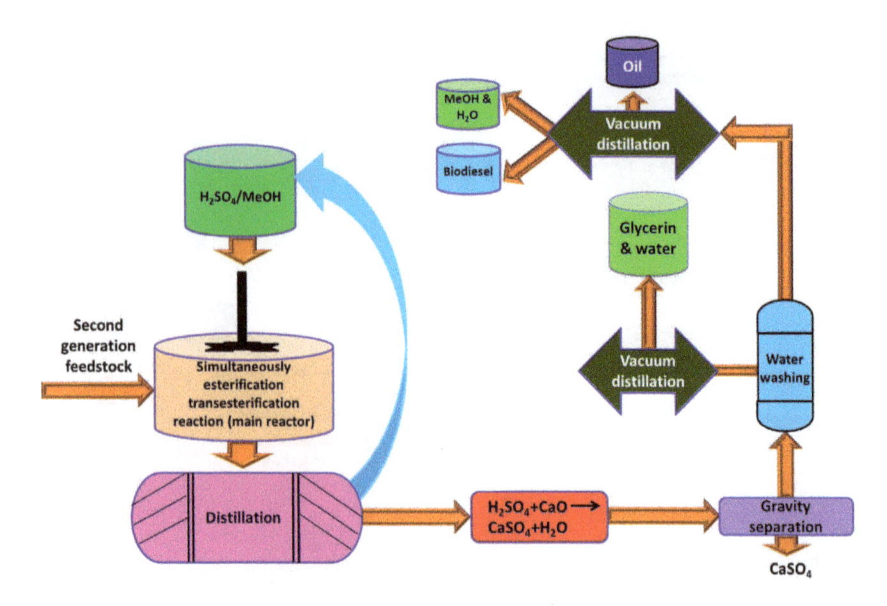

FIGURE 3.3 Process flow diagram for the production of biodiesel via acid-catalysed transesterification of second-generation feedstock. (Reproduced from Reference [46] with permission from Elsevier.)

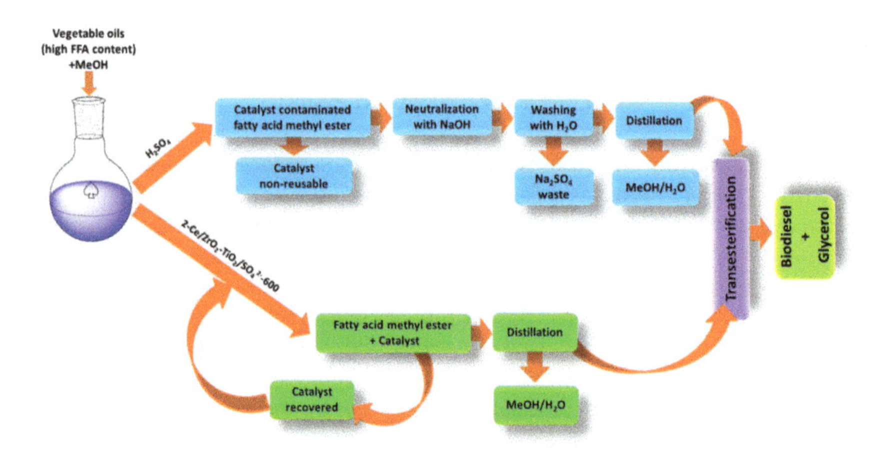

FIGURE 3.4 Comparison of the solid acid-catalysed esterification with that of homogeneous acid-catalysed one.

scale, Shirke Energy utilizes second-generation feedstock (jatropha oil and animal tallow) while employing conventional NaOH and KOH catalysts for methanolysis.[52]

In another approach, solid acid catalysts, $Ce/ZrO_2–TiO_2/SO_4^{2-}$, were employed[50] for the esterification of fatty acids present in second-generation feedstock followed by alkali-catalysed transesterification of the Vos, as shown in Figure 3.4. This procedure is advantageous over the process involving homogeneous acid catalyst as it does not require acid neutralization or a washing step. Esterified oil, in the second step, could be directly employed for homogeneous alkali-catalysed transesterification to obtain BD. Moreover, a solid acid catalyst could be recovered and reused in the next cycle for fatty acid esterification.

The major disadvantages of homogeneous (alkali or acidic) catalysis include the mandatory catalyst neutralization, removal, and purification of BD and glycerol, formed during the reaction. Glycerol, a valuable by-product, cannot be utilized for edible purposes due to the presence of catalysts, alcohol, and glycerides as impurities. Further, BD washing is essential to remove the partially dissolved salts to maintain the total metal content in BD to < 5 ppm. BD purification steps are not only cumbersome and time-consuming but also generate a huge quantity of industrial effluents that are often difficult to dispose of.

3.5.2 Heterogeneous Acid Catalysts

To address the issues associated with homogeneous catalysts, a varied range of solid acid catalysts has been utilized for VO transesterification. Solid acid catalysts are usually less reactive than their alkali equivalents and hence, usually demand high temperature, high pressure, and a higher alcohol-to-oil molar ratio to achieve adequate conversion levels. Despite their poor reactivity, the use of solid acids has several advantages over homogeneous counterparts, such as they (i) are less sensitive to FFA present in the feedstock, (ii) are capable of catalysing simultaneous esterification and transesterification, (iii) can eliminate the need for catalyst neutralization and BD washing, (iv) can be separated easily from the reaction medium, (v) are reusable, and (vi) can reduce reactor corrosion, which simplifies the reactor design and reduces its construction cost.[45]

An ideal solid acid (having Bronsted and/or Lewis centres) catalyst, to be employed for second-generation feedstock, must be able to catalyse simultaneous esterification and transesterification of fatty acids and triglycerides, respectively.[53] A variety of heterogeneous acid catalysts for biodiesel production has been reported in the literature,[54] such as sulphated oxides (SO_4^{2-}/ZrO_2, SO_4^{2-}/SnO_2, SO_4^{2-}/SiO_2, and SO_4^{2-}/TiO_2), native ($H_3PW_{12}O_{40}{\cdot}24H_2O$ and $H_4SiMo_{12}O_{40}{\cdot}13H_2O$) and supported heteropolyacids over SnO_2, SiO_2, TiO_2, Al_2O_3, and ZrO_2, zeolites (H–ZSM-5, Y, and Beta), ion exchange resins with sulphonic acid group (Nafion and Amberlyst), sulphated mesoporous silica (SO_3H-SBA-15, ZnO/SBA-15, and SO_3H-KIT-6), sulphated carbon (derived from D-glucose and cellulose), and diverse solid acid catalysts, such as $Fe(HSO_4)_3$ and WO_3/SnO_2.[53,55,56] A solid acid-catalysed commercial-scale BD production was established by Benefuel's Ense in association with NCL Pune, applying double-metal cyanide of Fe-Zn as a solid acid catalyst. The catalyst was able to perform simultaneous esterification and transesterification of a variety of feedstocks, including second-generation feedstocks (*viz.* yellow grease, palm fatty acid distillate, crude palm oil, and even a blend of degummed soybean oil).[43]

3.5.2.1 Sulphated Metal Oxides

Sulphuric and sulphonic acids have shown excellent activity towards esterification and transesterification as homogeneous catalysts. Hence, to prepare the solid acid

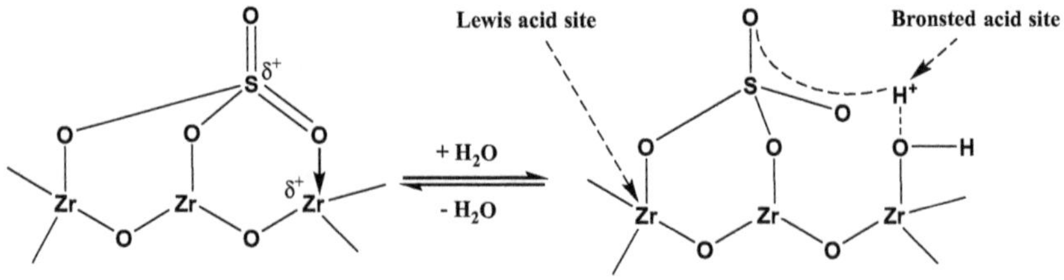

FIGURE 3.5 Illustration of Lewis and Bronsted acidic sites over SO_4^{2-}/ZrO_2 catalyst. (Reproduced from reference 61 with permission from the Royal Society of Chemistry.)

catalysts, sulphate or sulphonic acid groups were immobilized over a variety of matrices consisting of metal oxides. Sulphated metal oxides such as phenyl and propyl sulphonic acid-supported SBA-15, ZrO_2, SnO_2, SiO_2, and TiO_2 are frequently used as solid super acid catalysts in the literature, due to their capability in catalysing the esterification and transesterification of VOs with high FFA concentrations.[57–59] Metal oxides such as ZrO_2, SnO_2, SiO_2, and TiO_2 possess both Bronsted and Lewis acid sites as they can simultaneously protonate as well as coordinate to the pyridine molecule.[60] The surface modification of these oxides with sulphate groups can enhance the Bronsted acid sites which in turn resulted in improved catalytic activity. As shown in Figure 3.5, in a typical SO_4^{2-}/ZrO_2 unit, three oxygen atoms from S–O and S=O are bound to Zr atoms, while the fourth oxygen atom remains bound with S as an S=O bond. The S=O–Zr linkage could easily undergo reversible protonation by a water molecule to form an acidic proton in the Zr–OH group. Thus the addition of a small amount of water to the catalyst generates more Bronsted acid sites and may further enhance its catalytic activity.[61]

Owing to the presence of structural defects, mixed metal oxides are found to be a superior matrix for catalyst preparation, followed by distinct metal oxides. It is not known which of the metal ions acts as an active centre in mixed metal oxides.[62] Sulphate-functionalized ZrO_2-TiO_2/La^{3+} was also utilized as an efficient catalyst for simultaneous esterification and transesterification of VOs containing up to 60 wt% FFA. The role of La^{3+} was to fortify the interaction of SO_4^{2-} with ZrO_2-TiO_2 matrix to decrease the loss of SO_4^{2-} in the reaction media. Lanthanum, while improving the catalyst stability as a secondary component, remains silent during the catalytic procedure. The catalyst was recycled over five cycles without requiring any regeneration process and maintained a 90.2% FAME yield after the fifth cycle.[63,64] Some other sulphated mixed oxides employed for BD production, along with the reaction conditions, are described in Table 3.2.

3.5.2.2 Mesoporous Silica

Mesoporous SiO_2, due to its insolubility in organic media, tuneable pore size, and ease of surface modification, has proven to be a versatile support for the preparation of a variety of industrial catalysts.[65] To develop solid catalysts for biodiesel production, sulphonic acid, phenyl, and propyl sulphonic acid groups have been incorporated over various categories of silica supports, *viz.*, MCM-41, SBA-15,

TABLE 3.2

Activity of a Few Literature-Reported Heterogeneous Acidic Catalysts

		Reaction conditions						
	Feedstock	Catalyst amount (wt%)	Alcohol to oil molar ratio (m/m)	Reaction temp. (°C)	Reaction time (h)	FAAE Yield (%)	Reusability	References
Methanolysis								
PSH/UiO-66NO_2	Jatropha	4.0	25:1	70	4.0	97.57	5	[88]
K/TPA/Al_2O_3	Waste oil	10.0	9:1	65	1.25	>97.0	4	[76]
([Ps-im]HSO_4)/ SO_4^{2-}/ZrO_2-SiO_2	Ricebran oil	5.0	18:1	150	3.0	99.0	5	[89]
Mesoporous W/Ti/SiO_2	Waste cottonseed	5.0	30:1	65	4.0	98.0	4	[90]
Mesoporous sulphated zirconia	Soyabean oil	4.0	12:1	120	4.0	94.9	NR	[91]
Mesoporous polymer solid acid	Jatropha	6.0	50:1	160	8.0	94.0	4	[92]
Mesoporous sulphonated carbon	Acidic oils (21–41 wt% FFA)	16.6	20:1	120	NR	91.0	NR	[55]
Al-SBA-15	Jatropha	NR	12:1	180	24.0	99.0	NR	[67]
Ti/SBA-15	Jatropha	15.0	27:1	200	3.0	90.0	3	[54]
SO_4^{2-}/ZT/La^{3+}	Acid oil	5	15:1	200	2	96.24	5	[64]
SO_4^{2-}/SnO_2-TiO_2	Waste cooking	6	15:1	150	1.5	88.2	NR	[93]
Ethanolysis								
Amberlyst A26	Jatropha	15.0	35:1	55	20.0	36.31	NR	[94]

etc.[66,67] The pore size of mesoporous material can be tuned to reduce the internal mass transfer limitations, by employing templates such as trimethyl-, triethyl-, or triisopropyl-benzene as pore swelling agents.[68,69] To further enhance the catalyst stability/activity instead of pure silica, Ti- or Al-doped silica were used for the impregnation of active species. For example, Ti-doped SBA-15 was employed for the simultaneous esterification and transesterification of high FFA-containing VOs with methanol.[70] The catalyst was found to be efficient even in the presence of impurities that are frequently present in second-generation feedstock, *viz.*, up to 5 wt% water or 30 wt% FFA contents. Superior activity can be achieved in the case of mesoporous-based catalysts by incorporating more acidic sites, adjusting the pore diameter (5–20 nm) appropriately, and modifying the surface hydrophilic and hydrophobic characteristics. High-temperature requirements to achieve the acceptable BD yield and sulphate group leaching in the reaction medium are the main drawbacks of such catalysts.

3.5.2.3 Heteropolyacids

Heteropolyacids (HPAs), namely Keggin-type $H_3PMo_{12}O_{40}$, $H_3PW_{12}O_{40}$, $H_4SiW_{12}O_{40}$, and $H_4SiMo_{12}O_{40}$, are an additional category of acid catalysts that possess super acid sites ($pK_H > 12$) and flexible structure. However, in their natural form, they are not suitable as heterogeneous catalysts for BD production due to their high solubility in polar media.[71,72] To decrease the solubility of these acids during the reaction, they are often impregnated over a matrix with a high surface area. Proton exchange with heavier cations, such as Cs^+, into Keggin-type $Cs_xH_{(3-x)}PW_{12}O_{40}$ and $Cs_yH_{(4-y)}SiW_{12}O_{40}$ can improve their stability. Although acidic proton was lost, both acids were found to be active towards palmitic acid esterification and tributyrin transes terification.[73,74] To make the catalyst separation easier, Duan *et al.*,[75] reported the immobilization of $H_3PW_{12}O_{40}$ over Fe_2O_3 to prepare a magnetic catalyst for the esterification of palmitic acid under mild conditions. The water-resistance ability of the catalyst was further improved by incorporating nonyl chains over the catalyst surface. In another approach, to perform the simultaneous transesterification and esterification of WO, K^+ and $H_3PW_{12}O_{40}$ were simultaneously loaded over alumina. Under the optimized conditions for the reaction, a 97% FAME yield was obtained within 1.25 h of the reaction time. The catalyst activity was not adversely affected due to the presence of 4 wt% moisture and 7.8 wt% FFA contents in the feedstock.[76,77]

3.5.2.4 Miscellaneous Solid Acids

A range of additional solid acid catalysts have also been examined for BD production such as sulphonated carbonaceous material (carbon is mainly derived from biomass), supported tungsten oxides (WO_3/SnO_2,WO_3/ZrO_2), ferric hydrogen sulphate [$Fe(HSO_4)_3$], ferric zinc metal cyanide, and bifunctional catalysts, such as Mo-Mn/Al_2O_3-15 wt% MgO, ZnO-La_2O_3 having the advantages of acidic well as basic sites.[78,79–87] A few heterogeneous acidic catalysts used for the simultaneous transesterification and esterification of high-FFA-containing vegetable oils are compared in Table 3.2.

3.5.3 Heterogeneous Base Catalysts

Heterogeneous base catalysts are usually more active than their acidic counterparts, and hence, have been found to be suitable, in the literature, for the transesterification of first- as well as second-generation feedstocks.[30] The advantages of the solid base-catalysed reaction include: (i) relatively faster reaction rate than acid-catalysed transesterification, (ii) accomplishment of standard FAME yield even under mild reaction conditions, (iii) less energy-intensive process, and (iv) easy separation from the product and hence, a high possibility of catalyst reuse and regeneration.[46]

The Esterfip process, developed by the French Institute of Petroleum, utilizes Zn/Al mixed-oxide as a solid catalyst for industrial-scale BD production but requires relatively higher pressure (3–5 MPa) and temperature (483–523 K) compared to a conventional homogeneous catalysed process.[55] This process yielded high-purity BD and glycerol and hence does not require catalyst neutralization and intensive BD and glycerol purification steps.[95] The process targets making industrial BD synthesis cost-effective by reducing the crude glycerol purification cost. A variety of other solid base catalysts, for the transesterification reaction, are reported in the literature, *viz.*, alkali or alkaline earth oxides, alkali-doped mixed oxides, transition metal oxides, alkaline zeolites and clays such as hydrotalcite, and immobilized organic bases.[96,97]

3.5.3.1 Alkaline Earth Oxides

Alkaline earth oxides owing to their cost-effectiveness and ease of preparation were commonly used as solid catalysts for the transesterification of a diversity of feedstock. In such catalysts, oxide ions (O^{2-}) in metal-oxide (M–O) pairs, are responsible for the evolution of Lewis basic sites which were found to be more active towards transesterification than the corresponding Bronsted sites.[98] The strength of these sites could further be enhanced by generating coordination defects at corners and edges, or on high Miller index surfaces.[31] The order of activity for the alkaline earth oxides, employed for VO transesterification, was found to be: BaO > SrO > CaO > MgO.[99] Usually, single metal oxides either demonstrate low catalytic activity and/or get partially dissolved into the reaction media causing a significant homogeneous contribution in catalytic activity. The stability and reactivity of these oxides could be improved by preparing mixed oxides of alkali earth metals. The active centres in alkaline earth mixed oxides are either oxide ion or impregnated metal oxide or the defect created due to metal impregnation. For example, a nanocomposite of Sr/Al was prepared as $Sr_3Al_3O_6$ and employed as a heterogeneous catalyst for soybean oil transesterification to produce 95.7% FAME yield in 1 h of reaction duration.[100] Calcium oxide is non-toxic, easily available, and cost-effective, and hence, frequently studied in the literature either as a catalyst or matrix to impregnate other active metals while preparing the solid catalyst for the triglyceride transesterification. Furthermore, CaO can also be prepared from bio-waste consisting of $CaCO_3$, such as mollusc shells, egg shells, etc.[101] The native group–II metal oxides have been primarily employed for the transesterification of first-generation feedstock. To make them suitable for the transesterification of second-generation feedstock CaO or MgO have been impregnated with many transition metals. For example, CaO-ZrO_2 or CaO-La_2O_3 mixed oxides have been prepared as stable catalysts due to the formation of a homogeneous solid solution. Both these catalysts were found to be effective for WCO transesterification to produce BD.[84,87,100,102] An increase in Ca/Zr ratio was found to enhance the catalyst activity, however, with a compromise with catalyst stability as Ca ions were found to leach out from the catalyst surface. Teo *et al.*[9] employed CaO-NiO and CaO-Nd_2O_3 mixed oxides as heterogeneous catalysts for jatropha oil transesterification claiming a > 80% FAME yield. In a CaO-Nd_2O_3 binary system, a strong interaction between CaO and Nd_2O_3 was observed due to the transfer of electrons from Nd_2O_3 to CaO. Active sites of high basic strength formed in both catalysts due to the existence of an unbounded oxide anion. Table 3.3 illustrates the activity of a few alkaline earth oxide-based catalysts used for a variety of triglyceride transesterifications.

3.5.3.2 Alkali-Doped Metal Oxides

In another approach, alkali metals were doped in CaO, MgO, ZrO, etc. (Table 3.3) to improve the basicity as well as activity by creating the defects. Replacing alkali metal (M^+) for alkaline earth/transition metal (M^{n+}) can generate the defects in the regular structure in the form of O^-.[103] Kumar and Ali[104] prepared M/CaO (M = Li, Na, K) for the transesterification of WCO and found a better performance with Li/CaO to achieve a > 98% FAME yield in 45 min. The same catalyst was also found to be effective for VO transesterification, even containing 15 wt% water content. Alkali metal leaching was established as a major cause of the loss of catalytic activity and poor reusability. To improve the catalyst stability, Wang *et al.*[105] immobilized Li over silica to form Li_4SiO_4 catalyst for the production of biodiesel from soybean oil. The catalyst was found to be less basic than CaO but more stable as well as more tolerant towards air, moisture, and carbon dioxide. Similarly, K-impregnated SiO_2 was also employed as a heterogeneous catalyst for the transesterification of karanja (11.3 wt% FFA) and jatropha oil (5.6 wt% FFA) to obtain a > 97% FMAEs yield.[106]

Sodium silicate was also found to be an active catalyst for BD production from rapeseed and jatropha oils under conventional and microwave heating, giving a 98% FAME yield.[107] The loss of catalytic activity, during reusability experiments, was attributed to the Si–O–Si bond cleavage and sodium leaching on exposure to the moisture. In another approach, Li-, Na-, and K-doped ZrO_2 catalysts (Li/ZrO_2, Na/ZrO_2, and K/ZrO_2) were prepared by a wet chemical route for the simultaneous esterification and transesterification of second-generation feedstock (waste cooking oil) with methanol and ethanol. The kinetic parameters for VO transesterification were evaluated following the pseudo-first-order kinetic model. The Li/ZrO_2 catalyst was found to be more efficient towards methanolysis (E_a = 40.8 kJ mol^{-1}) than ethanolysis (E_a = 43.1 kJ mol^{-1}). This reusability study suggests that catalyst could be applied during nine reaction cycles without a significant drop in performance as a > 90% FAME yield was maintained.[38]

3.5.3.3 Transition Metal Oxides

Transition metal oxides (TMOs) are employed for catalysing a wide range of chemical reactions, even at the industrial level, due to their lower cost, ease of regeneration, and selective mode of action. Their catalytic activity is credited to the presence of partly filled *d*-orbitals which may further split under the effect of the O^{2-} ligand field.[62] TMOs could

TABLE 3.3

Comparison of Activity of a Few Alkaline Heterogeneous Catalysts

		Reaction conditions						
Catalyst	**Feedstock**	**Catalyst amount (wt%)**	**Alcohol to oil molar ratio (m/m)**	**Reaction temp. (°C)**	**Reaction time (h)**	**FAAE Yield (%)**	**Reusability**	**References**
Methanolysis								
7-SrO/CaO-H	Jatropha	4.77	27.6:1	65	1.50	99.71	5	[120]
20-CeO_2/Li/SBA-15	Waste cottonseed	10.0	40:1	65.0	4.0	98.0	5	[121]
Mn/ZnO	Mahua oil	8.0	7:1	50	0.83	97.0	5	[122]
K/animal bones	Jatropha	6.0	9:1	70	3.0	96.0	4	[123]
Mesoporous polymer solid acid	Jatropha	6.0	50:1	160	8.0	94.0	4	[93]
B. cepacia/ Fe_3O_4/lipase on mesoporous silica	Waste cooking	6.0	25:1	35	35.0	91.0	5	[124]
Mesoporous calcium titanate	Waste cooking oil	0.2	3:1	65	1.0	80.0	5	[125]
CaO-LaO	Jatropha	1.0	12:1	240	0.17	93.0	NR	[126]
Zn/CaO	Used cottonseed	5.0	9:1	65	0.75	98	7	[127]
CaO-La_2O_3	Jatropha	3.0	25:1	160	3.0	98.0	NR	[128]
Li_2ZrO_3	Waste oil	5.0	12:1	65	1.25	99.0	9	[38]
Li/ZrO_2	Waste cottonseed	5.0	12:1	65	1.25	99.0	9	[38]
CaO-Nd_2O_3, CaO-NiO	Jatropha	5	15:1	65	6.0	80,85	4	[9]
K_2SiO_3/AlSBA-15	Jatropha	3.0	9:1	65.0	0.5	95.0	4	[129]
KF/CaO-NiO	Waste cottonseed	5.0	15:1	65	4.0	>99.0	4	[130]
K/La-Mg	Used cottonseed	5.0	54:1	65	0.33	96.0	3	[131]
Zn/CaO	Waste cooking	5.0	9:1	65	0.75	99.0	3	[132]
Li/MgO	Mutton fat	5.0	12:1	65	0.66	>99.0	NR	[133]
K/CaO	Waste cottonseed	7.5	12:1	65	1.25	98.0	3	[134]

(continued)

TABLE 3.3 (Continued)

Comparison of Activity of a Few Alkaline Heterogeneous Catalysts

		Reaction conditions						
Catalyst	Feedstock	Catalyst amount (wt%)	Alcohol to oil molar ratio (m/m)	Reaction temp. (°C)	Reaction time (h)	FAAE Yield (%)	Reusability	References
$Cs\text{-}H_3PW_{12}O_{40}$	Used VO	3.0	40:1	260	0.66	92.0	NR	[135]
$CaO\text{-}ZrO_2$	Waste cooking	10.0	30:1	65	2.0	92.1	NR	[102]
MgO	Jatropha	4 .0	15:1	65	6.0	10 .0	NR	[136]
Sr/ZrO_2	Waste cooking	2.7	6:1	115.5	2.8	79.7	NR	[112]
CaMgO, CaZnO	Jatropha	4.0	15:1	65	6.0	83,81	6	[136]
$TiO_2\text{-}MgO$	Waste cooking	5.0	30:1	150	6.0	92.3	4	[137]
Li/CaO	Used cottonseed	5.0	12:1	65	0.75	>98.0	NR	[138]
$CaO\text{-}La_2O_3$	Waste oil	5.0	20:1	58	1.0	94.3	NR	[149]
Na/SiO_2	Jatropha	65.0	15:1	65	0.75	99.0	3	[140]
Ethanolysis								
Li/ZrO_2	Waste cottonseed	5	15:1	75	2.5	98	NR	[51]
Sr:Zr	Waste cottonseed	5	12:1	75	7	>99	4 (98)	[50]
Li/NiO	Waste cottonseed	5	12:1	65	3	98	7	[141]
Zr/CaO	Jatropha	5	21:1	75	7	> 99	N.R.	[142]
Li/CaO	Waste cottonseed	5	12:1	65	2.5	98	4	[143]

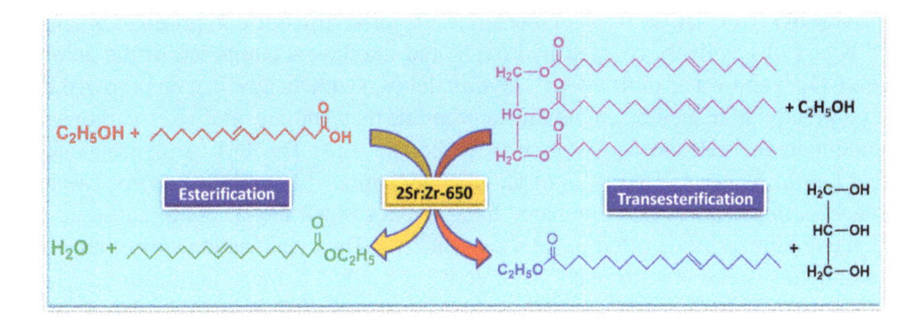

FIGURE 3.6 Simultaneous transesterification and esterification of waste cooking oil with ethanol in the presence of Sr/Zr catalyst.

demonstrate higher activity towards transesterification than solid acids, thus a variety of such catalysts (e.g., MnO, MoO_3, TiO_2, and ZrO_2) of varying Lewis base character have been utilized in biodiesel production from first- and second-generation feedstock.[108–110] Table 3.3 illustrates a few transition metal oxide-based heterogeneous catalysts employed for the transesterification of second-generation feedstock to produce BD. Among them, MnO and TiO_2 are soft bases, which have demonstrated simultaneous transesterification and esterification of VOs, having up to 15 wt% FFA contents. These catalysts, even in the presence of FFAs, do not form soap, which is the major drawback of the homogeneous alkali catalysts.[108] On the other hand, alkali and alkaline earth metal-impregnated zirconia, owing to the presence of acidic and basic properties, was successfully engaged for catalysing simultaneous esterification and transesterification of high fatty acid second-generation feedstock.[111] The interaction of zirconia with impregnated metal resulted in a distinct interaction to yield a catalyst with enhanced activity and selectivity. A series of alkaline earth-impregnated zirconia catalysts (Mg/ZrO_2, Ca/ZrO_2, Sr/ZrO_2, and Ba/ZrO_2) were prepared for biodiesel production by employing waste cooking oil as feedstock.[112] Among the catalysts, Sr/ZrO_2 was found to demonstrate better catalytic activity owing to the presence of more basic and acidic sites to facilitate concurrent transesterification and esterification, respectively. However, the FAMEs yield was found to be less than the acceptable limit of 96.5%, even at a higher reaction temperature of 115.5 °C. To improve the BD yield, Kaur and Ali[51] prepared a series of mixed oxides of Zr/M (M = Mg, Ca, Ba, and Sr) by the co-precipitation method and employed them as solid catalysts for the one-pot transesterification and esterification of WCO with up to 18% FFA and 2% moisture contents (Figure 3.6). The catalyst was effective even in utilizing ethanol as alcohol to produce fatty acid ethyl esters (FAEE). The catalyst was amenable to recovery from the reaction media by simple filtration and was reused over four runs without losing its activity to any significant extent. The catalyst was found to be stable as minimal metal leaching from the catalyst was observed in the reaction media.

3.5.3.4 Hydrotalcites

Hydrotalcites are naturally occurring layered double hydroxides (LDHs) with a general formula of $([M^{2+}_{1-x}M^{3+}_{x}(OH)_2]^{x+} (A^{n-})_{x/n} \cdot mH_2O)$, where $M^{2+} = Mg^{2+}$, Zn^{2+}, or Ni^{2+}; $M^{3+} = Al^{3+}$, Fe^{3+}, or Cr^{3+}; $A^{n-} = OH^-$, Cl^-, NO_3^-, CO_3^{2-}, or SO_4^{2-}; and $x \sim 0.1–0.5$ (e.g.,

$Mg_6Al_2(OH)_{16}CO_3 4H_2O$). They demonstrate a layered structure comprised of brucite-like layers of positively charged hydroxide and interlayers composed of the anions (e.g., CO_3^{2-}) and water molecules.[113] Hydrotalcites, synthetically, can be prepared by the co-precipitation method employing appropriate metal nitrates and alkali carbonates to maintain the pH and as a source of carbonate ions. The OH^- in hydrotalcites, according to literature reports, is of an alkaline nature.[114] Physically adsorbed water molecules in the pores of as-prepared hydrotalcites block the access of substrate to the active site and thus often make them inactive. Thus, calcination of as-prepared hydrotalcites is essential to make them active towards VO transesterification.[115] For instance, Zn-Al hydrotalcite was prepared by the co-precipitation method at 8.5 pH using Na_2CO_3 and NaOH as the precipitating agents.[116] Calcination of the prepared hydrotalcite at 400 °C incorporated the Al^{3+} into the lattice of ZnO to create the cationic vacancies. Surface Zn^{2+} compensated the vacancies to form M^{n+}–O^{2-} (M = Zn or Al) pairs along with isolated oxide ions. On the other hand, Mg-Al hydrotalcites have been found to be effective for the transesterification of second-generation feedstock, having 9.5 wt% FFA and 45 wt% moisture contents to obtain 99% conversion levels within 3 h at 200 °C.[117] Hydroxides or carbonates of Na or K were engaged during hydrotalcite preparation as a precipitating agent and their removal from the catalyst is a challenging task. The presence of Na or K ions always causes homogeneous involvement in catalytic activity.[118] To overcome this issue, NH_4OH or $(NH_4)_2CO_3$ could be employed as a precipitating agent during the preparation.[119]

Thus a variety of inorganic solid acid and alkali catalysts have been prepared in the literature, and employed for the transesterification/esterification of second-generation feedstock to produce BD.[30] However, in comparison to the virgin oil, limited reports are available for the application of solid catalysts for simultaneous transesterification and esterification of second-generation feedstock with high FFA and moisture contents. A few commonly occurring issues with the solid catalysts and their probable solutions are provided in Table 3.4.

3.6 Properties of the Biodiesel Produced from Second-Generation Feedstock

Most of the BD physicochemical properties, *viz.*, viscosity, freezing point, cold flow properties, CN, pour point, etc., are a function of the carbon chain length of fatty acid ester and the extent of unsaturation present in them. Hence, profiling of BD fatty acid ester is an important step while determining its physicochemical properties. Analysis of the BD sample by GC-MS technique not only quantifies the BD yield but also provides information on the percentage of individual fatty acid ester present. A typical gas chromatogram of the FAEE produced from waste cottonseed oil is provided in Figure 3.7. The presence of seven distinct peaks indicates the presence of seven different fatty acids esters in the sample. The area under each peak is directly proportional to the amount of fatty acid esters present in the mixture. With the help of MS analysis, these peaks have been correlated with the respective fatty acid esters.

The fatty acid composition of the BD sample investigated, based on GC-MS analysis, is provided in Table 3.5. The total FAEE contents in BD were found to be 99.61 wt%, consisting of total saturated and unsaturated fatty acid alkyl ester of 31.32 and

TABLE 3.4

Application of Solid Catalysts for Biodiesel Production: Issues and Resolutions

S. no.	Issues	Causes	Probable solutions
1.	High reaction temperature and pressure requirement	Weaker reactive catalytic sites or their partial deactivation	Catalyst active sites must be replaced with strong acidic or basic sites
2.	Mass transfer	Creation of three immiscible phases, *viz.*, oil, alcohol, and catalyst	Application of co-solvent (THF, n-hexane, DMSO) in which both, oil and alcohol, form a single phase and preparation of catalysts of high surface area
3.	Lesser catalytic activity	The difference in catalyst and substrate phase	Application of high surface area and pore size matrix for active site immobilization in high density
4.	Catalyst stability and its dissolution in the reaction media	Loosely bound active–active sites, and their higher solubility in alcohol or glycerol or BD	Application of novel and better catalyst synthesis methodologies to make the stronger interactions between the matrix and active sites
5.	Catalyst deactivation	Interaction of catalyst active sites with impurities/substrate/product	Catalyst washing with a suitable solvent and/or re-calcination at high temperature

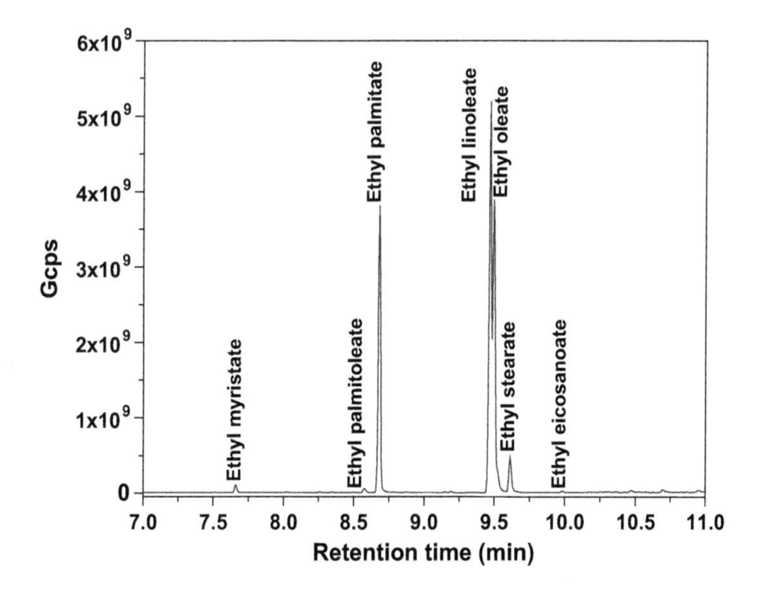

FIGURE 3.7 Gas chromatogram of biodiesel containing various fatty acid ethyl esters.

TABLE 3.5

Analysis of Fatty Acid Ethyl Esters Produced from Waste Cooking (Cottonseed) Oil

S. no.	Retention time (min)	Composition; Molecular formula	Corresponding acid	wt%
1	7.66	Myristic acid ethyl ester; $C_{16}H_{32}O_2$	Myristic acid (C14:0)	0.86
2	8.57	Palmitoleic acid ethyl ester; $C_{18}H_{34}O_2$	Palmitoleic acid (C16:1)	0.62
3	8.68	Palmitic acid ethyl ester; $C_{18}H_{36}O_2$	Palmitic acid (C16:0)	26.64
4	9.46	Linoleic acid ethyl ester; $C_{20}H_{36}O_2$	Linoleic acid (C18:2)	37.77
5	9.49	Oleic acid ethyl ester; $C_{20}H_{38}O_2$	Oleic acid (C18:1)	29.90
6	9.61	Stearic acid ethyl ester; $C_{20}H_{40}O_2$	Stearic acid (C18:0)	3.44
7	9.90	Eicosanoic acid ethyl ester;$C_{22}H_{44}O_2$	Eicosanoic acid (C20:0)	0.38

68.29 wt%, respectively. Higher unsaturated fatty acid esters indicate better low-temperature operability of the BD.[144]

For a BD sample, frequently studied physicochemical parameters along with prescribed test methods and corresponding ASTM D6751 and EN 14214 limits are provided in Table 3.6. Biodiesel produced from various feedstock sources must satisfy the prescribed specifications before their application in a diesel engine. The FAME content in the BD fuel must be > 96.5% and kinematic viscosity should be below 6 mm^2/s as it affects the operation of the fuel injector, particularly at low temperatures.[30] On the other hand, a low cetane number causes fuel knocking, which increases gaseous and particulate exhaust emissions due to incomplete combustion of biodiesel fuel. As indicated in Table 3.6, the BD produced from second-generation feedstock [such as jatropha oil (JO), karanja oil (KO), waste cooking oil (WO), and animal fats (AF)] meets the specified parameters and consequently would be safe to use in existing diesel engines in the form of a prescribed blend.[145–147]

3.7 Current Scenario

The second-generation feedstock for BD production has not yet adequately been explored and evaluated in the internal combustion engine.[30,148] To improve the technical and economic feasibility, further research is essential to develop novel technologies for BD production from second-generation feedstock. Indian National Policy on Biofuels (2018) highlighted the government's strategy to reduce the country's dependency on imported fossil-based fuels. The Ministry of Renewable Energy has set the target of 5% BD blending, which presently stands at < 0.1%, in petro-diesel by 2030. The total annual biodiesel production capacity of India was 650 million litres in 2019, against a demand for ~ 35,000 million litres to meet the target of 5% blending.[149] Indian biodiesel policy stresses the use of second-generation feedstock, such as jatropha and karanja oils, for BD production.[150] However, limited success has been attained so far due to the lack of feedstock availability. Some important milestones in BD utilization in India include the successful trial run of 5% BD-fuel on Delhi–Amritsar Shatabdi on December 31, 2002, by the Indian railway operator,

TABLE 3.6

Physicochemical Properties of the FAME and FAEE Prepared from Various Feedstocks

	FAME				FAEE				ASTM	
Parameters	**JO**	**KO**	**WO**	**AF**	**JO**	**KO**	**WO**	**EN14214**	**D-6751**	**Test method**
Ester content (%)	>99	>98	99%	>99	>97	>99	99%	≥96.5	≥96.5	^{1}H–NMR/GC
Flash point (°C)	110	160	120	150	178	120	114	110–170	100–170	ASTM D93
Pour point (°C)	1	5	2	4	2	4	1	-5 to10	–5 to 10	ASTM D2500
Water content (%)	–	0.034	0.25	–	–	0.045	0.27	–	0.05	ASTM D2709
Kinematic viscosity at 40 °C (cSt)	4.50	3.99	4.60	3.63	4.83	4.57	4.73	1.9–6.0	4–6.0	ASTM D445
Density at 31 °C (kg/mm^3)	870	880	867	0.86	880	880	870	860–900	878	ISI448 P:32
Iodine value (mg of I_2/g of sample)	75.9	86.5	78.9	–	87.8	86.5	80.1	<120	-	ASTM D 1510
Acid value (mg of KOH/g of sample)	0.4	0.43	0.3	0.1	0.5	0.46	0.4	≤0.5	0.50	ASTM D664
Saponification value (mg of KOH/g of sample)	180	187.37	180.5	–	182.32	194	181.23	–	–	ASTM D5558

Notes: JO= jatropha oil, KO= karanja oil, WO= waste oil, AF= animal fat.

the operation of a bio-jet fuelled civilian flight between Dehradun–Delhi by Spicejet on August 27, 2018, and the use of blended bio-jet fuel to fly military aircraft at the Republic Day Parade of 2019.[151]

3.8 Conclusions and Future Aspects

1. Biodiesel is a renewable, non-toxic, biodegradable alternative for fossil-based diesel fuel. The application of BD in conventional diesel engines can significantly decrease the emissions of greenhouse gases.
2. At the commercial level, BD is frequently produced *via* homogeneous alkali catalysed transesterification of first-generation feedstock, e.g., edible vegetable oils, with methanol, a highly toxic molecule, and a refinery residue. Such feedstock is costly and hence the BD produced is not cost-competitive with fossil-based fuels in India.
3. Being renewable and non-toxic, the application of ethanol for BD could be advantageous. However, it is less reactive than methanol and fatty acid ethyl ester is difficult to separate from glycerol.
4. Homogeneous alkali catalysts could be applied for the transesterification of costly refined edible oils having FFA and moisture contents of less than $<$ 0.5 and 0.3 wt%, respectively. Moreover, the cultivation of edible oils for BD production, in a country like India, may reduce the availability of agricultural land and could increase edible oil prices significantly.
5. To utilize second-generation feedstock (waste cooking oil, animal fats, and non-edible VOs, e.g. jatropha and karanja oil) for BD production, heterogeneous catalysts capable of catalysing simultaneous esterification and transesterification are required. Very few commercial BD production plants are employing heterogeneous catalysts for BD production.
6. Alkaline and acidic active sites over a variety of matrices, *viz.*, metal oxides, mixed metal oxides, zirconia, silica, etc., have been impregnated to perform the simultaneous esterification and transesterification of second-generation feedstock.
7. The catalyst deactivation, lower reusability, partial catalyst dissolution in reaction media, and mass transfer are a few important associated issues that are hindering the application of heterogeneous catalysts at a commercial scale.
8. Physicochemical properties of the BD produced from second-generation feedstock are found to comply with EN and ASTM standards and to make them suitable for commercial application in existing diesel engines.
9. To improve the second-generation feedstock availability there is a need to establish a collection chain for the waste cooking oils, fats, etc., from households, restaurants, industries, and slaughterhouses.
10. Based on this chapter it can be concluded that numerous research opportunities are available in the area of BD processing, economic feasibility, performance enhancement, and emission reduction. In the future, research should focus on the identification of suitable second-generation feedstocks for biodiesel production with high yield, the performance of second-generation BD

in existing diesel engines, and the evaluation of their emission characteristics. The reduction in BD production cost, to make it competitive with petro-diesel, without affecting the fuel quality, is one of the major challenges that needs to be addressed.

Acknowledgement

The authors are thankful to the TIET-VT Center of Excellence in Emerging Materials for the research funding.

REFERENCES

[1] Zahan, K.A., Kano, M. Biodiesel production from palm oil, its by-products, and mill effluent: a review. *Energies* 11 (2018): 2132.

[2] Singh, D., Bhoi, R., Ganesh, A., Mahajani, S. Synthesis of biodiesel from vegetable oil using supported metal oxide catalysts. *Energy Fuel* 28 (2014): 2743–2753.

[3] Asri, N.P., Machmudah, S., Wahyudiono, Budikarjono, S.K., Roesyadi, A., Goto, M. Palm oil transesterification in sub-and supercritical methanol with heterogeneous base catalyst. *Chemical Engineering and Processing: Process Intensification* 72 (2013): 63–67.

[4] Daud, N.M., Abdullah, S.R.S., Hasan, H.A., Yaakob, Z. Production of biodiesel and its wastewater treatment technologies: a review. *Process Safety and Environmental Protection* 94 (2015): 487–508.

[5] Saiful Islam, M., Ahmed, A.S., Islam, A., Aziz, S.A., Xian, L.C., Mridha, M. Study on emission and performance of diesel engine using castor biodiesel. *Journal of Chemistry* (2014): 1–8.

[6] Goering, C.E., Schwab, A.W., Daugherty, M.J., Pryde, E.H., Heakin, A.J. Fuel properties of Elven vegetable oils. *Transactions—American Society of Agricultural Engineers: General Edition* 85 (1982): 1472–1483.

[7] Larson, E.D. A review of life-cycle analysis studies on liquid biofuel systems for the transport sector. *Energy for Sustainable Development* 2 (2006): 109–126.

[8] Jothiramalingam, R., Wang, M.K. Review of recent developments in solid acid, base and enzyme catalysts for biodiesel production via transesterification. *Industrial & Engineering Chemistry Research* 48 (2009): 6162–6172.

[9] Teo, S.H., Rashid, U., Yap, Y.H.T. Biodiesel production from crude *Jatropha Curcas* oil using calcium based mixed oxide catalysts. *Fuel* 136 (2014): 244–252.

[10] Vyas, A.P., Verma, J.L., Subrahmanyam, N. A review on FAME production processes *Fuel* 89 (2010): 1–9.

[11] Wu, X., Leung, D.Y.C. Optimization of biodiesel production from camelina oil using orthogonal experiment. *Applied Energy* 88 (2011): 3615–3624.

[12] Leduc, S., Natarajan, K., Dotzauer, E., McCallum, I., Obersteiner, M. Optimizing biodiesel production in India. *Applied Energy* 86 (2009): 125–131.

[13] Gorji, A., Ghanei, R. A review on catalytic biodiesel production. *Journal of Biodiversity and Environmental Sciences* 5, 48–59.

[14] Talebian-Kiakalaieh, A., Amin, N.A.S., Mazaheri, H. A review on novel processes of biodiesel production from waste cooking oil. *Applied Energy* 104 (2013): 104, 683–710.

[15] McLaughlin, D.W. Land, food and biodiversity. *Conservation Biology* 25 (2011): 1117–1120.
[16] Anwar, M. Biodiesel feedstocks selection strategies based on economic, technical, and sustainable aspects. *Fuel* 283 (2021): 119204.
[17] Keera, S.T., El Sabagh, S.M., Taman, A.R. Castor oil biodiesel production and optimization. *Egyptian Journal of Petroleum* 27 (2018): 979–984.
[18] Negma, N.A., Sayed, G.H., Yehia, F.Z., Habiba, O.I., Mohamed, E.A. Biodiesel production from one-step heterogeneous catalyzed process of Castor oil and Jatropha oil using novel sulphonated phenyl silane montmorillonite catalyst. *Journal of Molecular Liquids* 234 (2017): 157–163.
[19] Karmakar, B., Hossain, A., Jha, B., Sagar, R., Halder, G. Factorial optimization of biodiesel synthesis from castor-karanja oil blend with methanol-isopropanol mixture through acid/base doped *Delonix regia* heterogeneous catalysis. *Fuel* 285 (2021): 119197.
[20] Yin, Z., Zhu, L., Li, S., Hu, T., Chu, R., Mo, F. Hu, D., Liu, C. Li, B. A comprehensive review on cultivation and harvesting of microalgae for biodiesel production: environmental pollution control and future directions. *Bioresource Technology* 301 (2020): 122804.
[21] Azadbakht, M. Ardebili, S.M.S. Rahmani, M. A study on biodiesel production using agricultural wastes and animal, fats. *Biomass Conversion and Biorefinery* (2021) https://doi.org/10.1007/s13399-021-01393-1.
[22] Sheldon, R.A. Green and sustainable manufacture of chemicals from biomass: state of the art. *Green Chemistry* 16 (2014): 950–963.
[23] Chung, K.H., Chang, D.R., Park, B.G. Removal of free fatty acid in waste frying oil by esterification with methanol on zeolite catalysts. *Bioresource Technology* 99 (2008): 7438–7443.
[24] Atadashi, I.M., Aroua, M.K,. Aziz, A.R. The effect of water on biodiesel production and refining technologies: a review. *Renewable and Sustainable Energy Reviews* 16 (2012): 3456–3470.
[25] Amin, A. Review of diesel production from renewable resources: catalysis, process kinetics and technologies, *Ain Shams Engineering Journal* 10 (2019): 821–839.
[26] Singh, D., Sharma, D., Soni, S.L., Sharma, S., Sharma, P.K., Jhalani, A. A review on feedstocks, production processes, and yield for different generations of biodiesel. *Fuel* 262 (2020): 116553
[27] Demirbas, A. Production of biodiesel fuels from linseed oil using methanol and ethanol in non-catalytic SCF conditions. *Biomass Bioenergy* 33 (2009) 113–118.
[28] Tran, K.D., Miller, M.R., Doe, C.Q. Recombineering Hunchback identifies two conserved domains required to maintain neuroblast competence and specify early-born neuronal identity. *Development* 137 (2010): 1421–1430.
[29] Aransiola, E.F., Ojumu, T.V., Oyekola, O.O., Madzimbamuto, T.F., Omoregbe, D.I.O. A review of current technology for biodiesel production: state of the art. *Biomass and Bioenergy* 61 (2014): 276–297.
[30] Atabani, A.E., Silitonga, A.S., Ong, H.C., Mahlia, T.M.I., Masjuki, H.H., Badruddin, I.A., Fayaz, H. Non-edible vegetable oils: a critical evaluation of oil extraction, fatty acid compositions, biodiesel production, characteristics, engine performance and emissions production. *Renewable and Sustainable Energy Reviews* 18 (2013): 211–245.
[31] Lee, A.F., Bennett, J.A., Manayil, J.C., Wilson, K. Heterogeneous catalysis for sustainable biodiesel production via esterification and transesterification. *Chemical Society Reviews* 43 (2014): 7887–7916.

[32] Allen, C.A.W., Watts, K.C., Ackman, R.C., Pegg, M.J. Predicting the viscosity of biodiesel fuels from their fatty acid ester composition. *Fuel* 87 (1999): 1329–1336.
[33] Knothea, G., Matheaus, A.C., Ryan, T.W. III. Cetane numbers of branched and straight-chain fatty esters determined in an ignition quality tester. *Fuel* 82 (2003): 971–975.
[34] Knothe, G., Steidley, K.R. Lubricity of components of biodiesel and petrodiesel. The origin of biodiesel lubricity. *Energy Fuels* 19 (2005): 1192–1200.
[35] Stamenkovic, O.S., Velickovic, A.V., Veljkovic, V.B. The production of biodiesel from vegetable oils by ethanolysis: current state and perspectives. *Fuel* 90 (2011): 3141–3155.
[36] Brunschwig, C., Moussavou, W., Blin, J. Use of bioethanol for biodiesel production. Progress in Energy and *Combustion Science* 38 (2012): 283–301.
[37] Srilatha, K., Lingaiah, N., Prasad, P.S.S. Devi, B.L.A.P., Prasad, R.B.N., Venkateswar, S. Influence of carbon chain length and unsaturation on the esterification activity of fatty acids on Nb_2O_5 catalyst. *Industrial & Engineering Chemistry Research* 48 (2009): 10816–10819.
[38] Kaur, N., Ali, A. Lithium zirconate as solid catalyst for simultaneous esterification andtransesterification of low quality triglycerides. *Applied Catalysis A: General* 489 (2015): 193–202.
[39] Lam, M.K., Lee, K.T., Mohamed, A.R. Homogeneous, heterogeneous and enzymatic catalysis for transesterification of high free fatty acid oil (waste cooking oil) to biodiesel: a review. *Biotechnology Advances* 28 (2010): 500–518.
[40] Katiyar, M., Ali, A. One-pot lipase entrapment within silica particles to prepare a stable and reusable biocatalyst for transesterification. *Journal of the American Oil Chemists' Society* 92 (2015): 623–632.
[41] Katiyar, M., Abida, K., Ali, A. *Candida rugosa* lipase immobilization over SBA-15 to prepare solid biocatalyst for cotton seed oil transesterification. *Materials Today: Proceedings* 36 (2021): 763–768.
[42] Quaysona, E., Amoah, J., Rachmadona, N., Hama, S., Yoshida, A., Kondo, A., Ogino, C. Biodiesel-mediated biodiesel production: a recombinant Fusarium heterosporum lipase-catalyzed transesterification of crude plant oils. *Fuel Processing Technology* 199 (2020): 106278.
[43] Lavekar, V. Bacteria can breakdown biodiesel waste into useful products. https://vigyanprasar.gov.in/isw/bacteria_breakdown_biodiesl_story.html (accessed 11 August 2021).
[44] Ma, X., Liu, F., Helian, Y., Li, C., Wu, Z., Li, H., Chu, H., Wang, Y., Wang, Y., Lu, W., Guo, M. Yu, M., Zhou. S. *Energy Conversion and Management* (2021): 113760.
[45] Hincapie G., Mondragon F., Lopez D. Conventional and in situ transesterification of castor seed oil for biodiesel production. *Fuel* 90 (2011): 1618–23.
[46] Helwani, Z., Othman, M.R., Aziz, N., Fernando, W.Z.N., Kim, J. Technologies for production of biodiesel focusing on green catalytic techniques: a review. *Fuel Processing Technolog* 90 (2009): 1502–1514.
[47] Kulkarni, M.G., Gopinath, R., Meher, L.C., Dalai, A.K. Solid acid catalyzed biodiesel production by simultaneous esterification and transesterification. *Green Chemistry* 8 (2006): 1056–1062.
[48] Meher, L.C., Kulkarni, M.C., Dalai, A.K., Naik, S.N. Transesterification of karanja (*Pongamia pinnata*) oil by solid basic catalysts. *European Journal of Lipid Science and Technology* 108 (2006): 389–397.

[49] Sharma, S., Saxena, V., Baranwal, A., Pandey, L.M. Engineered nanoporous materials mediated heterogeneous catalysts and their implications in biodiesel production. *Materials Science for Energy Technologies* 1 (2018): 11–21.

[50] Kaur, N., Ali, A. Preparation and application of Ce/ZrO-TiO_2/SO_2^{4-} as solid catalyst for the esterification of fatty acids. *Renewable Energy* 81 (2015) 421–431.

[51] Kaur, N. Ali, A. One-pot transesterification and esterification of waste cooking oil via ethanolysis using Sr:Zr mixed oxide as solid catalyst. *RSC Advances* 4 (2014): 43671.

[52] Syam, A.M., Yunus, R. Methanolysis of Jatropha oil in the presence of potassium hydroxide catalyst. *Journal of Applied Sciences* 17 (2009): 3161–3165.

[53] Thiruvengadaravi, K.V., Nandagopal, J., Baskaralingam, P., Bala, V.S.S., Sivanesan, S. Acid-catalyzed esterification of karanja (*Pongamia pinnata*) oil with high free fatty acids for biodiesel production. *Fuel* 98 (2012) 1–4.

[54] Su, F., Guo, Y. Advancements in solid acid catalysts for biodiesel production. *Green Chemistry* 16 (2014): 2934–2957.

[55] Konwar, L.J., Warna, J., Arvela, P.M., Mikkola, J.P. Reaction kinetics with catalyst deactivation in simultaneous esterification and transesterification of acidic oils to biodiesel over mesoporous sulphonated carbon catalyst. *Fuel* 166 (2016): 1–11.

[56] Chen, S.Y., Mochizuki, T., Abe, Y., Toba, M., Yoshimura, Y. Ti-incorporated SBA-15 mesoporous silica as an efficient and robust Lewis solid acid catalyst for production of high quality diesel fuels. *Applied Catalysis B: Environmental* 148–149 (2014): 345–346.

[57] Petchmala, A., Laosiripojana, N., Jongsomjit, B., Goto, M., Panpranot, J., Mekasuwandumrong, O., Shotipru, A. Transesterification of palm oil and esterification of palm fatty acid in near and super-critical methanol with SO_4-ZrO_2 catalysts. *Fuel* 89 (2010): 2387–2392.

[58] Jitputti, J., Kitiyanan, B., Rangsunvigit, P., Bunyakiat, K., Attanatho, L., Jenvanitpanjakul, P. Transesterification of crude palm kernel oil and crude coconut oil by different solid catalysts. *Chemical Engineering Journal* 116 (2006): 61–66.

[59] Chen X.R., Ju, Y.H., Mou, C.Y. Direct synthesis of mesoporous sulfated silica-zirconia catalysts with high catalytic activity for biodiesel via esterification. *The Journal of Physical Chemistry C* 111 (2007): 18731–18737.

[60] Alsalme, A.M., Wiper, P.V., Khimyak, Y.Z., Kozhevnikova, E.F., Kozhevikov, I.V. Solid acid catalysts based on $H_3PW_{12}O_{40}$ heteropoly acid: acid and catalytic properties at a gas-solid interface. *Journal of Catalysis* 276 (2010): 181–189.

[61] Arata, K. Organic syntheses catalyzed by superacidic metal oxides: sulfated zirconia and related compounds. *Green Chemistry* 11 (2009): 1719–1728.

[62] Gawande, M.B., Pandey, R.K., Jayaram, R.V. Role of mixed metal oxides in catalysis science-versatile applications in organic synthesis. *Catalysis Science & Technology* 2 (2012): 1113–1125.

[63] Li, Y., Zhang, X.D., Sun, L., Xu, M. Zhou, W.G., Liang, X.H. Solid superacid catalyzed fatty acid methyl esters production from acid oil. *Applied Energy* 87 (2010a): 2369–2373.

[64] Li, Y., Zhang, X.D., Sun, L., Zhang, J., Xu, H.P. Fatty acid methyl ester synthesis catalyzed by solid superacid catalyst SO_4^{2-}/ZrO_2–TiO_2/La^{3+}. *Applied Energy* 87 (2010b): 156–159.

[65] Sierra, L., Guth, J.L. Synthesis of mesoporous silica with tunable pore size from sodium silicate solutions and a polyethylene oxide surfactant. *Microporous and Mesoporous Materials* 27 (1999): 243–253.

[66] Viscardi, R., Barbarossa, V., Maggi, R., Pancrazzi, F. Effect of acidic MCM-41 mesoporous silica functionalized with sulfonic acid groups catalyst in conversion of methanol to dimethyl ether. *Energy Reports* 6 (2020): 49–55.

[67] Meloni, D., Perra, D., Monaci, R., Cutrufello, M.G., Rombi, E., Ferino, I. Transesterification of *Jatropha curcas* oil and soyabean oil over Al-SBA-15 catalysts. *Applied Catalysis B: Environmental* 184 (2016): 163–173.

[68] Chen, D., Li, Z., Wan, Y. Tu, X., Shi, Y., Chen, Z., Shen, W., Yu, C., Tu, B., Zhao, D. Anionic surfactant induced mesophase transformation to synthesize highly ordered largepore mesoporous silica structures. *Microporous and Mesoporous Materials* 16 (2006): 1511–1519.

[69] Cao, L., Man, T., Kruk, M. Synthesis of ultra-large-pore SBA-15 silica with twodimensional hexagonal structure using triisopropylbenzene as micelle expander. *Chemistry of Materials* 21 (2009): 1144–1153.

[70] Chen, S.Y., Mochizuki, T., Abe, Y., Toba, M., Yoshimura, Y. Ti-incorporated SBA-15 mesoporous silica as an efficient and robust Lewis solid acid catalyst for the production of high-quality biodiesel fuels. *Applied Catalysis B: Environmental* 148–149 (2014): 344–356.

[71] Mizuno, N., Misono, M. Heterogeneous catalysis. *Chemical Reviews* 98 (1998): 199–218.

[72] Kozhevnikov, I.V. Catalysis by heteropoly acids and multicomponent polyoxometalates in liquid-phase reactions. *Chemical Reviews* 98 (1998): 98, 171–198.

[73] Narasimharao, K., Brown, D., Lee, A.F., Newman, A.D., Siril, P.F., Tavener, S.J., Wilson, K.; Structure–activity relations in Cs-doped heteropolyacid catalysts for biodiesel production. *Journal of Catalysis* 248 (2007): 248, 226–234.

[74] Pesaresi, L., Brown, D.R., Lee, A.F., Montero, J.M. Williams, H., Wilson, K. Cs-doped $H_4SiW_{12}O_{40}$ catalysts for biodiesel applications. *Applied Catalysis A: General* 360 (2009): 50–58.

[75] Duan, X., Liu, Y., Zhao, Q. Wang, X., Li, S. Water-tolerant heteropolyacid on magnetic nanoparticles as efficient catalysts for esterification of free fatty acid. *RSC Advances* 3 (2013): 13748–13755.

[76] Singh, H., Ali, A. Potassium and 12-tungstophosphoric acid loaded alumina as heterogeneous catalyst for the esterification as well as transesterification of waste cooking oil in a single pot. *Asia-Pacific Journal of Chemical Engineering* 16 (2021): 2585.

[77] Faessler, P., Kolmetz, K., Seang, K.W., Lee, S.H. Advanced fractionation technology for the oleochemical industry. *Asia-Pacific Journal of Chemical Engineering* 2 (2007): 315–321.

[78] Sreeprasanth, P.S., Srivastava, R., Srinivas, D., Ratnasamy, P. Hydrophobic, solid acid catalysts for production of biofuels and lubricants. *Applied Catalysis A: General* 314 (2006): 148–159.

[79] Sivasamy, A., Cheah, K.Y., Fornasiero, P., Kemausuor, F., Zinoviev, S., Miertus, S. Catalytic applications in the production of biodiesel from vegetable oils. *ChemSusChem* 2 (2009): 278–300.

[80] Zong, M.H., Duan, Z.Q., Lou, W.Y., Smith, T.J., Wu, H. Preparation of a sugar catalyst and its use for highly efficient production of biodiesel. *Green Chemistry* 9 (2007): 434–437.

[81] Okamura, M., Takagaki, A., Toda, M., Kondo, J.N., Tatsumi, T., Domen, K., Hara, M., Hayashi, S. Acid-catalyzed reactions on flexible polycylic aromatic carbon in amorphous carbon. *Chemistry of Materials* 18 (2006): 3039–3045.

[82] Xie, W., Wang, T. Biodiesel production from soybean oil transesterification using tin oxidesupported WO_3 catalysts. *Fuel Processing Technology* 109 (2013): 150–155.
[83] Laosiripojana, N., Kiatkittipong, W., Sutthisripok, W., Assabumrungrat, S. Synthesis of methyl esters from relevant palm products in near-critical methanol with modifiedzirconia catalysts. *Bioresource Technology* 101 (2010): 8416–8423.
[84] Sheikh, R., Choi, M.S., Im, J.S., Park, Y.H. Study on the solid acid catalysts in biodiesel production from high acid value oil. *Journal of Industrial and Engineering Chemistry* 19 (2013): 1413–1419.
[85] Alhassan, F.H., Yunus, R., Rashid, U., Sirat, K., Islam, A., Lee, H.V., Yap, Y.H.T. Production of biodiesel from mixed waste vegetable oils using ferric hydrogen sulphate as an effective reusable heterogeneous solid acid catalyst. *Applied Catalysis A: General* 456 (2013) 182–187.
[86] Farooq, M., Ramli, A., Subbarao, D. Biodiesel production from waste cooking oil using bifunctional heterogeneous solid catalysts. *Journal of Cleaner Production* 59 (2013): 131–140.
[87] Yan, S., Kim, M., Salley, S.O., Ng, K.Y.S. Oil transesterification over calcium oxides modified with lanthanum. *Applied Catalysis A: General* 360 (2009): 163–170.
[88] Dai, Q., Yang, Z., Li, J., Cao, Y., Tang, H., Wei, X. Zirconium-based MOFs-loaded ionic liquid-catalyzed preparation of biodiesel from Jatropha oil. *Renewable Energy* 163 (2021): 1588–1594.
[89] Fan, M., Liu, H., Zhang, P. Ionic liquid on the acidic organic-inorganic hybrid mesoporous material with good acid-water resistance for biodiesel production. *Fuel* 215 (2018) 541–550.
[90] Kaur, M., Malhotra, R., Ali, A. Tungsten supported Ti/SiO_2 nanoflowers as reusable heterogeneous catalyst for biodiesel production. *Renewable Energy* 116 (2018): 116, 109–119.
[91] Luoa, Y., Mei, Z., Liu, N., Wang, H., Han, C., He, S. Synthesis of mesoporous sulfated zirconia nanoparticles with high surface area and their applies for biodiesel production as effective catalysts. *Catalysis Today* 298 (2017): 99–108.
[92] Pan, H., Li, H., Liu, X.F., Zhang, H. Yang, K.L., Huang, S., Yang, S. Mesoporous polymeric solid acid as efficient catalyst for transesterification of crude Jatropha curcas oil. *Fuel Processing Technology* 150 (2016): 50–57.
[93] Lam, M.K., Lee, K.T., Mohamed, A.R. Sulfated tin oxide as solid superacid catalyst for transesterification of waste cooking oil: an optimization study. *Applied Catalysis B: Environmental* 93 (2009): 134–139.
[94] Yadessa G. K., Kathrine (Trine) Hvoslef-Eideb, A., Marchetti, J.M. Optimization of the production of biofuel form Jatropha oil using a recyclable anion-exchange resin. *Fuel* 278 (2020): 118253.
[95] Bournay, L., Casanave, D., Delfort, B., Hillion, G., Chodorge, J.A. New heterogeneous process for biodiesel production: a way to improve the quality and value of crude glycerine produced by biodiesel plants. *Catalysis Today* 106 (2005): 190–192.
[96] Lee, D., Park, Y., Lee, K. Heterogeneous base catalysts for transesterification in biodiesel synthesis. *Catalysis Surveys from Asia* 13 (2009): 63–77.
[97] Pang, H., Yang, G., Li, L., Yu, J. Efficient transesterification over two-dimensional zeolites for sustainable biodiesel production. *Green Energy & Environment* 5 (2020): 405–413.
[98] Hattori, H. Heterogeneous basic catalysis. *Chemical Reviews* 95 (1995): 537–558.

[99] Yan, S., Lu, H., Liang, B. Supported CaO catalysts used in the transesterification of rapeseed oil for the purpose of biodiesel production. *Energy & Fuels* 22 (2008): 646–651.

[100] Rashtizadeh, E., Farzaneh, F., Talebpour, Z. Synthesis and characterization of $Sr_3Al_2O_6$ nanocomposite as catalyst for biodiesel production. *Bioresource Technology* 154 (2014): 32–37.

[101] Kouzu, M., Hidaka, J. Transesterification of vegetable oil into biodiesel catalyzed by CaO: a review. *Fuel* 93 (2012): 1–12.

[102] Dehkordi, A.M., Ghasemi, M. Transesterification of waste cooking oil to biodiesel using Ca and Zr mixed oxides as heterogeneous base catalysts. *Fuel Processing Technology* 79 (2012): 45–51.

[103] Montero, J., Wilson, K., Lee, A. Cs promoted triglyceride transesterification over MgO nanocatalysts. *Topic in Catalalysis* 53 (2010): 737–745.

[104] Kumar, D., Ali A. Nanocrystalline lithium ion impregnated calcium oxide as heterogeneous catalyst for transesterification of high moisture containing cotton seed oil. *Energy & Fuel* 24 (2010): 2091–7.

[105] Wang, J.X., Chen, K.T., Wu, J.S., Wang, P.H., Huang, S.T., Chen, C.C. Production of biodiesel through transesterification of soybean oil using lithium orthosilicate solid catalyst. *Fuel Processing Technology* 104 (2012): 167–173.

[106] Mutreja, V., Singh, S., Minhas, T.K., Ali, A. Nanocrystalline potassium impregnated SiO_2 as heterogeneous catalysts for the transesterification of karanja and jatropha oil. *RSC Advances* 5 (2015): 46890.

[107] Guo, F., Peng, Z.G,., Dai, J.Y., Xiu, Z.L. Calcined sodium silicate as solid base catalyst for biodiesel production. *Fuel Processing Technology* 91 (2010): 322–328.

[108] Gombotz, K., Parette, R., Austic, G., Kannan, D., Matson, J.V. MnO and TiO solid catalysts with low-grade feedstocks for biodiesel production. *Fuel* 92 (2012): 9–15.

[109] Xie, W., Zhao, L. Heterogeneous CaO-MoO_3-SBA-15 catalysts for biodiesel production from soybean oil, *Energy Conversion and Management* 79 (2014): 34–42.

[110] Kumar, D., Ali, A. Direct synthesis of fatty acid alkanolamides and fatty acid alkyl esters from high free fatty acid containing triglycerides as lubricity improvers using heterogeneous catalyst. *Fuel* 159 (2015) 845–853.

[111] Li, Y., He, D. Yuan, Y., Cheng, Z., Zhu, Q. Influence of acidic and basic properties of ZrO_2 based catalysts on isosynthesis. *Fuel* 81 (2002): 1611–1617.

[112] Omar, W.N.N.W., Amin, N.A.S. Biodiesel production from waste cooking oil over alkaline modified zirconia catalyst. *Fuel Processing Technology* 92 (2011): 2397–2405.

[113] Cosimo, J.I.D., Diez, V.K., Xu, M., Iglesia, E., Apesteguia, C.R. Structure and surface and catalytic properties of Mg-Al basic oxides. *Journal of Catalysis* 178 (1998): 499–510.

[114] Choudary, B.M., Kantam, M.L., Reddy, V., Rao, K.K., Figueras, F. Henry reactions catalysed by modified Mg–Al hydrotalcite: an efficient reusable solid base for selective synthesis of β-nitroalkanols. *Green Chemistry* (1999): 187–189.

[115] Winter, F., Xia, X.Y., Hereijgers, B.P.C., Bitter, J.H., Dillen A.J.V., Muhler, M., Jong, K.P. On the nature and accessibility of the bronsted-base sites in activated hydrotalcite catalysts. *Journal of Physical Chemistry B* 110 (2006): 9211–9218.

[116] Liu, Q., Wang, B., Wang, C., Tian, Z., Qu, W., Maa, H., Xua, R. Basicities and transesterification activities of Zn–Al hydrotalcites-derived solid bases. *Green Chemistry* 16 (2014):16, 2604–2613.

[117] Barakos, N., Pasias, S., Papayannakos, N. Transesterification of triglycerides in high and low quality oil feeds over an HT_2 hydrotalcite catalyst. *Bioresource Technology* 99 (2008): 5037–5042.
[118] Fraile, J.M., Garcıa, N., Mayoral, J.A., Pires, E., Roldan, L. The influence of alkaline metals on the strong basicity of Mg–Al mixed oxides: the case of transesterification reactions. *Applied Catalysis A: General* 364 (2009): 87–94.
[119] Cantrell, D.G., Gillie, L.J., Lee, A.F., Wilson, K. Structure-reactivity correlations in MgAl hydrotalcite catalysts for biodiesel synthesis. *Applied Catalysis A: General* 287 (2005): 183–190.
[120] Palitsakun, S., Koonkuer, K., Topool, B., Seubsai, A., Sudsakorn, K. Transesterification of Jatropha oil to biodiesel using SrO catalysts modified with CaO from waste eggshell. *Catalysis Communications* 149 (2021): 106233.
[121] Malhotra, R., Ali, A. Lithium-doped ceria supported SBA-15 as mesoporous solid reusable and heterogeneous catalyst for biodiesel production via simultaneous esterification and transesterification of waste cottonseed oil. *Renewable Energy* 119 (2018): 32–44.
[122] Baskar, G., Gurugulladevi, A., Nishanthini, T., Aiswarya R., Tamilarasan, K. Optimization and kinetics of biodiesel production from Mahua oil using manganese doped zinc oxide nanocatalyst. *Renewable Energy* 103 (2017): 641–646.
[123] Nisar, J., Razaq, R., Farooq, M., Iqbal, M., Khan, R.A. Sayed, M., Shah, A., Rahman, I. Enhanced biodiesel production from Jatropha oil using calcined waste animal bones as catalyst. *Renewable Energy* 101 (2017): 111–119.
[124] Karimi, M. Immobilization of lipase onto mesoporous magnetic nanoparticles for enzymatic synthesis of biodiesel. *Biocatalysis and Agricultural Biotechnology* 8 (2016): 182–188.
[125] Yahya, N.Y., Ngadi, N., Jusoh, M., Abdul Halim, N.A. Characterization and parametric study of mesoporous calcium titanate catalyst for transesterification of waste cooking oil into biodiesel. *Energy Conversion and Management* 129 (2016): 275–283
[126] Teo, S.H., Goto, M., Taufiq-Yap, Y.H. Biodiesel production from Jatropha curcas L. oil with Ca and La mixed oxide catalyst in near supercritical methanol conditions. *Journal of Supercritical Fluids* 104 (2015): 243–250.
[127] Kumar D., Ali, A. A solid catalyst composition and process for its preparation (Patent Application No: 3358/DEL/2013)
[128] Lee, H.V., Juan, J.C., Taufiq-Yap, Y.H. Preparation and application of binary acidebase $CaOeLa_2O_3$ catalyst for biodiesel production. *Renewable Energy* 74 (2015): 124–132.
[129] Wu, H., Zhanga, J., Liub, Y., QinWei, J. Biodiesel production from Jatropha oil using mesoporous molecular sieves supporting K_2SiO_3 as catalysts for transesterification. *Fuel Processing Technology* 119 (2014): 114–120.
[130] Kaur, M., Ali, A. Potassium fluoride impregnated CaO/NiO: an efficient heterogeneous catalyst for transesterification of waste cottonseed oil. *European Journal of Lipid Science and Technology* 116 (2014): 80–88.
[131] Mutreja, V., Singh, S., Ali, A. Potassium impregnated nanocrystalline mixed oxides of La and Mg as heterogeneous catalysts for transesterification. *Renewable Energy* 62 (2014): 226–233.
[132] Kumar D., Ali, A. Transesterification of low-quality triglycerides over a Zn/CaO heterogeneous catalyst: kinetics and reusability studies. *Energy Fuels* 27 (2013): 3758–3768.

[133] Kaur, N., Ali, A. Lithium ions-supported magnesium oxide as nano-sized solid catalyst for biodiesel preparation from mutton fat, *Energy Sources, Part A*, 35 (2013): 184–192.

[134] Kumar D., Ali, A. Nanocrystalline K-CaO for the transesterification of a variety of feedstocks: structure, kinetics and catalytic properties. *Biomass & Bioenergy* 46 (2012): 459–468.

[135] Shin, H., An, S., Sheikh, R., Ho Park, Y., Bae, S. Transesterification of used vegetable oils with a Cs-doped heteropolyacid catalyst in supercritical methanol. *Fuel* 96 (2012) 572–578.

[136] Yap, Y.H.T., Lee, H.V., Hussein, M.Z., Yunus, R. Calcium-based mixed oxide catalysts for methanolysis of Jatropha curcas oil to biodiesel. *Biomass & Bioenergy* 35 (2011): 827–834.

[137] Wen, Z., Yu, X., Tu, S.T., Yan, J., Dahlquist, E. Synthesis of biodiesel from vegetable oil with methanol catalyzed by Li-doped magnesium oxide catalysts. *Applied Energy* 87 (2010a): 743–748.

[138] Kaur, M., Kumar, D., Ali, A. Physicochemical properties of the biodiesel prepared from jatropha and used cottonseed oils in the presence of Li/CaO solid catalyst. *Proceedings of Conference on Advances in Chemical Engineering*. Macmillan Publishers India Ltd (2011): 315–323.

[139] Yan, S., Salley, S.O., Ng, K.Y.S. Simultaneous transesterification and esterification of unrefined or waste oils over ZnO-La_2O_3 catalysts. *Applied Catalysis A: General* 353 (2009): 203–212.

[140] Akbar, E., Binitha, N., Yaakob, Z., Kamarudin, S.K., Salimon, J. Preparation of Na doped SiO_2 solid catalysts by the sol-gel method for the production of biodiesel from jatropha oil. *Green Chemistry* 11 (2009): 1862–1866.

[141] Kaur, M., Ali, A. An efficient and reusable Li/NiO heterogeneous catalyst for ethanolysis of waste cottonseed oil. *European Journal of Lipid Science and Technology* 117 (2015): 550–560.

[142] Kaur, N., Ali, A. Kinetics and reusability of Zr/CaO as heterogeneous catalyst for the ethanolysis and methanolysis of *Jatropha crucas* oil. *Fuel Processing Technology* 119 (2014): 173–184.

[143] Kaur, M., Ali, A. Lithium ion impregnated calcium oxide as nano catalyst for the biodiesel production from karanja and jatropha oils. *Renewable Energy* 36 (2011): 2866–2871.

[144] Lotero, E., Liu, Y., Lopez, D.E. Suwannakarn, K., Bruce, D.A., Goodwin, J.G. Synthesis of biodiesel via acid catalysis. *Industrial & Engineering Chemistry Research* 44 (2005): 5353–5363.

[145] Baiju, B., Naik, M.K., Das, L.M. A comparative evaluation of compression ignition engine characteristics using methyl and ethyl esters of Karanja oil. *Renewable Energy* 34 (2009): 1616–1621.

[146] Verma, R., Sharma, D.K., Bisen, P.S. Determination of free fatty acid composition in *Jatropha* crude oil and suitability as biodiesel feedstock. *Current Alternative Energy* 3 (2019): 59–64.

[147] Moser, B.R. Biodiesel production, properties, and feedstocks. *In Vitro Cellular & Developmental Biology - Plant* 45 (2009): 229–266.

[148] Eisentraut, A. Sustainable production of second-generation biofuels. © OECD/IEA (2010).

[149] https://energy.economictimes.indiatimes.com/news/oil-and-gas/indias-bio-diesel-blending-for-road-transport-to-remain-muted-in-2019-us-dept-of-agriculture/70868387 (accessed 10 August 2021).
[150] https://economictimes.indiatimes.com/small-biz/productline/power-generation/national-policy-on-biofuels-2018-here-are-key-things-you-should-know/articleshow/71922729.cms?from=mdr (accessed 10 August 2021).
[151] www.gktoday.in/current-affairs/indias-biofuel-flight-fly-dehradun-delhi (accessed 10 August 2021).

4

Role of Catalysts and Their Mechanisms in Biodiesel Synthesis with Reference to Nanomaterials

Steven Lim,* **Sim Jia Huey, Shuit Siew Hoong, and Pang Yean Ling**

CONTENTS

4.1 Introduction: Background and Latest Developments

Biodiesel has been lauded as one of the best alternative fuels to replace mineral diesel with the aim of decarbonizing the transportation and energy generation sectors. It can be produced from numerous sources, which include vegetable oils, animal fats and waste oil, as long as they contain high fatty acid and triglyceride contents. The combustion of fuel using biodiesel instead of mineral diesel has proven to be more sustainable due to its feedstock renewability, biodegradability, cleaner emissions, non-toxicity and high availability. Most importantly, it is readily blended into the existing energy infrastructure without any engine modifications, which saves huge capital cost compared

* *Corresponding Author* stevenlim@utar.edu.my

DOI: 10.1201/9781003120858-4

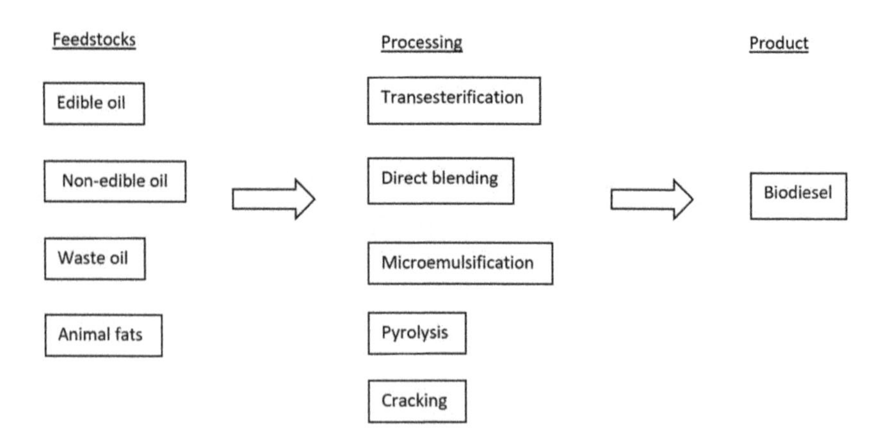

FIGURE 4.1 Various synthesis pathways for biodiesel from different feedstocks.

to other renewable energies (Lim and Lee, 2010). There are several pathways to synthesize biodiesel from a variety of feedstocks, as shown in Figure 4.1. Among these processes, transesterification is the most popular method due to the higher quality and compatibility of the biodiesel produced, which meets the ASTM D6751 and EN 14214 standards (Demirbas et al., 2003). Over the years, extensive research has been devoted to improving the biodiesel synthesis methods with the aim of further enhancing the process viability and economic feasibility. Several processing methods for biodiesel synthesis have been developed in order to cater to different generations of biodiesel, as summarized in Figure 4.2 (Lim et al. 2019). Among these methods, the catalytic heterogeneous biodiesel production process has received considerable attention due to several advantages such as tuneable catalytic properties, easier catalyst separation, higher reusability and reduction of waste generation compared to its homogeneous catalytic process counterpart. However, the heterogeneous catalytic process is still beleaguered by several challenges which restrict its commercialization potential and development. Its main drawbacks are comprised of lower reaction kinetics due to higher mass transfer resistance, higher process intensity for optimum biodiesel yield, expensive catalyst synthesis methods and the use of hazardous chemicals during the synthesis process (Lim et al., 2020). Consequently, there is a need for continuous research and development to synthesize better and more efficient catalysts for biodiesel production. In this context, the application of nanomaterials as a high-performing catalyst in biodiesel production offers attractive catalytic properties to overcome the conventional challenges in the heterogeneous catalytic process.

The primary objective of introducing nanomaterial as a catalyst for biodiesel synthesis either directly as the active site itself or indirectly as a catalyst support is to reduce the mass transfer resistance due to the heterogeneous nature of the biodiesel synthesis reaction. Nanomaterial is defined as a material with the majority of its particles having dimensions between 1–100 nm. This small particle size allows nanomaterials to exhibit distinguishing characteristics such as increased surface area, mechanical

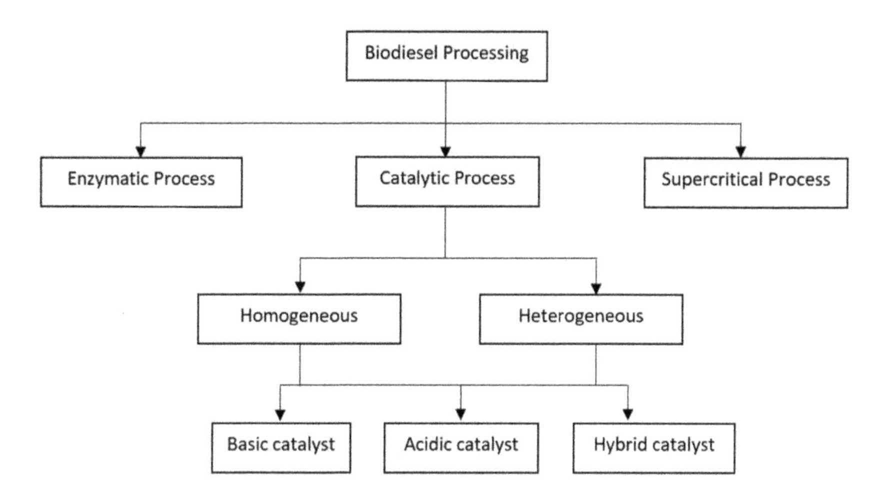

FIGURE 4.2 Different transesterification reaction pathways for biodiesel processing.

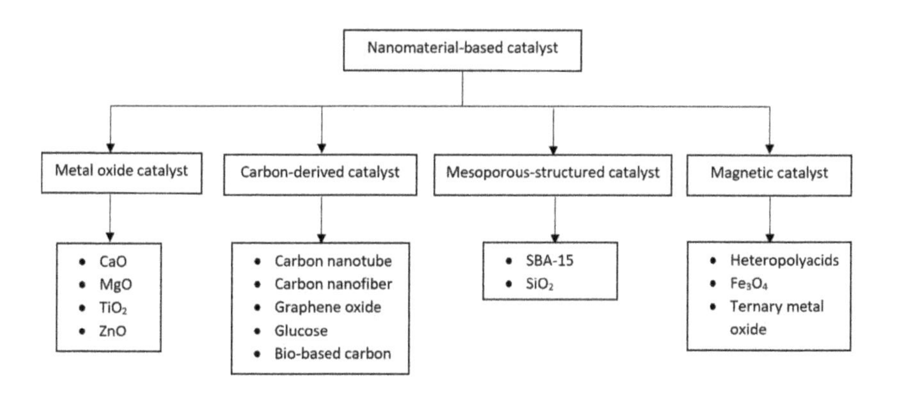

FIGURE 4.3 Different types of nanomaterial-based catalysts for biodiesel production.

strength, electrical conductivity and chemical reactivity (Tang et al., 2018). Figure 4.3 presents several types of nanomaterial-based catalyst which have been developed for the production of biodiesel. Most catalysts exhibit an ideal mesoporous structure with a pore size ranging from 2 nm to 50 nm (Gardy et al., 2019). This allows the larger reactant molecules to enter the pores easily for the reaction to take place with minimum steric hindrance while at the same time possessing a larger surface area to minimize the mass transfer resistance. In view of the current development of this type of nanocatalyst, this chapter is dedicated to elucidating the importance of its role and the mechanism in biodiesel synthesis which is still very rarely discussed in detail in the literature.

4.2 Role of Nanocatalysts in Biodiesel Production

Chemical catalysis technologies have been commonly used to produce biodiesel either in the laboratory or the production plant in the industry. The catalysis technologies for biodiesel production have undergone various phases of development and improvement, with the ultimate objective of developing cost-effective catalyst features with excellent catalytic performance. The catalytic performances include a rapid reaction rate with low activation energy, high conversion and selectivity.

4.2.1 Chemical Catalysis

In recent years, researchers have been focusing on converting waste resources into valuable solid catalysts for biodiesel production. Although the developed catalyst is cheap, with nearly 100% biodiesel yield achieved after 5–6 hours of reaction time, its reaction rate is still considered too low for commercialization. Heterogeneous catalyst in the liquid phase of a reaction mixture always exerts high mass transfer resistance to the liquid reactants to be smoothly adsorbed, reacted and desorbed on the active site. To minimize the mass transfer limitations due to the heterogeneous phase of the reaction, the development and application of nano-sized catalyst (less than 100 nm) in the reaction is a wise selection. Similar to other nano-sized materials, nanocatalyst is well known for its large surface area relative to the volume of catalyst, leading to a high number of active sites available for the simultaneous attachment of reactants prior to the reaction. Many research works currently focus on the study and understanding of the dependency of intrinsic catalytic performance with size reduction. Based on this knowledge, the effective size, porosity and structure of the nanocatalyst can be designed and synthesized.

4.2.2 Heterogeneous Nanocatalysts in Biodiesel Production

Chemical catalysis has been popular in chemical processes due to its superiority in reducing the activation energy drastically by providing many affinity platforms for all the reactive species to react instantly, thus accelerating the reaction rate per unit time. To further enhance the reaction rate and catalytic activity of the conventional catalyst, huge numbers of affinity platforms or active sites on the catalyst surface would become the determining factor. Therefore, the high surface area to volume ratio of the catalyst is one of the desirable features for chemical reaction due to the high availability of active sites for reaction. Heterogeneous nanocatalysts are effective in reducing the reaction time compared to conventional catalysts in micro or macro sizes. When a smaller surface material is developed into a larger surface material (both macro and micro sizes), its surface area tends to expand. However, its surface area to volume ratio is always less than the nano-sized material. Extremely small nano-sized material (less than 100 nm) is able to offer a tremendous surface area to volume ratio. Hence, less nanocatalyst is required for the reaction, thus enhancing the reaction efficiency (Banković–Ilić et al., 2017). In addition, the surface composition, spatial distribution, and chemical and thermal stability of nanocomponents can be regulated in a precise way to achieve 100% selectivity while simultaneously preserving a longer catalyst lifetime with lower energy consumption during the reaction (Kung and Kung, 2007). The application of nanocatalyst in chemical industries is favourable and cost-effective as it is a greener catalyst that generates minimum chemical wastes by optimizing the utilization of feedstock and it also possesses high energy efficiency (Philippot and Serp, 2013). Nano-sized catalyst has been proven to be a potential substitute for conventional catalysts to effectively increase the reaction rate to a satisfactory level. Among the nano-sized heterogeneous catalysts, CaO-based compounds are one of the most promising catalysts to produce biodiesel as the catalysts only require mild operating conditions to achieve a high biodiesel yield (Kesić et al., 2016). Nano-CaO catalysts have been synthesized and used in various forms, such as neat CaO catalysts, doped CaO catalyst (CaO carrier was doped with other compounds), and loaded form (CaO immobilization onto a different carrier).

Waste or natural source-derived CaO nanoparticles have been synthesized and modified by different researchers to improve the conversion rate of the transesterification reaction. Compared to commercial CaO, CaO nanocatalyst can exhibit up to 1.54 times greater surface area based on a study by Badnor et al. (2018). A CaO nanocatalyst with 97.63 m^2/g showed 18.81% higher conversion than the commercial CaO catalyst (74.64 m^2/g surface area). The small crystallite and particle size of nanocatalyst resulted in a large surface area to promote the reaction rate. The study also discovered that the reaction catalysed by 4% nanocatalyst at a moderate temperature of 60 °C and 6:1 methanol-to-oil ratio was completed in 80 min. Hebbar et al. (2018) also investigated the effectiveness of nano-sized CaO catalyst and discovered that a high biodiesel yield of 96.2% was achieved at a moderate temperature of 65 °C, 6:1 methanol-to-oil ratio and 1.5 wt% of catalyst loading. The activation energy for the reaction catalysed by CaO nanocatalyst was recorded at 35.99 kJ/mol. Hsiao et al. (2011) also reported a moderate biodiesel yield of 71.1% with CaO nanocatalyst compared to a larger size of CaO with only a 2.3% biodiesel yield with a 30 min reaction time. The CaO nanocatalyst was highly effective in biodiesel production compared to the larger size of CaO. Based on research carried out by Banković–Ilić et al. (2017), the rate of transesterification catalysed by CaO nanocatalyst was about 30 times higher than for a CaO microcatalyst within the 30 min reaction time. In addition to the CaO nanocatalyst's advantage in accelerating the reaction rate, it also helps to reduce the biodiesel production cost with a lower catalyst dosage required, simple and easy recovery of biodiesel products and permitting at least eight cycles of catalyst reusability without a significant drop in the biodiesel yield at moderate operating conditions. From the morphological analysis, fresh CaO nanocatalyst consists of numerous crystallites with well-defined edges. These well-defined edges are converted to polycrystals with less defined edges upon being recycled for reused. The surface micrograph also shows that CaO material is an amorphous structure in nature with high porosity.

The CaO nanocatalyst impregnated with K_2CO_3 was able to enhance the reaction rate at a moderate temperature of 65 °C. Degimenbasi et al. (2015) claimed that this was mainly due to the high surface area to volume ratio offered by CaO nanoparticles. In addition, the reaction catalysed by KF nanocatalyst supported on CaO (particle sizes of 30–100 nm) achieved a more than 96% biodiesel yield. Wen et al. (2010) reported that the high specific surface area and large pore size of nanocatalyst encouraged a higher frequency of intimate contacts between the reactants at the active sites, which drastically improved the efficiency of the transesterification reaction. The new crystal phase of $KCaF_3$ also contributes to the stability of the catalyst. Seffati et al. (2019) modified CaO nanocatalyst with magnetic properties of $CuFe_2O_4$ to ease the product separation from the catalyst. A high biodiesel yield of 94.52% was achieved at 3% nanocatalysts loading and at a moderate temperature of 70 °C. The $CuFe_2O_4$-doped CaO nanocatalyst offered a higher porosity advantage and thus a greater number of active sites were expected. The porous structure of KF/CaO-Fe_3O_4 nanocatalyst was reported to exhibit 50 nm particle diameter and 20.08 m^2/g surface area. The stability of ZnO/$BiFeO_3$ nanocatalyst after being reused was highly improved, where 95.42% and 95.02% of biodiesel yields were achieved in the first and second cycles, respectively.

4.2.3 Recent Prospects

In recent years, nanocatalysis application in chemical industries and research studies in chemical reaction has undergone an explosive growth. The important role played by nanocatalysis in chemical industries can be revealed by the increasing numbers of nanocatalysis-related patents, technologies and products in the market. With a thorough understanding of the nanocatalyst's intrinsic catalytic performance in relation to its properties (particle size and structure, porosity, surface area) accompanied with an optimum nanocatalyst synthesis protocol, the development of an excellent catalytic activity, cost-effective and environmentally benign nanocatalyst can be created. These will represent a significant milestone for the biodiesel industry.

4.3 Synthesis of Nanocatalysts for Biodiesel Production

The synthesis methods for nanocatalysts for biodiesel production can be divided into three main categories: doping via co-precipitation, wet impregnation and chemical functionalization or immobilization.

4.3.1 Doping via the Co-Precipitation Method

Doping via the co-precipitation method was one of the most common methods in the synthesis of nanocatalysts for biodiesel production. The nanocatalysts synthesized using this method are metal-based catalysts such as Zn-doped CaO (Rajendran and Gurunathan, 2020), Co-doped ZnO (Borah et al., 2019), Fe-doped ZnO (Baskar and Soumiya, 2016), and Mn-doped ZnO (Baskar et al., 2017). One of the most important criteria for doping via the co-precipitation method is that the metal used must present in the ionic form. Therefore, the starting material or the precursor for the metals reported in the literature were usually in nitrate (Rajendran and Gurunathan, 2020), acetate (Baskar et al., 2017), sulphate (Baskar & Soumiya, 2016) or chloride (Safakish et al., 2020) ionic solution. Firstly, pre-determined amounts of the metals in their respective ionic solutions (nitrate, acetate, chloride or sulphate) were mixed and dissolved homogeneously in distilled water to form the cation solution. The mixture was then continuously stirred at room temperature (Baskar et al., 2017; Baskar and Soumiya, 2016; Borah et al., 2019; Rajendran and Gurunathan, 2020). Furthermore, it was reported that ultrasonication (at a frequency of 57 kHz) coupled with mixing can be used to enhance homogeneous dispersion and mixing of the solution (Baskar and Soumiya, 2016). Next, the pre-calculated amount of anion solution such as Na_2CO_3 (Rajendran and Gurunathan, 2020), NaOH (Borah et al., 2019; Dehghani and Haghighi, 2020) or 8–25% of NH_3 aqueous solution (Naveenkumar and Baskar, 2019; Rahmani Vahid et al., 2018) was added dropwise into the cation solution until the formation of a precipitate. The pH of the solution was maintained in the range of 10–12 (Baskar and Soumiya, 2016; Dehghani and Haghighi, 2020; Rahmani Vahid et al., 2018; Seffati et al., 2019). The precipitate was then aged overnight (Singh et al., 2020). Next, the precipitate was filtered and washed several times with ethanol and distilled water (Borah et al., 2019). After that, it was dried at a temperature of 80–120 °C for at least 1 h (Baskar et al., 2017; Baskar and Soumiya, 2016; Borah et al., 2019; Dehghani and Haghighi, 2020; Rahmani Vahid et al., 2018; Rajendran and Gurunathan, 2020). Finally, the dried precipitate was then activated by a calcination process at a temperature range of 350–900 °C (Baskar et al., 2017; Baskar and Soumiya, 2016; Borah et al., 2019; Dehghani and Haghighi, 2020; Naveenkumar and Baskar, 2019; Rahmani Vahid et al., 2018). The synthesis route for doping via the co-precipitation method is summarized in Figure 4.4.

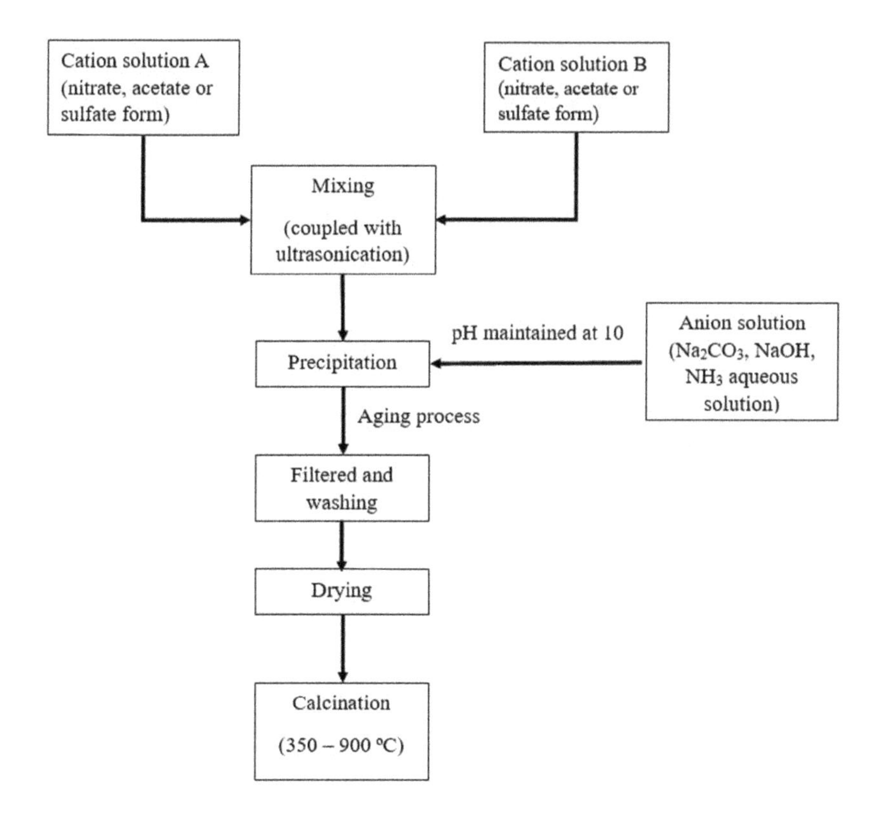

FIGURE 4.4 The synthesis route of doping via the co-precipitation method.

Instead of being directly applied as catalyst in a transesterification reaction, the nanocatalysts synthesized by doping via the co-precipitation method could be dispersed onto catalyst supports such as MCM-41 (Dehghani and Haghighi, 2019). Several advantages could be offered via the introduction of these nanocatalysts into the support, such as better controllability of the particle pore size, improvement of catalyst morphology and acidity as well as enhancing the adsorption and desorption of non-polar and polar molecules, respectively (Dehghani and Haghighi, 2020).

4.3.2 Wet Impregnation

Impregnation is the method to introduce or incorporate an active phase into the catalyst support. Unlike the co-precipitation method, the reactants for the wet impregnation method are in the opposite physical state. The catalyst support is in solid form, while the active phase is in aqueous form. The active phases that were impregnated into support and served as catalysts in biodiesel production included potassium (K) (Ambat et al., 2019; Seffati et al. 2019; Wen et al., 2010), sodium (Na) (Ibrahim et al., 2020), sulphate (SO_4^{2-}) (Safakish et al., 2020), gold (Au) (Bet-Moushoul et al., 2016), calcium (Ca) (Feyzi and Norouzi, 2016), copper (Cu) (Esmaeili et al., 2019), magnesium (Mg) (De and Boxi, 2020), strontium (Sr) (Feyzi and Khajavi, 2014) and zinc (Zn) (Kataria et al., 2017). The reported catalyst supports in biodiesel production were zeolite (ZSM-5) (Feyzi and Khajavi, 2014), activated carbon powder (Esmaeili et al., 2019), carbon nanotubes (Ibrahim et al., 2020) and metal oxides (Alaei et al, 2020; Alaei et al., 2018; Amani et al., 2019; Ambat et al., 2019; Bet-Moushoul et al., 2016; De and Boxi, 2020; Kataria et al., 2017; Safakish et al., 2020; Wen et al., 2010). First, a pre-determined amount of catalyst support (solid form) was mixed or dispersed in the aqueous solution of active phase. The agitation duration and temperature of the mixture ranged between 0.5 to 12 h and room temperature to 80 °C, respectively (Alaei et al., 2020; De and Boxi, 2020; Safakish et al., 2020; Wen et al., 2010). It was reported that ultrasonication treatment for 1 h could be applied to the mixture prior to the agitation process (Ibrahim et al., 2020). Next, the mixture was filtered and dried overnight in a temperature ranging between 50 to 120 °C. Finally, the dried catalyst sample was subjected to a calcination process. The calcination temperature and duration were in the range of 400–800 °C and 2–4 h (Alaei et al., 2018; Amani et al., 2019; Ambat et al., 2019; De and Boxi, 2020; Feyzi and Khajavi, 2014; Feyzi and Norouzi, 2016; Ibrahim et al., 2020; Kataria et al., 2017; Safakish et al., 2020; Wen et al., 2010), respectively, depending on the types of catalysts produced. The synthesis route of the wet impregnation method is shown in Figure 4.5.

4.3.3 Chemical Functionalization or Immobilization

The catalysts prepared via the wet impregnation method encountered the problem of active phase leaching into the reaction media. This problem can be overcome by chemically bonding the active phase to the catalyst support (Shuit et al., 2013). The catalyst supports that were reported to be chemically functionalized or immobilized with the active phase and served as nanocatalysts for biodiesel production included polymer-, carbon- or metal-based supports. The examples of polymer-based and carbon-based catalyst supports were poly(4-divinylbenzene-co-vinylbenzyl chloride) (PDVC)

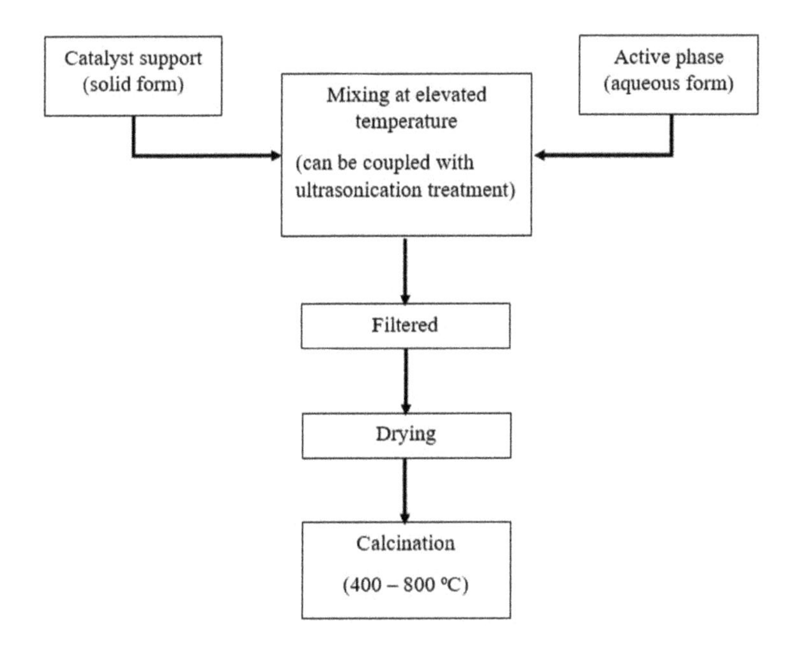

FIGURE 4.5 The synthesis route of the wet impregnation method.

(Negm et al., 2019) and multiwalled carbon nanotubes (MWCNTs) (Shuit et al., 2015), respectively. Meanwhile, the metal-based supports could be boehmite (Hosseini et al., 2019) or iron oxide (Fe_3O_4) (Jambulingam et al., 2019; Raita et al., 2015). The active phases that were reported to be grafted on the catalyst support and served as catalyst for biodiesel production included phosphotungstic acid (PTA) (Negm et al., 2019), sulphonic group (SO_3H) (Shuit et al., 2015), basic ionic liquid (Hosseini et al., 2019) and lipase enzyme (Jambulingam et al., 2019; Raita et al., 2015).

To produce phosphotungstic acid-functionalized PDVC (PTA-PDVC), the cross-linked PDVC was firstly synthesized via an emulsion copolymerization reaction. The pre-determined amount of cetrimonium bromide (CTAB) was dispersed in distilled water under nitrogen flow for 30 min. Subsequently, 4-divinylbenzene (DVB) and 4-vinylbenzyl chloride (VBC) monomers were added into the CTAB solution. The mixture was bubbled with nitrogen gas for 30 min to remove dissolved oxygen before heating to 75 °C. After heating for 15 min, the polymerization initiator, dibenzoyl peroxide (BOP), was charged into the mixture. The mixture was periodically analysed by thin-layer chromatography (TLC) to ensure a complete polymerization reaction. The yellowish precipitate resulting from the polymerization reaction was thoroughly washed with water/methanol solution and then dried under vacuum (60 °C) (Negm et al., 2019; Jeevanantham et al., 2018). Prior to functionalization with PTA, the PDVC was aminated by ethylenediamine (EDA) or 1–8,diaminooctane (DAO) via a quaternization reaction. The pre-calculated amount of PDVC in toluene solution was stirred under nitrogen flow for 30 min. Next, a dilute solution of EDA or DAO (in toluene) was added into the PDVC solution and stirred for 2 h at 0 °C. Then, the temperature of the mixture was gradually increased to 85 °C and maintained for 10

h. The yellowish precipitate (NH_2-PDVC) was recovered via filtration, washed thoroughly with water/acetone solution and dried overnight under vacuum (40 °C). The functionalization of NH_2-PDVC with PTA was achieved by quaternization (a reaction which can produce a compound with uneven alkyl chain length). The catalyst support of NH_2-PDVC was dispersed in 80% v/v of a water–acetonitrile mixture. The solution was stirred at room temperature for 3 h. Next, the pre-determined amount of PTA was added into the solution and the mixture was heated to 75 °C for 8 h with continuous stirring. Upon completion, the precipitate was filtered and subjected to Soxhlet water washing for 8 h. The purpose of Soxhlet water washing was to remove the unreacted PTA. Lastly, the washed catalyst was dried overnight under vacuum (40 °C) (Negm et al., 2019).

MWCNTs can be activated via a sulphonation process. Sulphonation is a process of functionalizing the catalyst support with SO_3H group. Various sulphonation methods, such as thermal treatment with concentrated sulphuric acid, *in situ* polymerization of acetic anhydride and sulphuric acid, thermal decomposition of ammonium sulphate and *in situ* polymerization of poly(sodium4-styrenesulphonate) (Shuit and Tan, 2014) have been reported to be feasible in grafting the MWCNTs with the SO_3H group. Prior to the sulphonation process, MWCNTs were subjected to acid coupling with ultrasonication treatment to purify them and the carboxyl groups were attached to the sidewalls of the MWCNTs (Shuit et al., 2013).

Basic ionic liquid was prepared by mixing (2-chloroethyl)-tri-methyl ammonium chloride (CCH) with KOH in ethanol. The mixture was subjected to ultrasonication for 2 h followed by the separation of precipitate KCl. The basic ionic liquid catalyst, CCH, was formed after the removal of ethanol. For the immobilization of CCH on nanoboehmite, pre-determined amounts of CCH and nanoboehmite were mixed with toluene and stirred for 24 h at 100 °C. Upon completion, the mixture was filtered, washed with ethanol and dried at 50 °C for 6 h. CCH was attached to the nanoboehmite via a hydrogen bond (Hosseini et al., 2019).

For the immobilization of lipase on Fe_3O_4, Fe_3O_4 nanoparticles were initially dispersed in potassium phosphate buffer solution containing NaCl. The pH of the solution was maintained at 6. Then, the suspension was charged into 1-ethyl-3-(3-dimethylaminopropyl) carbodiimide (EDC) solution with the aid of ultrasonication (35 kHz) for 15 min. Next, lipase enzyme was added into the mixture and the suspension was ultrasonicated at 30 kHz for 30 min. The biocatalyst was then recovered via magnetic attraction and washed several times with buffer solution until no lipase was detected in the supernatant. The lipase-Fe_3O_4 was air dried and stored at a low temperature of 4 °C (Jambulingam et al., 2019; Raita et al., 2015).

4.4 Mechanism of Nanomaterials as Catalysts in Biodiesel Synthesis

In order to fully establish the understanding of the role of nanomaterial-based catalyst in biodiesel synthesis, exploration of its mechanism of action is paramount, especially when comparing with a non-catalytic process. The mechanism of nanomaterials as catalyst in biodiesel synthesis will entail the electron mobility, molecular structural transformation, formation of intermediates, reaction pathways, mass transfer, kinetics and thermodynamic data in detail. This information will pave the way for a more thorough scientific investigation to further improve the efficiency and reactivity of the nanomaterial-based catalyst in biodiesel synthesis starting at the molecular level. The relevant kinetics and thermodynamics data will also be crucial for the potential future up-scaling of the process into the commercial scale (Wong et al., 2020). It is also possible to overcome the common challenges faced by the implementation of nanomaterial-based catalyst in biodiesel production which include the high spatial hindrance, low reusability due to leaching issue and low stability of the catalyst by proper dissecting of the relevant mechanism pathways. Unfortunately, relevant studies which focus on the mechanism study of nanomaterial-based catalyst for biodiesel synthesis are still very much lacking. Therefore, this section serves to provide a compilation of the latest mechanism studies conducted in this area and highlights several of the key findings related to the two main catalytic pathways, which are solid base and acidic catalysts.

4.4.1 Solid Base-Catalysed Reaction

Generally, a solid base-catalysed transesterification reaction for biodiesel synthesis is preferred if the feedstock used has a low content of free fatty acids and water. This is because it usually demonstrates a higher catalytic efficiency at a lower process severity compared to its acidic catalyst counterpart. Basic earth oxide catalysts such as CaO possess reactive ion pairs [M^{+2}-O^{-2}] which give rise to its basicity and catalytic activity during the transesterification reaction. Due to the nature of the coupled acid–base reaction pair, the reactants will be acting as the acid towards the basic catalyst (Hattori, 1995). Liu et al. (2008) studied the mechanism of CaO as a basic catalyst for the transesterification of soybean oil into biodiesel. They summarized the mechanism into seven steps as shown in Figure 4.6, where R represents different long chain alkyl groups. For basic catalysts such as CaO, the surface oxides (O^{2-}) will accept the protons (H^+) from the dissociation of water molecules to form surface hydroxyl ions (step 1). With the addition of methanol, it will combine with the generated H^+ ions to form strong basic methoxide ions which are responsible for the main reaction activity (steps 2 and 3). During the transesterification process, the methoxide ions will react with the glyceride molecules through nucleophilic attack on the carbonyl carbon to form a tetrahedral intermediate (step 4). Afterwards, the tetrahedral intermediate can undergo two possible pathways to generate methoxide anion. It can be protonated by the H^+ ions from the surface of the CaO catalyst (step 5) or from the methanol molecules themselves (step 6). Finally, the electrons in the unstable tetrahedral intermediate will rearrange themselves to form the metastable fatty acid methyl ester and glycerol molecules (step 7). Steps 4 to 7 are repeated thrice for triglyceride, diglyceride and monoglyceride, and the overall transesterification reaction is combined to form step 8. Due to the basic nature of the catalyst, the presence of water exceeding a certain limit will induce hydrolysis of the fatty acid methyl esters to free the fatty acid molecules (step 9). Subsequently, they will react with the basic catalyst to form soap, which is the primary cause of catalyst deactivation (Kouzu and Hidaka, 2012). In contrast to this mechanism, Kouzu and Hidaka (2012) suggested that the formation of the basic methoxide ions was not the rate-determining step in the reaction. Their study deduced that the original surface activity of the CaO catalyst may play a larger part in the transesterification process directly. In addition, there was also a possibility that the

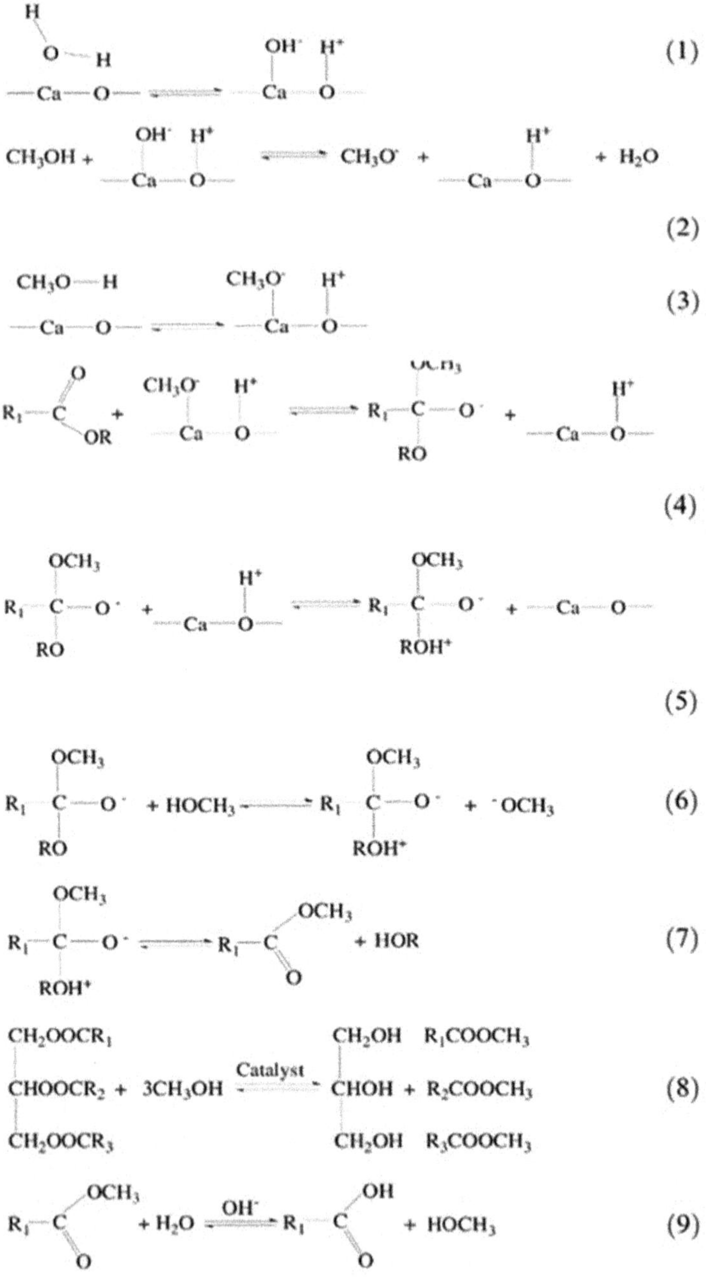

$$\text{H–O···H (adsorbed on) –Ca–O–} \rightleftharpoons \text{–Ca(OH}^-\text{)–O(H}^+\text{)–} \quad (1)$$

$$CH_3OH + \text{–Ca(OH}^-\text{)–O(H}^+\text{)–} \rightleftharpoons CH_3O^- + \text{–Ca–O(H}^+\text{)–} + H_2O \quad (2)$$

$$CH_3O\text{–H (adsorbed on) –Ca–O–} \rightleftharpoons \text{–Ca(CH}_3O^-\text{)–O(H}^+\text{)–} \quad (3)$$

$$R_1\text{–C(=O)–OR} + \text{–Ca(CH}_3O^-\text{)–O(H}^+\text{)–} \rightleftharpoons R_1\text{–C(OCH}_3\text{)(RO)–O}^- + \text{–Ca–O(H}^+\text{)–} \quad (4)$$

$$R_1\text{–C(OCH}_3\text{)(RO)–O}^- + \text{–Ca–O(H}^+\text{)–} \rightleftharpoons R_1\text{–C(OCH}_3\text{)(ROH}^+\text{)–O}^- + \text{–Ca–O–} \quad (5)$$

$$R_1\text{–C(OCH}_3\text{)(RO)–O}^- + HOCH_3 \rightarrow R_1\text{–C(OCH}_3\text{)(ROH}^+\text{)–O}^- + {}^-OCH_3 \quad (6)$$

$$R_1\text{–C(OCH}_3\text{)(ROH}^+\text{)–O}^- \rightleftharpoons R_1\text{–C(=O)–OCH}_3 + HOR \quad (7)$$

$$\begin{matrix} CH_2OOCR_1 \\ CHOOCR_2 \\ CH_2OOCR_3 \end{matrix} + 3CH_3OH \xrightarrow{\text{Catalyst}} \begin{matrix} CH_2OH \\ CHOH \\ CH_2OH \end{matrix} + \begin{matrix} R_1COOCH_3 \\ R_2COOCH_3 \\ R_3COOCH_3 \end{matrix} \quad (8)$$

$$R_1\text{–C(=O)–OCH}_3 + H_2O \xrightarrow{OH^-} R_1\text{–C(=O)–OH} + HOCH_3 \quad (9)$$

FIGURE 4.6 Mechanisms of base-catalysed transesterification reactions. (Adapted from Liu et al., 2008, with permission.)

chemically reactive calcium glyceroxide (Ca-Gly) compounds which formed from the glycerolysis after sufficient concentration of glycerol molecules was produced could also act as the chemically active sites for the transesterification process. However, it was less reactive compared to CaO catalyst. They also proposed that the reaction kinetics catalysed by the basic catalyst generally followed the Eley-Rideal kinetic model. According to Al-Sakkari et al. (2017), the Eley-Rideal kinetic model could be applied to reactions controlled by surface activity, such as CaO-catalysed transesterification. It is important to ascertain the rate-determining step associated with the mass transfer resistance during the kinetic study in order to validate the proposed mechanism. The mass transfer resistances involved could be distinguished into the bulk resistance for the transfer of the reactants to the catalyst surface, surface resistance for the reaction on the catalyst surface and internal diffusion resistance in the pores of the catalyst. The internal diffusion resistance was particularly important for highly porous catalyst formed from the nanomaterial. It is also noteworthy that the soluble active sites leached from the catalyst can also promote the reaction for biodiesel synthesis. The leaching of CaO nanocatalyst to the reaction medium occurs mainly due to the combined effects of high polarity alcohol groups and the glycerolysis with glycerol which exists as the by-products in the transesterification reaction (Kouzu and Hidaka, 2012). Furthermore, basic nanocatalyst derived from CaO has a higher tendency to be deactivated by the presence of free fatty acid through the neutralization process. Excessive catalyst leaching into the reaction mixture could affect the accuracy of the Eley-Rideal model. Thus, other more suitable homogeneous models, such as first order or second order, could be applied instead (Dhawane et al., 2021).

4.4.2 Solid Acidic-Catalysed Reaction

In contrary to base-catalysed biodiesel synthesis, solid acidic catalysts can withstand lower quality feedstock such as animal fats and waste cooking oil which contain high amounts of water and free fatty acids. Its acidic nature in the form of Bronsted and Lewis acids allows this type of catalyst to promote both transesterification and esterification reactions simultaneously. However, its catalytic efficiency is highly dependent on the accessibility of reactants to the active sites in order for the reaction to proceed. In order to improve the reactants' accessibility, synthesis of the solid acidic catalyst from nanomaterial has been focused on increasing the amount of active sites, pore sizes, surface hydrophobicity, reduction of mass transfer resistance to the catalyst surface and minimizing deactivation of the catalyst active sites. Soltani et al. (2017) synthesized a mesoporous nano-ZnO catalyst sulphonated by H_2SO_4 for the esterification of palm fatty acid distillate (PFAD) to produce fatty acid methyl esters. In their study, they proposed that the esterification reaction commenced when the PFAD molecules diffused to the surface of the catalyst from the bulk medium. Then, these molecules would absorb into the pores of the catalyst to access the active sulphonated sites. Under the high pressure and temperature in the autoclave reactor, hydrogen atoms generated from the cleavage protonated the PFAD molecules before being converted into fatty acid methyl esters by the SO_3H functional groups. Lopez et al. (2008) conducted a study on the simultaneous esterification and transesterification of a tricaprylin and caprylic acid mixture using a sulphonated zirconia catalyst. They opined that the kinetic rate for the esterification reaction was higher compared to the transesterification reaction due to the bulkier triglyceride molecules. However, the conversion of triglycerides to esters could also be enhanced via hydrolysis followed by rapid esterification. Thus, the presence of water molecules could in fact be beneficial to the reaction as long as it was not excessive. From the kinetic study, they concluded that the esterification reaction catalysed by the solid sulphonated nano-zirconia catalyst obeyed a pseudo second-order model. In addition, it was not diffusion-controlled and the mass transfer limitation was not obvious. The advantages of a simultaneous esterification and transesterification reaction were further studied by Raia et al. (2017) with the similar sulphonated zirconia catalyst in *Jatropha curcas* L. oil. The high acidity of the catalyst was contributed by the sulphate groups attached to the nanocatalyst in the amorphous solid structure. After the synthesis process, the sulphate groups would transform into two S=O covalent double bonds on the catalyst surface and pores. This gave rise to the strong Bronsted and Lewis acid characteristics which promoted the simultaneous reaction. A three-stage mechanism was proposed with the esterification of free fatty acids proceeding first due to its higher kinetic rate. This was followed by the transesterification of the bulkier triglyceride molecules with the production of di- and mono-glyceride as intermediates. Two side reactions might occur which would compete with the transesterification process. Due to the requirement for a higher temperature, the triglyceride molecules could undergo cracking to breakdown into shorter chain fatty acids. Subsequently, esterification would take place to convert them into esters. The presence of water molecules could also induce hydrolysis of the triglyceride molecules to produce more free fatty acids for esterification. These reactions would proceed to equilibrium at the final stage. The general mechanism for both the esterification and transesterification reactions was deduced based on the Bronsted and Lewis acid sites on the sulphated nano-zirconia catalyst, as shown in Figure 4.7. These acidic active sites promoted the nucleophilic attack of the hydroxyl groups from the alcohol reactant to the carbonyl oxygen group present on the fatty acid chains. At the Bronsted acid sites, protonation of the carbonyl oxygen group would occur. Meanwhile, Lewis acid sites would produce a reactive complex intermediate when the carbonyl oxygen groups were attached to them. This reactive intermediate would then undergo tetrahedral rearrangement to produce esters and water molecules (Raia et al., 2017).

4.5 Conclusions

The development of biodiesel has reached a maturity stage as it is poised to displace part of the fossil fuel energy demand and decarbonize the transportation sector. However, its commercial production is still beleaguered by several challenges including low reactivity, low yield, high costs and the huge generation of waste products. Consequently, intensive research and development are on-going to ensure that biodiesel synthesis becomes more sustainable and economically competitive. The application of nanomaterials as solid catalysts in biodiesel production has the potential to address several of these shortcomings. Its distinctive characteristics include high surface area, tuneable surface properties and high stability. Recently, more studies have been conducted on the synthesis of bio-based nanomaterials as catalyst support for biodiesel production (Lim et al., 2020; Wong et al., 2020). This

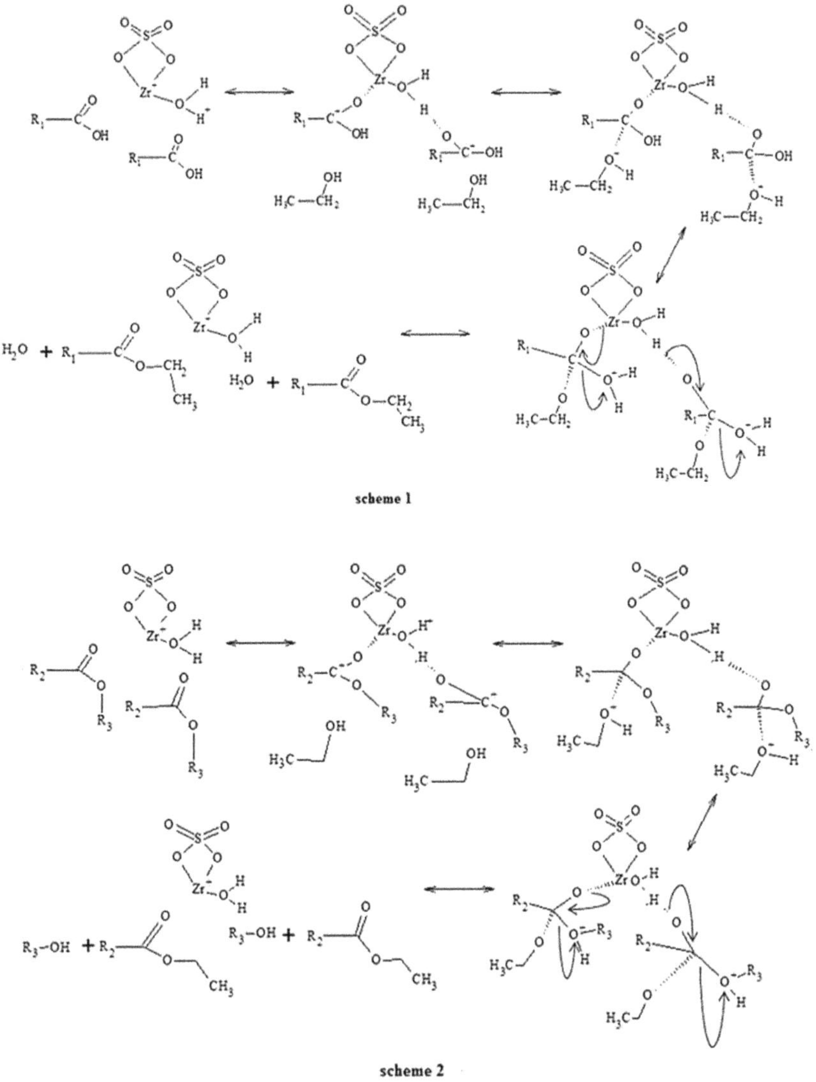

scheme 1

scheme 2

FIGURE 4.7 General mechanisms for simultaneous esterification (Scheme 1) and transesterification (Scheme 2) reactions catalysed by solid acid catalysts. (Adapted from Raia et al., 2017, with permission.)

could drive down the cost of the catalyst as well as ensuring the sustainability of the process due to its biodegradable nature. However, there remains a technical gap to be addressed in order to translate the superior laboratory results to larger commercial-scale production in the industry. Furthermore, the synthesis of the nanomaterial as a catalyst for biodiesel production still has several issues to be overcome. The fundamental knowledge pertaining to the characteristics of nanomaterials and their relation to conversion reactions is still very much lacking. The synthesis parameters are yet to be optimized and more discoveries are required on the effectiveness of the different chemical reagents towards the reactivity of the catalyst. More critically, the leaching problem associated with catalyst deactivation needs to be further improved by the investigation of stronger attachment and bonding of the active groups for the reaction. As such, it is important to understand its role, synthesis pathways and mechanism in biodiesel production as elucidated in this chapter for further advancement of its scientific research. It is believed that nanomaterials could play a hugely beneficial role in sustainable biodiesel synthesis with more in-depth and fundamental studies being performed in the future.

REFERENCES

Alaei, S., Haghighi, M., Rahmanivahid, B., Shokrani, R., and Naghavi, H. 2020. Conventional vs. Hybrid methods for dispersion of MgO over magnetic Mg–Fe mixed oxides nanocatalyst in biofuel production from vegetable oil. *Renewable Energy 154*: 1188–203.

Alaei, S., Haghighi, M., Toghiani, J., and Rahmani Vahid, B. 2018. Magnetic and reusable $MgO/MgFe_2O_4$ nanocatalyst for biodiesel production from sunflower oil: influence of fuel ratio in combustion synthesis on catalytic properties and performance. *Industrial Crops and Products* 117: 322–32.

Al-Sakkari, E.G., El-Sheltawy, S.T., Attiab, N.K., and Mostafa, S.R. 2017. Kinetic study of soybean oil methanolysis using cement kiln dust as a heterogeneous catalyst for biodiesel production. *Applied Catalyst B* 206: 146–157.

Amani, T., Haghighi, M., and Rahmanivahid, B. 2019. Microwave-assisted combustion design of magnetic Mg–Fe spinel for Mgo-based nanocatalyst used in biodiesel production: influence of heating-approach and fuel ratio. *Journal of Industrial and Engineering Chemistry* 80: 43–52.

Ambat, I., Srivastava, V., Haapaniemi, E., and Sillanpää, M. 2019. Nano-magnetic potassium impregnated ceria as catalyst for the biodiesel production. *Renewable Energy* 139: 1428–36.

Badnore, A.U., Jadhav, N.L., Pinjari, D.V., and Pandit, A.B. 2018. Efficacy of newly developed nano-crystalline calcium oxide for biodiesel production. *Chemical Engineering Process* 133: 312–319.

Banković-Ilić, I.B., Miladinović, M.R., Stamenković, O.S., and Veljković, V.B. 2017. Application of nano CaO–based catalysts in biodiesel synthesis. *Renewable and Sustainable Energy Reviews* 72: 746–760.

Baskar, G., Gurugulladevi, A., Nishanthini, T., Aiswarya, R., and Tamilarasan, K. 2017. Optimization and kinetics of biodiesel production from mahua oil using manganese doped zinc oxide nanocatalyst. *Renewable Energy* 103: 641–46.

Baskar, G., and Soumiya, S. 2016. Production of biodiesel from castor oil using iron (ii) doped zinc oxide nanocatalyst. *Renewable Energy* 98: 101–07.

Bet-Moushoul, E., Farhadi, K., Mansourpanah, Y., Nikbakht, A.M., Molaei, R., and Forough, M. 2016. Application of CaO-based/au nanoparticles as heterogeneous nanocatalysts in biodiesel production. *Fuel* 164: 119–27.

Borah, M.J., Devi, A., Borah, R., and Deka, D. 2019. Synthesis and application of co doped zno as heterogeneous nanocatalyst for biodiesel production from non-edible oil. *Renewable Energy* 133: 512–19.

De, A., and Boxi, S.S. 2020. Application of cu impregnated TiO2 as a heterogeneous nanocatalyst for the production of biodiesel from palm oil. *Fuel* 265: 117019.

Degirmenbasi, N., Coskun, S., Boz, N., and Kalyon, D.M. 2015. Biodiesel synthesis from canola oil via heterogeneous catalysis using functionalized CaO nanoparticles. *Fuel* 153: 620–627.

Dehghani, S., and Haghighi, M. 2019. Sono-dispersed mgo over cerium-doped MCM-41 nanocatalyst for biodiesel production from acidic sunflower oil: surface evolution by altering si/ce molar ratios. *Waste Management* 95: 584–92.

Dehghani, S., and Haghighi, M. 2020. Sono-enhanced dispersion of Cao over Zr-doped MCM-41 bifunctional nanocatalyst with various Si/Zr ratios for conversion of waste cooking oil to biodiesel. *Renewable Energy* 153: 801–12.

Demirbas, A. 2003. Biodiesel fuels from vegetable oils via catalytic and non-catalytic supercritical alcohol transesterifications and other methods: a survey. *Energy Conversion and Management* 44: 2093–2109.

Dhawane, S.H., Al-Sakkari, E.G., Kumar, T., and Halder, G. 2021. Comprehensive elucidation of the apparent kinetics and mass transfer resistances for biodiesel production via in-house developed carbonaceous catalyst. *Chemical Engineering Research and Design* 165: 192–206.

Esmaeili, H., Seffati, K., Honarvar, B., and Esfandiari, N. 2019. Ac/$CuFe_2O_4$@Cao as a novel nanocatalyst to produce biodiesel from chicken fat. *Renewable Energy* 147, 25–34.

Feyzi, M., and Khajavi, G. 2014. Investigation of biodiesel production using modified strontium nanocatalysts supported on the ZSM-5 zeolite. *Industrial Crops and Products* 58: 298–304.

Feyzi, M., and Norouzi, L. 2016. Preparation and kinetic study of magnetic Ca/Fe_3O_4@SiO_2 nanocatalysts for biodiesel production. *Renewable Energy* 94: 579–86.

Gardy, J., Rehan, M., Hassanpour, A., Lai, X., and Nizami, A.S. 2019. Advances in nano-catalysts based biodiesel production from non-food feedstocks. *Journal of Environmental Management* 249: 109316.

Hattori, H. 1995. Heterogeneous basic catalysis. *Chemical Reviews* 95 (3): 537–558.

Hebbar, H.R.H., Math, M.C., and Yatish, K.V. 2018. Optimization and kinetic study of CaO nano-particles catalysed biodiesel production from *Bombax coiba* oil. *Energy* 143: 25–34.

Hosseini, S., Moradi, G.R., and Bahrami, K. 2019. Synthesis of a novel stabilized basic ionic liquid through immobilization on boehmite nanoparticles: a robust nanocatalyst for biodiesel production from soybean oil. *Renewable Energy* 138: 70–78.

Hsiao, M.C., Lin, C.C., and Chang, Y.H. 2011. Microwave irradiation assisted transesterification of soybean oil to biodiesel catalyzed by nanopowder calcium oxide. *Fuel* 90: 1963–1967.

Ibrahim, M.L., Nik Abdul Khalil, N.N.A., Islam, A., Rashid, U., Ibrahim, S.F., Sinar Mashuri, S.I., and Taufiq-Yap, Y.H. 2020. Preparation of Na_2O supported cnts nanocatalyst for efficient biodiesel production from waste-oil. *Energy Conversion and Management* 205: 112445.

Jambulingam, R., Shalma, M., and Shankar, V. 2019. Biodiesel production using lipase immobilised functionalized magnetic nanocatalyst from oleaginous fungal lipid. *Journal of Cleaner Production* 215: 245–58.

Jeevanantham, S., Hosimin, S., Vengatesan, S., and Sozhan, G. 2018. Quaternized poly(styrene-co-vinylbenzyl chloride) anion exchange membranes: role of different ammonium cations on structural, morphological, thermal and physio-chemical properties. *New Journal of Chemistry* 42(1): 380–87.

Kataria, J., Mohapatra, S., and Kundu, K. 2017. Biodiesel production from frying oil using zinc-doped calcium oxide as heterogeneous catalysts. *Energy Sources, Part A: Recovery, Utilization, and Environmental Effects* 39: 1–6.

Kesić, Ž., Lukić, I., Zdujić, M., Mojović, L., and Skala, D. 2016. Calcium oxide based catalysts for biodiesel production: a review. *Chemical Industry and Chemical Engineering Quarterly* 22: 391–408.

Kouzu, M., and Hidaka, J. 2012. Transesterification of vegetable oil into biodiesel catalyze by CaO: a review. *Fuel* 93: 1–12.

Kung, H., and Kung, M. 2007. Nanotechnology and Heterogeneous Catalysis. In: B. Zhou, S. Han, R. Raja and G. Somorjai, eds. *Nanotechnology in Catalysis.* New York: Springer Science+Business Media: 1–11.

Lim, S., and Lee, K.T. 2010. Recent trends, opportunities and challenges of biodiesel in Malaysia: an overview. *Renewable and Sustainable Energy Reviews* 14(3): 938–954.

Lim, S., Yap, C.Y., Pang, Y.L., and Wong, K.H. 2019. Biodiesel synthesis from oil palm empty fruit bunch biochar derived heterogeneous solid catalyst using 4-benzenediazonium sulfonate. *Journal of Hazardous Materials* 390: 121532.

Lim, S., Pang, Y.L., Shuit, S.H., Wong, K.H., and Leong, C.K. 2020. Synthesis and characterization of monk fruit seed (*Siraitia grosvenorii*)-based heterogeneous acid catalyst for biodiesel production through esterification process. *International Journal of Energy Research* 44(12): 9454–9465.

Liu, X., He, H., Wang, Y., Zhu, S., and Piao, X. 2008. Transesterification of soybean oil to biodiesel using CaO as a solid base catalyst. *Fuel* 87 (2): 216–221.

López, D.E., Goodwin, J.G., Bruce, D.A., and Furuta, S. 2008. Esterification and transesterification using modified-zirconia catalysts. *Applied Catalysis A: General* 339 (1): 76–83.

Naveenkumar, R., and Baskar, G. 2019. Biodiesel production from calophyllum inophyllum oil using zinc doped calcium oxide (plaster of paris) nanocatalyst. *Bioresource Technology* 280: 493–96.

Negm, N.A., Betiha, M.A., Alhumaimess, M.S., Hassan, H.M.A., and Rabie, A.M. 2019. Clean transesterification process for biodiesel production using heterogeneous polymer-heteropoly acid nanocatalyst. *Journal of Cleaner Production* 238: 117854.

Philippot, K., and Serp, P. 2013. Concepts in Nanocatalysis. In: P. Serp and K. Philippot, eds. *Nanomaterials in Catalysis.* Weinheim, Germany: Wiley–VCH Verlag GmbH & Co. KGaA: 1–54.

Rahmani Vahid, B., Haghighi, M., Toghiani, J., and Alaei, S. 2018. Hybrid-coprecipitation vs. Combustion synthesis of Mg-Al spinel based nanocatalyst for efficient biodiesel production. *Energy Conversion and Management 160*: 220–29.

Raia, R.Z., Sabino da Silva, L., Marcucci, S.M.P., and Arroyo, P.A. 2017. Biodiesel production from *Jatropha curcas* L. oil by simultaneous esterification and transesterification using sulphated zirconia. *Catalysis Today* 289: 105–114.

Raita, M., Arnthong, J., Champreda, V., and Laosiripojana, N. 2015. Modification of magnetic nanoparticle lipase designs for biodiesel production from palm oil. *Fuel Processing Technology* 134: 189–97.

Rajendran, N., and Gurunathan, D.B. 2020. Optimization and techno-economic analysis of biodiesel production from calophyllum inophyllum oil using heterogeneous nanocatalyst. *Bioresource Technology* 315: 123852.

Seffati, K., Honarvar, B., Esmaeili, H., and Esfandiari, N. 2019. Enhanced biodiesel production from chicken fat using CaO/CuFe_2O_4 nanocatalyst and its combination with diesel to improve fuel properties. *Fuel* 235: 1238–1244.

Safakish, E., Nayebzadeh, H., Saghatoleslami, N., and Kazemifard, S. 2020. Comprehensive assessment of the preparation CaO/CuFe_2O_4 a separable magnetic nanocatalyst for biodiesel production from algae. *Algal Research* 49: 101949.

Seffati, K., Honarvar, B., Esmaeili, H., and Esfandiari, N. 2019. Enhanced biodiesel production from chicken fat using cao/cufe2o4 nanocatalyst and its combination with diesel to improve fuel properties. *Fuel* 235: 1238–44.

Shuit, S.H., Ng, E.P., and Tan, S.H. 2015. A facile and acid-free approach towards the preparation of sulphonated multi-walled carbon nanotubes as a strong protonic acid catalyst for biodiesel production. *Journal of the Taiwan Institute of Chemical Engineers* 52: 100–08.

Shuit, S.H., and Tan, S.H. 2014. Feasibility study of various sulphonation methods for transforming carbon nanotubes into catalysts for the esterification of palm fatty acid distillate. *Energy Conversion and Management* 88: 1283–89.

Shuit, S.H., Yee, K.F., Lee, K.T., Subhash, B., and Tan, S.H. 2013. Evolution towards the utilisation of functionalised carbon nanotubes as a new generation catalyst support in biodiesel production: an overview. *RSC Advances* 3(24): 9070–94.

Singh, S., Mukherjee, D., Dinda, S., Ghosal, S., and Chakrabarty, J. 2020. Synthesis of coo nio promoted sulfated ZrO2 super-acid oleophilic catalyst via co-precipitation impregnation route for biodiesel production. *Renewable Energy* 158: 656–67.

Soltani, S., Rashid, U., Al-Resayes, S.I., and Nehdi, I.A. 2017. Sulfonated mesoporous ZnO catalyst for methyl esters production. *Journal of Cleaner Production* 144: 82–491.

Tang, Z.E., Lim, S., Pang, Y.L., Ong, H.C., and Lee, K.T. 2018. Synthesis of biomass as heterogeneous catalyst for application in biodiesel production: state of the art and fundamental review. *Renewable and Sustainable Energy Reviews* 92: 235–253.

Wen, L., Wang, Y., Lu, D., Hu, S., and Han, H. 2010. Preparation of KF/CaO nanocatalyst and its application in biodiesel production from chinese tallow seed oil. *Fuel 89*(9): 2267–71.

Wong, W.Y., Lim, S., Pang, Y.L., Shuit, S.H., Chen, W.H., and Lee, K.T. 2020. Synthesis of renewable heterogeneous acid catalyst from oil palm empty fruit bunch for glycerol-free biodiesel production. *Science of the Total Environment* 727: 138534.

5

Metal Oxide/Sulphide-Based Nanocatalysts in Biodiesel Synthesis

Juan S. Villarreal and José R. Mora*

CONTENTS

5.1 Introduction

The search for renewable energies as alternatives to fossil fuels has grown greatly due to the limited reserves of petroleum, instability of its price and its negative environmental impact (Guo, Song, and Buhain, 2015). In this regard, biodiesel has been demonstrated to be an appropriate alternative to conventional diesel and its production

* *Corresponding Author* jrmora@usfq.edu.ec

DOI: 10.1201/9781003120858-5

has increased significantly in the last decade (Ogunkunle and Ahmed, 2019). Biodiesel production is performed by applying transesterification and esterification reactions (Ambat, Srivastava, and Sillanpää 2018). The most accepted mechanism for the transesterification process is shown in Figure 5.1, being a reaction with three steps. First, triglycerides (TG) are converted to diglycerides (DG), then to monoglycerides (MG) and finally to glycerol (Figure 5.1) (Chua et al., 2019). The mixture of fatty acids methyl esters (FAMEs) produced in the reaction is the biodiesel.

Usually, strong bases and acids are widely employed as homogeneous catalysts (HomC), obtaining good reaction yields with low cost (Marinković et al., 2016). Nevertheless, homogeneous catalysis produces some problems such as difficulty in separation or non-recoverability (Chua et al., 2019), corrosion of equipment (Wang et al., 2020) and the generation of high amounts of wastewater in the purification step (Marinković et al., 2016). All of these issues increase the production cost and, if catalyst is recovered, the process would require a lot of energy, aqueous quench and neutralization (Lee et al., 2014). This situation is the main motivation for the search for alternative processes for biodiesel production and the use of heterogeneous catalysts as metallic oxides (MOs) and metallic sulphides appears to be a viable way. They are environmentally friendly because less wastewater is generated and it is possible with the use of this type of material for more efficient separation of the biodiesel (Singh Chouhan and Sarma, 2011).

Divalent metal oxides with covalent characteristics facilitate transesterification and, for this reason, alkaline earth metals are frequently used by researchers (Singh Chouhan and Sarma, 2011), where CaO is the most studied catalyst (Marinković

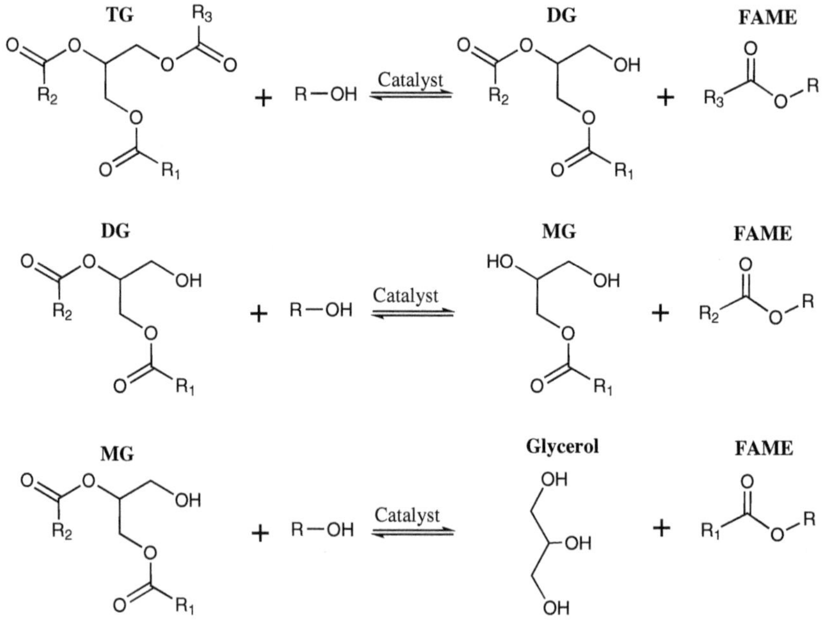

FIGURE 5.1 Reaction mechanism of the transesterification process.

et al., 2016). MgO and SrO are widely used also (Singh Chouhan and Sarma, 2011). In addition, researchers have developed and studied bifunctional catalysts. These types of catalysts take advantage of acid and base properties at the same time (Wang et al., 2020).

5.2 Metal Oxide-Based Nanocatalyst Preparation

There are several methods for obtaining nanoparticles, which can be divided into chemical, physical and biological methods. Chemical methods are the most commonly used, especially for nanoparticles that will be used later in catalysis processes. Worldwide handled methods used for magnetic nanoparticle preparation, which will be discussed further, are shown in Figure 5.2, with the chemical methods being the most popular (90%) (Ali et al., 2016). For this reason, the most commonly used techniques in this category are discussed further in detail.

5.2.1 Chemical Precipitation/Co-Precipitation Methods

The chemical precipitation method is one of the most used techniques to obtain MO nanoparticles and one of the most quoted in academic articles (Cruz et al., 2018). It consists of the transformation of metal precursors, such as chlorides, nitrates, carbonates, sulphates or oxalate salts (Singh and Ashok, 2016), into precipitates using

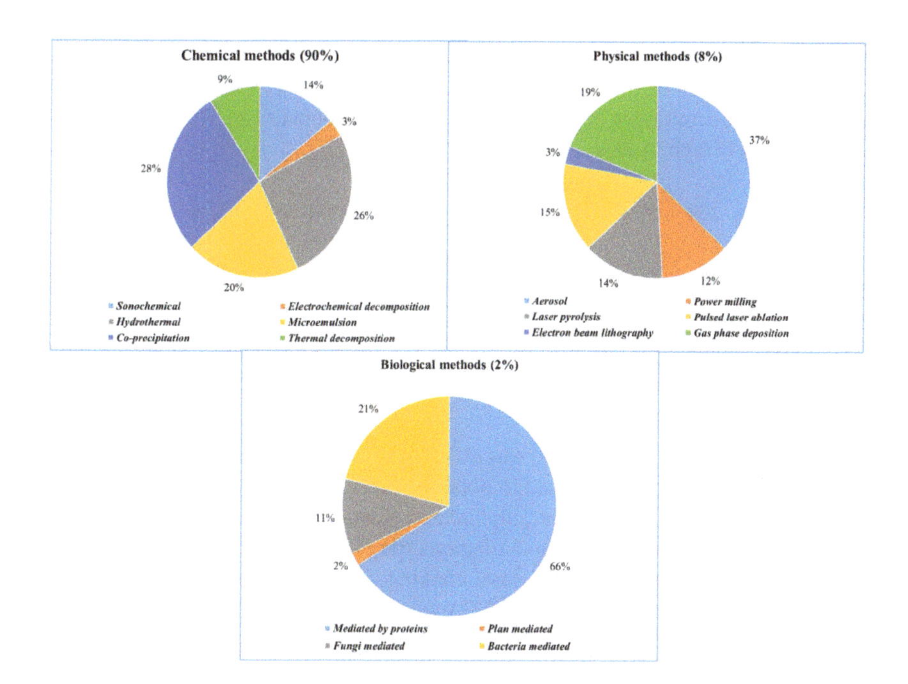

FIGURE 5.2 Worldwide handled methods used for the magnetic nanoparticle preparation.

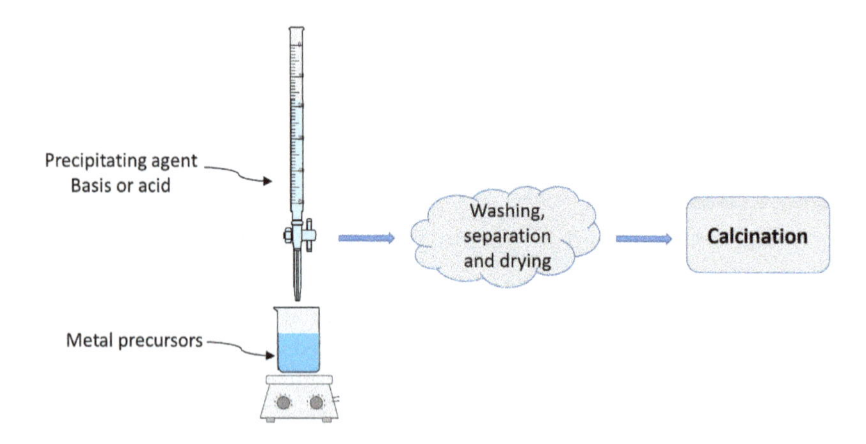

SCHEME 5.1 Graphical representation of the general procedure for the precipitation method.

an acid or basic media. Generally, metal precursors are stirred in a basic solution, such as NaOH, KOH, NH_3OH, $N(CH_3)_4OH$ (Nawaz et al., 2019), $NaHCO_3$ or Na_2CO_3, until supersaturated conditions are reached (Scheme 5.1).

Small particles are formed by nucleation and further, nuclei begin to grow by diffusion of solute into the crystal surface and aggregation (Nawaz et al., 2019). Species formed in the reaction are mostly in the form of hydroxides (Ashik, Kudo, and Hayashi, 2018) and they are insoluble under supersaturated conditions. Then, hydroxides are transformed into oxides by annealing treatment (Rashid et al., 2018). Nevertheless, dispersion and narrow particle size distribution are difficult to achieve because nucleation and growth occur simultaneously (Nawaz et al., 2019). For example, precipitation of $ZnSO_4$ with sodium carbonate can be represented as Reaction (5.1) (Singh, Ashok 2016):

$$2ZnSO_4 + 2Na_2CO_3 + H_2O \longrightarrow 2Zn_2(CO_3)(OH)_2 + 2Na_2SO_4 + CO_2\uparrow \quad (5.1)$$

where the basic carbonate of zinc is the precipitate that will be calcined to produce the metal oxide (Reaction 5.2):

$$2Zn_2(CO_3)(OH)_2 \longrightarrow 2ZnO + H_2O\uparrow + CO_2\uparrow \quad (5.2)$$

As well, waste-iron-filling (WIF) was used to produce $FeCl_2$ in the presence of HCl. $FeCl_2$ was precipitated and calcined to obtain hematite, which was later sulphonated. The catalyst contained mainly a sulphonated aluminium–hematite complex. The material calcined at 900 °C reached a nanometric size and its catalytic activity was studied in the transesterification of waste cooking oil (WCO) with methanol. A yield of 92% was achieved with a MeOH:WCO molar ratio of 12:1, 6 wt% of catalyst and 3 h of reaction at a temperature of 80 °C (Ajala et al., 2020).

Magnetic nanoparticles (MNPs) are a popular option in transesterification reactions and the co-precipitation method is widely used to obtain them (Huang, Lu, and Yang,

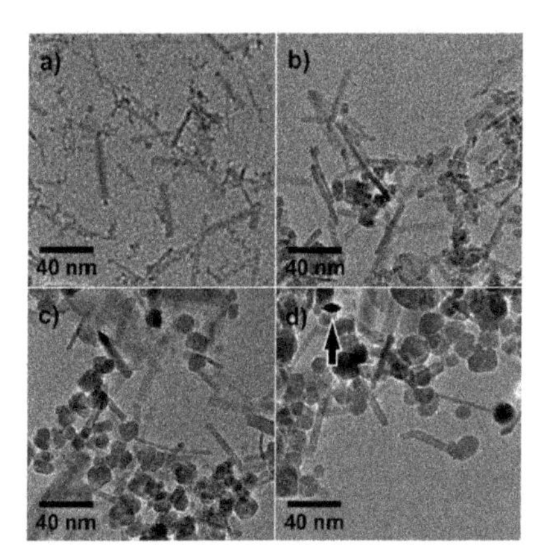

FIGURE 5.3 TEM images obtained for magnetic nanoparticles at different base-to-salt molar ratios: (a) ratio = 2.1, (b) ratio = 2.2, (c) ratio = 2.3 and (d) ratio = 2.4. (Reprinted with permission from Ahn et al. 2012. Copyright © 2012 American Chemical Society.)

2019). Magnetite nanoparticles for example, are obtained by co-precipitating ferric and ferrous salt precursors (Reaction 5.3) (Singh, Ashok 2016):

$$Fe^{2+} + 2Fe^{3+} + 8OH^{-} \longrightarrow Fe_3O_4 + 4H_2O \qquad (5.3)$$

Ahn et al. demonstrated that magnetite formation by the co-precipitation method occurs in several steps and highlighted the importance of base concentration. They observed that the precursor's chloride ions are replaced by hydroxide ions to form akaganeite through continuous ammonium addition. Akagaenite formed goethite, which quickly was transformed to hematite and magnetite (Figure 5.3), where even arrow-shaped particles were formed as intermediaries between goethite and magnetite (Ahn et al., 2012). The influence of the impregnation of two different super bases was analysed on the reaction of soybean oil with methanol. The catalyst impregnated with 1,5,7-triazabicyclo[4,4,0]dec-5-ene reached a reaction yield of 96% at the first cycle, while the catalyst impregnated with 1,1,3,3-tetramethylguanidine obtained only 11% (Santos et al., 2015).

5.2.2 Solvothermal Method

Solvothermal synthesis is a method in which metal precursors are mixed in a solvent at high temperatures (100–250 °C) and moderate or high vapour pressures (0.3–4 MPa) (Soytaş, Oğuz, and Menceloğlu, 2019). A sealed container, usually made of stainless steel and Teflon-lined, is employed, as can be observed in Scheme 5.2. At these conditions, the properties of the solvent change, allowing greater reactivity of the precursor. As the temperature increases, the dielectric constant decreases (Nunes, Pimentel, and Santos, 2019), which facilitates the formation of the nuclei and the crystalline growth around it, forming particles with well-defined shapes and morphologies (Na, Zhang, and Somorjai, 2014).

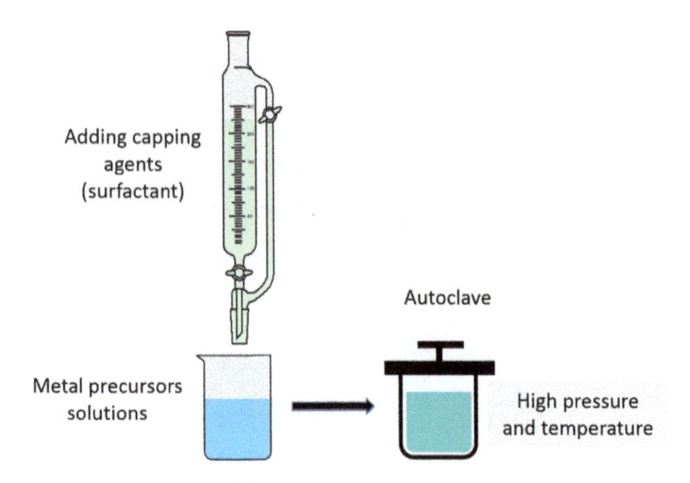

SCHEME 5.2 Schematic representation of the hydrothermal method.

Solvents such as water, ethanol, octadecene, ethylene glycol and diethylene glycol are commonly used (Chang, He, and Pan, 2020). In most cases, a surfactant or capping agent is placed to improve the properties of the product (Cruz et al., 2018) and to prevent agglomeration.

The influence of surfactants on the solvothermal synthesis and catalytic activity of Na-doped ZrO_2 has been studied (Lara-García, Romero-Ibarra, and Pfeiffer, 2014). Zirconium acetate was dissolved and stirred in a NaOH solution. Then, ethanol together with different surfactants were added. The solution was transferred to a Teflon-lined stainless steel autoclave to react for 24 hours at a temperature of 150 °C. The resulting powder was calcined at 500 °C. The catalytic activity was tested in a batch reaction performed with a methanol-to-soybean oil molar ratio of 30:1 and 3 wt% of catalyst for 3 hours at 65 °C, reaching a yield of 93%.

5.2.3 Sol–Gel Method

The sol–gel method consists of using metal precursors to be hydrolysed, condensed, dried and treated thermally to obtain nanoparticles with high stability, surface area, porosity and dispersion, with an average size of between 10 and 100 nm (Nawaz et al., 2019) (Scheme 5.3). Sol is the colloidal dispersion formed from the hydrolysis and condensation of the precursors (Wabeke et al., 2014), commonly alkoxides and alkoxysilanes, while gel is formed after aggregation, aging and drying as a three-dimensional network structure (Rane et al., 2018).

Size-related properties can be affected by parameters such as pH, stirring, drying conditions, reaction time and especially the rate of hydrolysis and condensation which are governed by the solvent-to-precursor ratio. The hydrolysis reaction can be performed by a catalytic/non-catalytic way in acidic/basic medium, with the solvent being the oxygen atoms supplier. In general terms, the reaction can be defined as (Ashik, Kudo, and Hayashi, 2018):

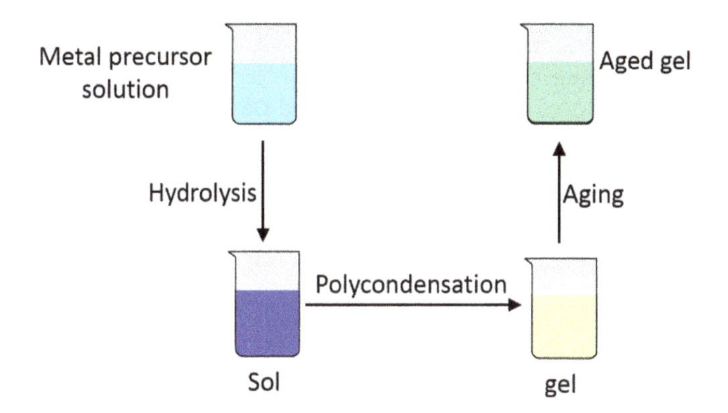

SCHEME 5.3 General representation of the sol–gel techniques.

$$M(OR)_x + mH_2O \longrightarrow M(OR)_{x-m}(OH)_m + mROH \tag{5.4}$$

Then, the condensation occurs as described in Reaction (5.5):

$$2M(OR)_{x-m}(OH)_m \longrightarrow (OH)_{m-1}(OR)_{x-m}\text{-M-O-M}(OR)_{x-m}(OH)_{m-1} + H_2O \tag{5.5}$$

$$2M(OR)_{x-m}(OH)_m \longrightarrow (OH)_{m-1}(OR)_{x-m}\text{-M-O-M}(OR)_{x-m-1} + ROH \tag{5.6}$$

The sol–gel method is considered a cost-effective process, and pure amorphous phases, with different sizes and shapes can be obtained (Nawaz et al., 2019), such as nano-alloys and three-dimensional nanostructured materials (Ashik, Kudo, and Hayashi, 2018).

La_2O_3-ZrO_2 catalysts were produced by sol–gel synthesis, where zirconium n-butoxide and lanthanum acetylacetonate were subjected to a gelling reaction in the presence of water and butanol. The gel was aged, air dried and calcined. The amount of lanthanum oxide was optimized to obtain the highest yield in the canola oil transesterification reaction (Salinas et al., 2017). Wang et al. developed a bifunctional catalyst based on Ca and B by the sol–gel method. They compared the catalytic activity of different boron materials in the transesterification reaction of jatropha oil, and the highest reaction yield of 96% was achieved using the calcium-boron catalyst calcined at 700 °C (Wang et al., 2018).

5.2.4 Impregnation Method

In this technique, a metallic precursor is impregnated in a porous support material *via* capillary absorption (Rashid et al., 2018) and later, precursor is thermally treated (Mehradabi et al., 2017). The precursors are generally inorganic salts such as chlorides, sulphates, nitrates and carbonates. Nitrates do not disperse well because they are redistributed during the heat treatment (Sietsma et al., 2006).

Supports can be inorganic or organic, and their polar character will define the type of solvent to be employed (Munnik, De Jongh, and De Jong, 2015). An ideal support must have strong interactions with the active phase to avoid particle migration and agglomeration without causing isolation (Haukka, Lakomaa, and Suntola, 1998), and a structure that allows accessibility to the active sites to enhance interaction during reaction is recommended (Sietsma et al., 2006). One of the disadvantages of the impregnation method is that control of the size of the particles is complicated, and so the size and narrow distribution pore size of the support are of critical importance (Mehradabi et al., 2017).

Seffati et al. produced a catalyst with a $CuFe_2O_4$ spinel-type active phase into an activated carbon as support. Copper (II) chloride and iron (II) chloride were co-impregnated into activated carbon, and the nanomaterial was magnetically separated and dried. In addition, a CaO layer was impregnated into the nanomaterial. A total of 3% of this catalyst was used for the transesterification reaction of a chicken fat oil with methanol, and a yield of 95.6% was reached. The experimental conditions include a molar methanol-to-oil ratio of 12:1 for 4 h at 65 °C (Seffati et al., 2020).

5.2.5 Combustion Synthesis Method

Combustion synthesis is a method that takes advantage of the exothermicity of a sequence of reactions to generate MO nanoparticles. Solution combustion synthesis (SCS) is the most commonly used method, and so it is discussed further (Singh et al., 2010). SCS consists of heating a solution containing the metal precursor or oxidizer and a soluble fuel with low decomposition temperature until its ignition temperature (Aruna and Mukasyan, 2008).

The most common precursors are metal nitrates, sulphates and carbonates. Nitrates are the most used since they have a low decomposition temperature and good oxidizing characteristics due to their NO_3 group (Varma et al., 2016). With regard to the fuels, the most commonly used are urea and glycine, due to their compatibility with nitrates, low cost and high exothermicity (Aruna and Mukasyan, 2008). Different types of solvents are used, such as water, alcohols, kerosene and formaldehydes (Varma et al., 2016). The reaction of metallic nitrates with urea can be described as follows (Li et al. 2015):

$$M(NO_3)_v + (\tfrac{5}{6}v\phi)CO(NH_2)_2 + \tfrac{15}{6}v(\phi-1)O_2 \rightarrow MO_{v/2} + (\tfrac{5}{6}v\phi)CO_2 + (\tfrac{5}{3}v\phi)H_2O + (\tfrac{3+5\phi}{6})N_2 \quad (5.7)$$

where ϕ is the fuel/oxidizer ratio and v is the oxidation state of the metal M. When the coefficient ϕ is equal to 1, there is a precise stoichiometry and the precursor will not require an external oxidant like oxygen that is present in the reaction (Christy, Umadevi, and Sagadevan, 2020). The advantage of this method is the gas release, which produces a lot of pores and prevents agglomeration of the particles (Aruna and Mukasyan, 2008).

An $NiFe_2O_4$ catalyst was prepared by SCS using four different fuels: urea, glycine, sucrose and glycerol. The precursors were $Ni(NO_3)_2$ and $Fe(NO_3)_3$, and the fuel/

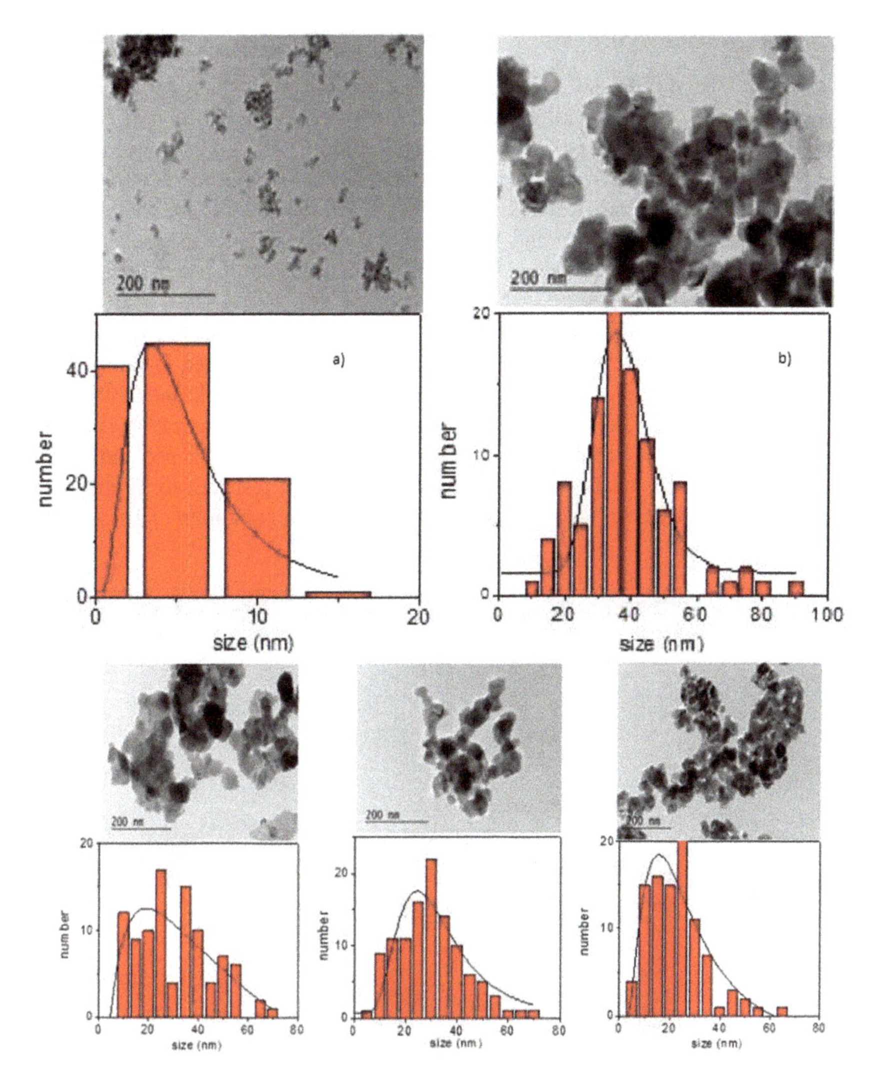

FIGURE 5.4 TEM images obtained for the $NiFe_2O_4$ nanocatalyst in different conditions. (Reprinted with permission from Lazarova et al. 2017. Copyright © 2017 Elsevier B.V. All rights reserved.)

oxidant ratio was 1:1. It was found that the temperature of the heat treatment was important in the average particle size and the nature of the fuel specially affected the particle size distribution, as shown in Figure 5.4 (Lazarova et al., 2017).

Vahid et al. performed a comparative study between the activities of $MgAl_2O_4$ catalysts synthesized by various co-precipitation methods and the combustion synthesis method. The catalyst obtained by the combustion synthesis method presented the lowest surface area and pore size, but the highest catalytic activity and stability in the transesterification reaction of sunflower oil with methanol, obtaining a conversion rate of 95.7% (Vahid et al., 2018).

5.2.6 Microwave Synthesis

Microwave-assisted synthesis consists of using electromagnetic energy to decompose metal precursors (Ashik, Kudo, and Hayashi, 2018). Generally, polar molecules are capable of absorbing this type of energy (Gupta et al., 2018) and its movements are displayed in a continuous orientation and reorientation to align their dipole to the external field (Dahiya, Tomer, and Duhan, 2018). These movements and friction cause the generation of heat (Mirzaei and Neri, 2016). The advantage of using this type of radiation is that the heat is produced inside the precursor solution, which distributes it better towards the entire volume.

Microwave assistance on the synthesis of ZnO nanoparticles has been reported in the literature. A zinc nitrate solution was prepared in the presence of ammonium, which was irradiated by two different microwave powers, 510W (A1) and 680W (A2). In addition, a zinc acetate solution with hydrazine was stirred and irradiated by 150 W microwave power for 15 minutes (B). Nanorod-shaped, needle-shaped and flower-shaped particles were obtained from tests B, A1, and A2, respectively (Figure 5.5) (Hasanpoor, Aliofkhazraei, and Delavari, 2015). Also, the preparation of zinc–copper

FIGURE 5.5 TEM images obtained for the ZnO nanoparticle synthesis. (a) Produced from B solution, (b) irradiated for 15 min at 540 watts microwave power to precursor A1 and (c) irradiated for 15 min at 540 watts microwave power to precursor A2. (Reprinted with permission from Hasanpoor, Aliofkhazraei, and Delavari 2015. Copyright © 2015 The Authors. Published by Elsevier Ltd.)

aluminate catalysts by the microwave-assisted combustion method has been reported. It achieved a 99.1% yield in oleic acid esterification using a catalyst prepared with a fuel mixture of urea–ammonium acetate and a microwave irradiation power of 900 W (Hashemzehi et al., 2019).

5.3 Nanoparticle Characterization

Nanomaterials still present a challenge regarding their characterization. Due to their scale, they cannot be analysed, as their bulk counterparts and more refined techniques are required to have a better material appreciation (Mayeen et al., 2018). Great care must be taken since the mechanical characteristics may differ widely (Mourdikoudis, Pallares, and K. Thanh, 2018). Heterogeneity of composition and size, and changes in the surface and shape characteristics over time can lead to misinterpretations (Kumar and Kumar Dixit, 2017). Also, it is necessary to take into account that the electronic properties of the particles are different at this scale due to the wave-particle duality of the electrons (Mohan Bhagyaraj and Oluwafemi, 2018). For these reasons, readers must be very rigorous in choosing techniques to characterize nanoparticles.

5.3.1 Scanning Electron Microscopy (SEM) and Transmission Electron Microscopy (TEM)

There are two types of electron microscopy that are widely used: scanning electron microscopy (SEM) and transmission electron microscopy (TEM) (Ealias and Saravnakumar, 2017). An electron beam is launched to the sample and then, backscattered electrons (Mayeen et al. 2018) and transmission electrons are detected (Mohan Bhagyaraj and Oluwafemi 2018) in SEM and TEM, respectively. SEM allows to obtain a visualization of the surface of the particles with a 3D reference, so it is possible to estimate its size, size distribution, morphology and agglomerations (Ali et al., 2016). On the other hand, TEM provides the possibility to visualize the contour, shape, size and also density of particles, which is appreciable by the colour intensity (Mohan Bhagyaraj and Oluwafemi, 2018).

Field emission SEM or FE-SEM and TEM images of K-catalysts supported in calcium aluminate can be observed in Figures 5.6 and 5.7. The target material was prepared by SCS with microwave assistance and plasma treatment (KCAC-P) and its catalytic activity was compared with the untreated counterpart KCAC (KOH/$Ca_{12}Al_{14}O_{33}$). FE-SEM allowed the observation of different forms of nanoplate- and nanorod-type particles, and the dimensions of the structures as well as the average particle size were estimated by surface particle distribution size (SPDS). The authors emphasize that a greater interaction between potassium and support was detected in TEM and the size of the particles was sharply reduced. KCAC-P proved to have better stability, larger surface area and higher catalytic activity in the transesterification of canola oil with methanol, even after many recycles (Nayebzadeh, Haghighi, and Saghatoleslami, 2019).

FIGURE 5.6 Histogram representation for the surface particle size of synthesized KOH/$Ca_{12}Al_{14}O_{33}$-C nanoparticles through a hybrid microwave-combustion-plasma methodology. (Reprinted with permission from Nayebzadeh, Haghighi, and Saghatoleslami 2019. Copyright © 2019 Elsevier Ltd. All rights reserved.)

5.3.2 X-Ray Diffraction

X-ray diffraction (XRD) uses X-rays to determine the crystallinity or amorphousness of a material. The rays are diffracted by electrons or neutrons, depending on the technique, and this provides information about the position of the atoms (Ealias and Saravnakumar, 2017) and produces a diffractogram which reveals what phases are present in the material, and also, the lattice parameters and size of the crystals can be determined (Mourdikoudis, Pallares, and K. Thanh, 2018).

Amani, Haghighi and Rahmavid prepared an MgO catalyst dispersed by the impregnation method on $MgFe_2O_4$ spinel prepared by SCS. They analysed the influence of the heat source for the combustion of the precursors, furnace (MF) or microwave (MW), and the change in the molar ratio of the fuel with respect to the nitrates of the precursors (1:1–2:1) on the catalytic activity of the solids. XRD patterns of all prepared materials were analysed regarding the ICDD reference patterns (the International

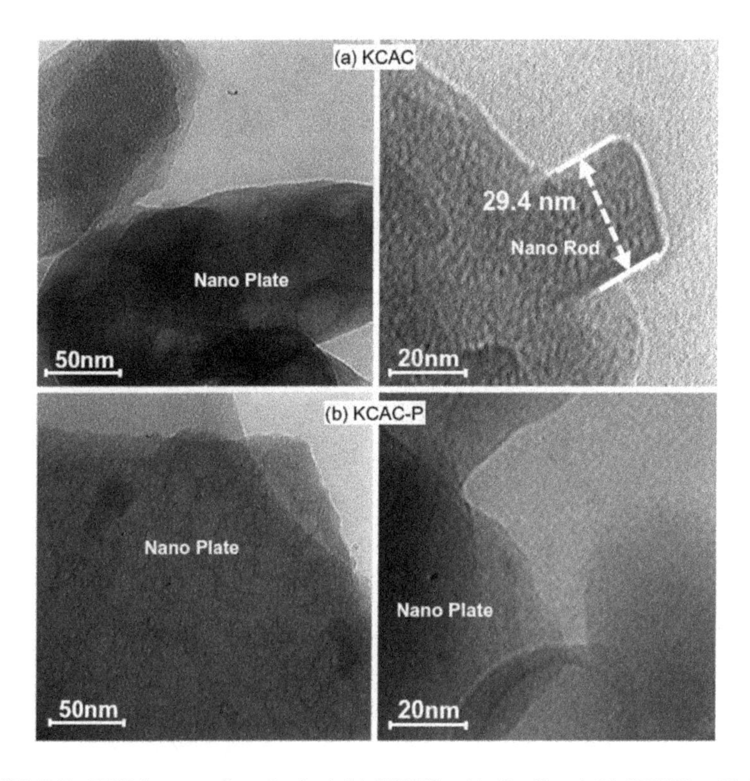

FIGURE 5.7 TEM images of synthesized (a) $KOH/Ca_{12}Al_{14}O_{33}$-C and (b) $KOH/Ca_{12}Al_{14}O_{33}$-P nanoparticles. (Reprinted with permission from Nayebzadeh, Haghighi, and Saghatoleslami (2019). Copyright © 2019 Elsevier Ltd. All rights reserved.)

Centre for Diffraction Data, formerly called JCPDS) are shown in Figure 5.8. In conclusion, hematite and maghemite were not produced, and the traditional furnace heating method formed a more crystalline material related to the intensity of the peaks (Amani, Haghighi, and Rahmanivahid, 2019).

5.3.3 Fourier Transform Infrared Spectroscopy (FTIR)

FTIR is a technique that uses electromagnetic radiation in a region between 4000–400 cm^{-1} to obtain spectra and information about the different functional groups from a sample (Mourdikoudis, Pallares, and K. Thanh, 2018). Each band of the spectrum represents a type of bond or interaction that is related to surface properties and molecular structures (Ali et al., 2016), so it is suitable to provide information about the interactions of the catalyst with the stabilizing agents (Mourdikoudis, Pallares, and K. Thanh, 2018).

In recent research, a mixed CaO-MgO oxide was prepared from eggshells to catalyse the biodiesel production from waste edible oil. The material was characterized with FTIR (Figure 5.9) and presented a combination of the spectra of its individual oxides, therefore a successful combination was achieved (Foroutan et al., 2020).

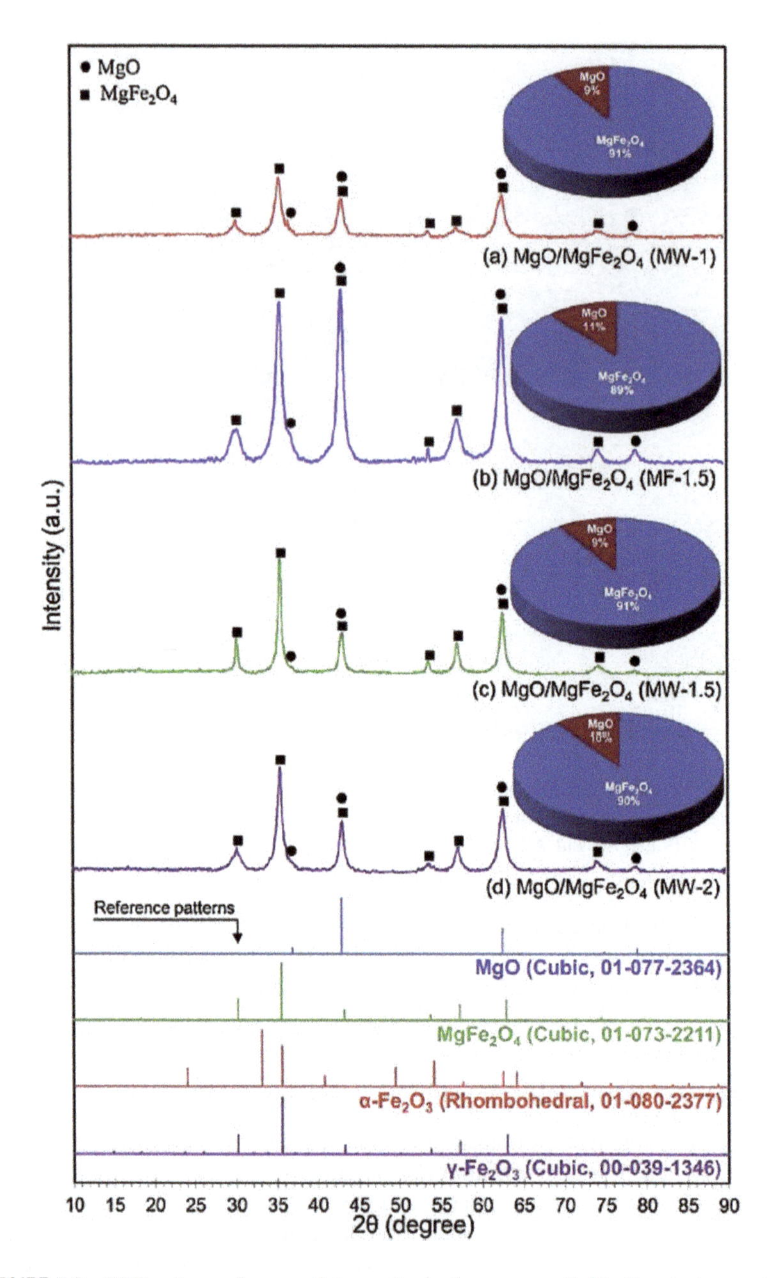

FIGURE 5.8 XRD patterns of nanoparticles synthesized over magnetic Mg-Fe spinel supports: (a) microwave heat source and 1:1 ratio, (b) furnace heat source and 1.5:1 ratio, (c) microwave heat source and 1.5:1 ratio and (d) microwave heat source and 2:1 ratio. (Reprinted with permission from Amani, Haghighi, and Rahmanivahid (2019). Copyright © 2019 The Korean Society of Industrial and Engineering Chemistry. Published by Elsevier B.V. All rights reserved.)

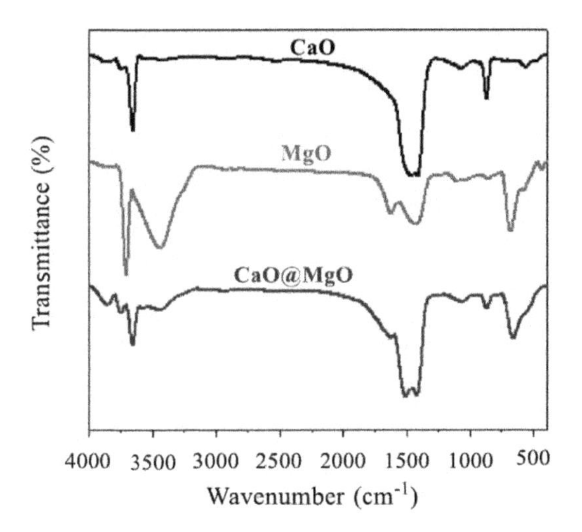

FIGURE 5.9 FTIR spectrum for the CaO-MgO materials prepared from eggshells and individual oxides. (Reprinted with permission from Foroutan et al. 2020. Copyright © 2020 Elsevier Ltd. All rights reserved.)

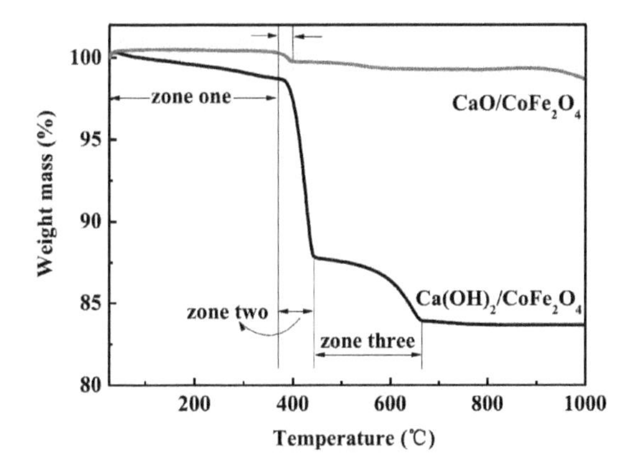

FIGURE 5.10 TGA curves of CaO catalysts impregnated in spinel-type magnetic supports. (Reprinted with permission from P. Zhang et al. 2014. Copyright © 2014 Elsevier B.V. All rights reserved.)

5.3.4 Thermo-Gravimetric Analysis (TGA)

TGA uses temperature variation to predict the composition of the nanomaterial due to the different degradation temperatures of present substances (Mourdikoudis, Pallares, and K. Thanh 2018). Then, it is possible to determine if there is any kind of impurity or even to discover the thermal stability of the nanomaterial (Abraham et al., 2018).

Recently, several CaO catalysts impregnated into spinel-type magnetic supports have been produced and used for soybean oil transesterification. The TGA technique was used to evaluate the decomposition of the precursor. In Figure 5.10, it is observed that the precursor has three significant mass losses up to 665 °C, attributed to different forms of water loss within the material. Thus, the calcination temperature must exceed 665 °C to obtain a thermal-stable catalyst and it can be corroborated with the red TGA curve (P. Zhang et al., 2014).

5.3.5 BET and Temperature Programmed Desorption (TPD)

The Brunauer-Emmett-Teller (BET) method consists of using a gas, usually nitrogen, to fill the empty space of materials due to the physical principle of gas adsorption on solid surfaces (Ealias and Saravnakumar, 2017). It is possible to determine the specific surface area (SSA), the volume-specific surface area (VSSA) and subsequently the pore size of the nanomaterial by obtaining the isotherm of adsorbed and desorbed gas (Kumar and Kumar Dixit, 2017).

The TPD technique also uses the principle of gas adsorption on solids. However, this technique does not use inert gases and detects adsorption-interaction signals with temperature variation, which are indicators of active sites and, depending on the temperature range that is encountered, it can be determined if the sites are strong or weak. CO_2 and NH_3 are used to detect basic and acidic active sites, respectively.

B-ZSM-5 (Boron-Zeolite Socony Mobil-5) and MoO_3/B-ZSM-5 catalysts were tested in the oleic acid esterification. BET analysis showed that the addition of MoO_3 to the silica B-ZSM-5 catalyst decreased the surface area of the catalyst from 376 to 248 $m^2.g^{-1}$ (Mohebbi, Rostamizadeh, and Kahforoushan, 2020). In addition, a NH_3-TPD was performed, and it was observed that B-ZSM-5 presented peaks at lower temperatures, so the acidic strength of its active sites was higher, while MoO_3/B-ZSM-5 presented higher signals, which represented a major concentration of acid sites. Finally, MoO_3/B-ZSM-5 showed higher catalytic activity in the esterification reaction due to the higher concentration of its acid sites, even with lower surface area and acid strength (Mohebbi, Rostamizadeh, and Kahforoushan, 2020).

5.3.6 Energy-Dispersive and Photoelectron X-Ray Spectroscopies

There are two methods commonly used to determine the elemental composition: EDS (energy-dispersive X-ray spectroscopy) and XPS (X-ray photoelectron spectroscopy). EDS consists of shooting an electron beam that collides inelastically with the electrons (secondary electrons) of atoms in a sample. Secondary electrons leave an empty space that is filled by L electrons. The energetic jump emission to occupy the vacancy is measured in X-ray wavelength and is characteristic for each element (Hodoroaba, 2020).

On the other hand, XPS is a technique that uses X-ray radiation to excite core-level electrons that are displaced. The energy difference between the rays and the kinetic energy of the electrons leaked from the sample is distinctive to each element, and a spectrum can be obtained with respect to the energy difference or even the binding energy that exceeds the electron when it separates from the material (Shard, 2020). The EDS technique is more related to the bulk composition, while XPS is related to the surface composition, and can also provide information of oxidation states (Mourdikoudis, Pallares, and K. Thanh, 2018).

Zhang et al. created a CeO_2-CaO material to catalyse the transesterification reaction of soybean oil with methanol. Materials with different amounts of CaO were characterized with EDS and XPS. It was found that the weight percentages of O, Ca and Ce were 24.32%, 40.85% and 34.83%, respectively, using EDS of CeO_2-CaO60 (60% of CaO). In the XPS spectrum the orbitals of the valence electrons of each element, Ce oxidation states and its composition ratio were determined (N. Zhang, Xue, and Hu, 2018).

5.4 Performance and Advantage of Nanocatalysts in Biodiesel Production

Characteristics such as greater catalytic activity, environmental friendliness and higher energy and cost efficiency, are indispensable in the catalysis field (Chaturvedi, Dave, and Shah, 2012). Nanocatalysts are materials that seem to achieve these qualities, because they act as heterogeneous materials, but their small size means that they have certain similar properties to HomC (Polshettiwar and Asefa, 2013).

Nanocatalysts tend to be more active and have a higher energy density because they have fewer coordinated atoms around them (Verma, Shukla, and Sinha, 2019). In addition, structural characteristics can be controlled more easily due to the high confinement of electrons. This characteristic has been widely analysed, since many of the shapes and morphologies can present higher catalytic activities even when it involves the same material (Polshettiwar and Asefa, 2013). Finally, the most important characteristic is that nanocatalysts have a much higher surface area-to-volume ratio (Chaturvedi, Dave, and Shah, 2012). Then, the number of surface or active atoms is similar to the total of atoms, therefore, the number of active sites becomes much more optimal and accessible, which improves the reaction rate (H. Liu et al., 2017). However, catalytic activity finds a critical point in the nanoscale, as can be observed in Figure 5.11. Too small particles do not provide a sufficient number of active sites to catalyse the reaction optimally (W. Liu, 2005) and additionally, an increase in the surface energy may produce agglomeration of particles to stabilize the catalyst. For this reason, stabilizing agents are usually added (Verma, Shukla, and Sinha, 2019).

5.5 Nanocatalyst Reusability and Leaching Analysis

The reusability of the catalyst is highly related to its stability, preparation, and recoverability, since reusability depends on how easily the catalyst can be separated and how active the catalyst remains through the reaction time (Nayebzadeh, Haghighi, and Saghatoleslami, 2019).

The stability and activity of the catalyst, after multiple uses, can be affected by several factors. Fouling, attrition, sintering, poisoning, and finally, leaching, are discussed in this section since they are problems inherent to liquid reactions. Leaching is the direct or indirect dissolution of the active phase in the reaction medium (Sádaba et al., 2015) which, apart from causing an irreversible decrease in the catalytic activity (Otor

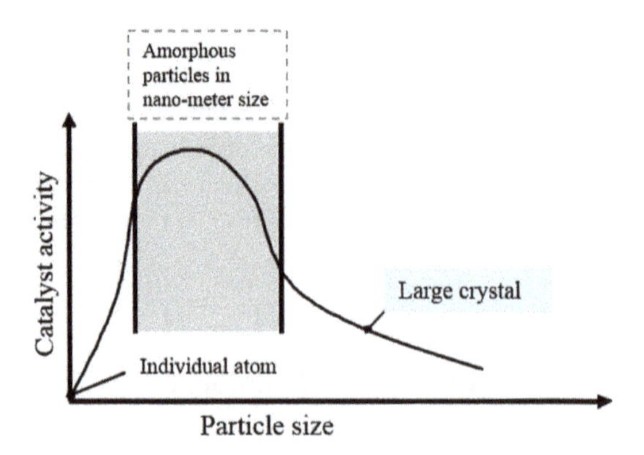

FIGURE 5.11 Conceptual illustrative scheme for the changes in the catalyst activity with the particle size.

et al., 2020), may also produce a lower-quality biodiesel by the incorporation of the catalyst cationic fraction (Chua et al., 2019). This is of great relevance, since many studies do not determine the amount of metal ions contained in the fuel (Di Serio et al., 2010).

CaO has been one of the most used catalysts due to its high availability, low production cost and high catalytic activity (Sharma et al., 2018), however, it has been observed that Ca^{2+} is leached in large quantities into biodiesel (Lukić et al., 2013). Also, it has been noted that catalysts containing alkaline metals are usually more susceptible to leaching (Kwong and Yung, 2015). The ASTM D6751 and EN 14214 standards state that there should not be a concentration greater than 5 ppm of Na, K, Ca or Mg within biodiesel. Na^+ or K^+ ions can be leached due to their presence in the bases used to obtain hydrotalcite-type precursors in some synthesis methods. This is one of the reasons why ammonium hydroxides and organic bases are sometimes preferably used (Lee et al., 2014) and new methods are trying to improve binding and adsorption of the active phase into supported materials (Polshettiwar et al., 2011). Other metals such as Fe, Mn, Co, Ni, Cu, Pb and Zn have been found to promote oxidative degradation of biodiesel and some may even decrease its induction period (Sánchez et al., 2015).

There are two ways to determine whether leaching has occurred in the reaction, and in many cases it is necessary to do more than one test. One method is the analysis of metal traces in the reaction medium. Atomic spectroscopy, emission spectroscopy or mass spectrometry coupled with inductively coupled plasma (ICP-AES, ICP-OES and ICP-MS), flame atomic absorption spectrometry (FAAS), X-ray fluorescence or XPS, can be performed depending on the concentration and nature of the leached material (Sánchez et al., 2015). The second method focuses on the evaluation of the changes occurring in the solid. XRD can be implemented to observe phase changes and even changes in peak intensities. Also, EDS can be implemented to observe if the proportion or content of an element has been decreased, which is quite useful with mixed oxides (Sádaba et al., 2015).

Furthermore, a reusability test can be performed, where reaction cycles are set and it is observed how many cycles catalyst remains active. Ideally, a catalyst should operate continuously, however, solids can be washed or thermally treated after each cycle. Although catalyst activity may depend on more factors, the reusability test can give an indication of the leaching of the solid.

Tang et al. developed a bromoctane-modified CaO catalyst which was tested in the transesterification reaction of soybean oil with methanol. The reaction yield reached a value of 99.5% using a methanol-to-oil molar ratio of 15:1 and 5 wt% catalyst for 3 h at 65 °C. The catalyst presented a high resistance to deactivation, maintaining yields greater than 95% after 15 cycles. However, according to the results of the ICP-MS, the concentration of leached Ca^{2+} was 26 ppm after the first cycle (Tang et al., 2013). In another study, CaO was modified with pectin in the reaction of soybean oil with methanol. The catalyst proved to be efficient because the reaction reached a 99% yield with a methanol-to-oil molar ratio of 7:1 and 1 wt% catalyst for 4 h at 65 °C. The reusability of the catalyst reached three cycles without regeneration with a yield greater than 90% and the amount of leached Ca^{2+} was just 0.7 ppm (Acosta et al., 2020).

5.6 Metal Sulphides

The traditional method of obtaining biodiesel is the transesterification of triglycerides into FAMEs or FAEEs (fatty acid ethyl esters). But, another type of diesel, with similar properties to the fossil fuel, is obtained from the hydrogenation of vegetable oils, called second-generation biodiesel or bio-hydrogenated diesel (BHD).

Traditional or first-generation biodiesel has been shown to have low resistance to oxidation, presence of glycerol by-products (Yoosuk et al., 2019), sometimes low thermal stability (Heriyanto et al., 2018), and troublesome secondary saponification reaction (Toba et al., 2011). In the other hand, production of hydrocarbons through hydrodeoxygenation, decarbonization and decarboxylation of vegetable oils with hydrogen lies in the diesel chain length range, appears to control the problems previously mentioned due to the release of heteroatoms such as oxygen (Tapia et al. 2017), and additionally, hydrocarbons apparently have a higher cetane number and heating value.

Unfortunately, this process requires higher energy conditions even in the presence of catalysts, and second-generation biodiesel has been reported to have poor cold flow properties (Ravi et al., 2020). Also, it is necessary to control the selectivity of the reactions, because an undesirable cracking reaction as olefins formation can decrease the oxidation stability of diesel (Tapia et al., 2017).

For the hydrodeoxygenation process (HDO), metal sulphide catalysts are usually used. Ni-Mo and Co-Mo sulphide catalysts have been studied in the reaction of oleic acid and palmitic acid (Yoosuk et al., 2019). Ammonium tetrathiomolybdate and $Ni(NO_3)_2.6H_2O$ or $Co(NO_3)_2.6H_2O$ were mixed in a reactor pressurized with H_2 at 28 bar and 350 °C for 60 min to synthesize the catalysts. Then, oleic acid or palmitic acid were submitted to react with pressurized H_2 in the presence of the sulphide catalyst. The authors suggested the three main reactions occurring in the system:

$$C_{17}H_{33}COOH + 4H_2 \xrightarrow{cat} C_{18}H_{38} + 2H_2O \quad \textit{Hydrodeoxygenation} \tag{5.8}$$
(oleic acid)

$$C_{17}H_{33}COOH + 2H_2 \xrightarrow{cat} C_{17}H_{36} + CO + 2H_2O \quad \textit{Decarbonylation} \tag{5.9}$$

$$C_{17}H_{33}COOH + H_2 \xrightarrow{cat} C_{17}H_{36} + CO_2 \quad \textit{Decarboxylation} \tag{5.10}$$

The mechanism of palmitic acid is very similar. Ni-Mo catalyst with Ni/(Ni+Mo) = 0.2 promoted HDO reaction and presented the highest catalytic activity in all cases. Oleic acid reached a 100% conversion rate, 78.8% n-C_{18} selectivity and 70.3% n-C_{18} yield. Palmitic acid reached a 95.2% conversion rate, 78.5% n-C_{16} selectivity and 65.6% n-C_{16} yield (Yoosuk et al., 2019).

Heriyanto et al. tested the efficiency of an alumina-supported Ni-Mo sulphate catalyst for the hydrogenation reaction of waste cooking oil by varying the reaction parameters such as temperature and pressure. The catalyst was sulphated with sulphur powder in an autoclave during the same reaction in the presence of hydrogen, where the gases were purged after the first stage, and then fresh hydrogen was added. A yield of 98.93% was obtained with stages of 1 hour, at 30 bars and 300 °C in the first stage, and 400 °C in the second stage (Heriyanto et al., 2018).

The authors suggested the transformation of triglycerides to free fatty acids from hydrogenation:

$$\text{Triglyceride}(C_{17}H_{33})_3 \xrightarrow{3H_2} \text{Triglyceride}(C_{17}H_{35})_3 \xrightarrow{3H_2} 3C_{17}H_{35}COOH + C_3H_8 \tag{5.11}$$

5.7 Conclusion

- The use of MO nanoparticles as catalyst for biodiesel production has increased widely in the last 20 years due to their high catalytic activity, reusability and efficient separation from the reaction system.
- Chemical methods are the most commonly used for the preparation of MO nanoparticles, with co-precipitation being the most commonly used technique, and based on its experimental simplicity this is a technique that could be used to prepare catalysts even at an industrial level.

- The exhaustive characterization of nanoparticles plays a pivotal role in the understanding of the relationship of the chemical structure/physical characteristics and the catalytic activity of the material, with the most used techniques being: TEM, SEM, XRD, FTIR, TGA, BET, TPD and XPS.
- Leaching and reusability of the catalyst is crucial for industrial scaling. Therefore, the material must be easily separated from the reaction and a small quantity of metals in biodiesel is mandatory to comply with the ASTM fuel norms.
- The use of metal sulphides for the production of second-generation biodiesel represents a relevant challenge for the scientific community focused on improvement of the experimental conditions used for the reaction process.

REFERENCES

Abraham, Jiji, Arif P. Mohammed, M.P. Ajith Kumar, Soney C. George, and Sabu Thomas. 2018. "Thermoanalytical Techniques of Nanomaterials." In Sneha M, and Samuel O. (eds.) *Characterization of Nanomaterials*, 213–36. https://doi.org/10.1016/B978-0-08-101973-3.00008-0.

Acosta, Paula I., Roberta R. Campedelli, Elder L. Correa, Heitor A.G. Bazani, Elvis N. Nishida, Bruno S. Souza, and José R. Mora. 2020. "Efficient Production of Biodiesel by Using a Highly Active Calcium Oxide Prepared in Presence of Pectin as Heterogeneous Catalyst." *Fuel* 271 (March): 117651. https://doi.org/10.1016/j.fuel.2020.117651.

Ahn, Taebin, Jong Hun Kim, Hee-Man Yang, Jeong Woo Lee, and Jong-Duk Kim. 2012. "Formation Pathways of Magnetite Nanoparticles by Coprecipitation Method." *Journal of Physical Chemistry C* 116: 6069–76. https://doi.org/10.1021/jp211843g.

Ajala, E. Olawale, Mary Adejoke Ajala, Ibrahim K. Ayinla, Ayomide David Sonusi, and S. E. Fanodun. 2020. "Nano-synthesis of Solid Acid Catalysts from Waste-Iron-Filling for Biodiesel Production Using High Free Fatty Acid Waste Cooking Oil." *Scientific Reports* 10: 13256. https://doi.org/10.1038/s41598-020-70025-x.

Ali, Attarad, Hira Zafar, Muhammad Zia, Abdul Rehman Phull, and Altaf Hussain. 2016. "Synthesis, Characterization, Applications, and Challenges of Iron Oxide Nanoparticles." *Nanotechnology, Science and Applications* 9: 49–67.

Amani, Tohid, Mohammad Haghighi, and Behgam Rahmanivahid. 2019. "Microwave-Assisted Combustion Design of Magnetic Mg – Fe Spinel for MgO-Based Nanocatalyst Used in Biodiesel Production: Influence of Heating-Approach and Fuel Ratio." *Journal of Industrial and Engineering Chemistry* 80: 43–52. https://doi.org/10.1016/j.jiec.2019.07.029.

Ambat, Indu, Varsha Srivastava, and Mika Sillanpää. 2018. "Recent Advancement in Biodiesel Production Methodologies Using Various Feedstock: A Review." *Renewable and Sustainable Energy Reviews* 90 (February 2017): 356–69. https://doi.org/10.1016/j.rser.2018.03.069.

Aruna, Singanahally T., and Alexander S. Mukasyan. 2008. "Combustion Synthesis and Nanomaterials." *Current Opinion in Solid State & Materials Science* 12: 44–50. https://doi.org/10.1016/j.cossms.2008.12.002.

Ashik, U. P. M., Shinji Kudo, and Jun-ichiro Hayashi. 2018. "An Overview of Metal Oxide Nanostructures." In *Synthesis of Inorganic Nanomaterials*, edited by Sneha Mohan Bhagyaraj, Oluwatobi Samuel Oluwafemi, Nandakumar Kalarikkal, and Sabu

Thomas, 19–57. Woodhead Publishing—Elsevier. https://doi.org/10.1016/B978-0-08-101975-7.00002-6.

Chang, Chih-Hung, Yujuan He, and Changqing Pan. 2020. "Aqueous Methods for the Synthesis of Colloidal Metal Oxide Nanoparticles at Ambient Pressure." In *Colloidal Metal Oxide Nanoparticles*, edited by Sabu Thomas, Anu Tresa Sunny, and Prajitha Velayudhan, 41–66. Elsevier. https://doi.org/10.1016/B978-0-12-813357-6.00004-8.

Chaturvedi, Shalini, Pragnesh N. Dave, and Nisha K. Shah. 2012. "Applications of Nano-Catalyst in New Era." *Journal of Saudi Chemical Society* 16 (3): 307–25. https://doi.org/10.1016/j.jscs.2011.01.015.

Christy, A. Jegatha, Mahalingam Umadevi, and Suresh Sagadevan. 2020. "Solution Combustion Synthesis of Metal Oxide Nanoparticles for Membrane Technology." In Jegatha (ed.) *Metal Oxide Powder Technologies*, 329–46. INC. https://doi.org/10.1016/B978-0-12-817505-7.00016-6.

Chua, Song Yuan, Loshinie Periasamy, Celine Ming Hui Goh, Yie Hua Tan, Nabisab Mujawar Mubarak, Jibrail Kansedo, Mohammad Khalid, Rashmi Walkevar, and E.C. Abdullah. 2019. "Biodiesel Synthesis Using Natural Solid Catalyst Derived from Biomass Waste—A Review." *Journal of Industrial and Engineering Chemistry* 81: 41–60. https://doi.org/10.1016/j.jiec.2019.09.022.

Cruz, Inés F., Cristina Freire, João P. Araújo, Clara Pereira, and André M. Pereira. 2018. "Multifunctional Ferrite Nanoparticles: From Current Trends Toward the Future." In *Magnetic Nanostructured Materials*, edited by Ahmed A. El-Gendy, José M. Barandiarán, and Ravi L. Hadimani, 59–116. Elsevier. https://doi.org/10.1016/B978-0-12-813904-2.00003-6.

Dahiya, Manjeet S., Vijay K. Tomer, and Surender Duhan. 2018. "Metal–Ferrite Nanocomposites for Targeted Drug Delivery." In *Applications of Nanocomposite Materials in Drug Delivery*, edited by Inamuddin, Abdullah M. Asiri, and Ali Mohammad, 739–62. Elsevier. https://doi.org/10.1016/B978-0-12-813 741-3.00032-7.

Ealias, Anu Mary, and M. P. Saravanakumar. 2017. "A Review on the Classification, Characterisation, Synthesis of Nanoparticles and Their Applications." *IOP Conference Series: Materials Science and Engineering* 263 (3). https://doi.org/10.1088/1757-899X/263/3/032019.

Foroutan, Rauf, Reza Mohammadi, Hossein Esmaeili, Fatemeh Mirzaee Bektashi, and Sajad Tamjidi. 2020. "Transesterification of Waste Edible Oils to Biodiesel Using Calcium Oxide @ Magnesium Oxide Nanocatalyst." *Waste Management* 105: 373–83. https://doi.org/10.1016/j.wasman.2020.02.032.

Guo, Mingxin, Weiping Song, and Jeremy Buhain. 2015. "Bioenergy and Biofuels: History, Status, and Perspective." *Renewable and Sustainable Energy Reviews* 42: 712–25. https://doi.org/10.1016/j.rser.2014.10.013.

Gupta, Divya, Deepika Jamwal, Dolly Rana, and Akash Katoch. 2018. *Microwave Synthesized Nanocomposites for Enhancing Oral Bioavailability of Drugs. Applications of Nanocomposite Materials in Drug Delivery*. Elsevier. https://doi.org/10.1016/B978-0-12-813741-3.00027-3.

Hasanpoor, M., M. Aliofkhazraei, and H. Delavari. 2015. "Microwave-Assisted Synthesis of Zinc Oxide Nanoparticles." *Procedia Materials Science* 11: 320–25. https://doi.org/10.1016/j.mspro.2015.11.101.

Hashemzehi, Mojgan, Vahid Pirouzfar, Hamed Nayebzadeh, and Afshar Alihosseini. 2019. "Effect of Synthesizing Conditions on the Activity of Zinc-Copper Aluminate

Nanocatalyst Prepared by Microwave Combustion Method Used in the Esterification Reaction." *Fuel* 263: 116422. https://doi.org/10.1016/j.fuel.2019.116422.

Haukka, Suvi, Eeva-Liisa Lakomaa, and Tuomo Suntola. 1998. "Adsoprtion Controlled Preparation of Heterogeneous Catalysts." *Surface Science and Catalysis* 120: 715–50.

Heriyanto, Heri, Sri Djangkung Murti Sumbogo, Septina Is Heriyanti, Inayatu Sholehah, and Ayi Rahmawati. 2018. "Synthesis of Green Diesel from Waste Cooking Oil through Hydrodeoxygenation Technology with NiMo/γ-Al_2O_3 Catalysts." *MATEC Web of Conferences* 156: 1–6. https://doi.org/10.1051/matecconf/201815603032

Hodoroaba, Vasile-Dan. 2020. "Energy-Dispersive X-Ray Spectroscopy (EDS)." In *Characterization of Nanoparticles*, edited by Vasile-dan Hodoroaba, Wolfgang E.S. Unger, and Alexander G. Shard, 397–417. Elsevier. https://doi.org/10.1016/B978-0-12-814182-3.00021-3.

Huang, Guoming, Chun-Hua Lu, and Huang-Hao Yang. 2019. "Magnetic Nanomaterials for Magnetic Bioanalysis." In *Novel Nanomaterials for Biomedical, Environmental and Energy Applications*, edited by Xiaoru Wang and Xi Chen, 89–109. Elsevier. https://doi.org/10.1016/B978-0-12-814497-8.00003-5.

Kumar, Ajeet, and Chandra Kumar Dixit. 2017. "Methods for Characterization of Nanoparticles." In *Advances in Nanomedicine for the Delivery of Therapeutic Nucleic Acids*, edited by Surendra Nimesh, Ramesh Chandra, and Nidhi Gupta, 43–58. Woodhead Publishing—Elsevier. https://doi.org/10.1016/B978-0-08-100 557-6.00003-1.

Kwong, Tsz-Lung, and Ka-Fu Yung. 2015. "Heterogeneous Alkaline Earth Metal-Transition Metal Bimetallic Catalysts for Synthesis of Biodiesel from Low Grade Unrefined Feedstock." *RSC Advances* 5: 83748–56. https://doi.org/10.1039/C5R A13819A.

Lara-García, Hugo A., Issis C. Romero-Ibarra, and Heriberto Pfeiffer. 2014. "Hierarchical Na-Doped Cubic ZrO2 Synthesis by a Simple Hydrothermal Route and Its Application in Biodiesel Production." *Journal of Solid State Chemistry* 218: 213–20. https://doi.org/10.1016/j.jssc.2014.06.040.

Lazarova, Tsvetomila, Milena Georgieva, Dimitar Tzankov, Dimitrinka Voykova, Lyubomir Aleksandrov, Zara Cherkezova-zheleva, and Daniela Kovacheva. 2017. "Influence of the Type of Fuel Used for the Solution Combustion Synthesis on the Structure, Morphology and Magnetic Properties of Nanosized NiFe2O4." *Journal of Alloys and Compounds* 700: 272–83. https://doi.org/10.1016/j.jallcom.2017.01.055.

Lee, Adam F., James A. Bennett, Jinesh C. Manayil, and Karen Wilson. 2014. "Heterogeneous Catalysis for Sustainable Biodiesel Production via Esterification and Transesterification." *Chemical Society Reviews* 43: 7887–7916. https://doi.org/10.1039/c4cs00189c.

Li, Fa-tang, Jingrun Ran, Mietek Jaroniec, and Shi Zhang Qiao. 2015. "Solution Combustion Synthesis of Metal Oxide Nanomaterials for Energy Storage and Conversion." *Nanoscale* 7: 17590–610. https://doi.org/10.1039/c5nr05299h.

Liu, Haichao, Jing Guan, Xindong Mu, Guoqiang Xu, Xicheng Wang, and Xiufang Chen. 2017. "Nanocatalysis." In *Encyclopedia of Physical Organic Chemistry*, edited by Zerong Wang, Uta Wille, and Eusebio Juaristi, 3697–3773. John Wiley & Sons. https://doi.org/10.1002/9781118468586.epoc5009.

Liu, Wei. 2005. "Catalyst Technology Development from Macro-, Micro- Down to Nano-Scale." *China Particuology* 3 (6): 383–94.

Lukić, Ivana, Svetolik Maksimović, Miodrag Zdujć, Hui Liu, Jugoslav Krstić, and Skala Dejan. 2013. "Kinetics of Sunflower and Used Vegetable Oil Methanolysis Catalyzed by CaO.ZnO." *Fuel* 113: 367–78. https://doi.org/10.1016/j.fuel.2013.05.093.

Marinković, Dalibor M., Miroslav V. Stanković, Ana V. Veličković, Jelena M. Avramović, Marija R. Miladinović, Olivera O. Stamenković, Vlada B. Veljković, and Dušan M. Jovanović. 2016. "Calcium Oxide as a Promising Heterogeneous Catalyst for Biodiesel Production: Current State and Perspectives." *Renewable and Sustainable Energy Reviews* 56: 1387–1408. https://doi.org/10.1016/j.rser.2015.12.007.

Mayeen, Anshida, Leyana K. Shaji, Anju K Nair, and Nandakumar Kalarikkal. 2018. "Morphological Characterization of Nanomaterials." In *Characterization of Nanomaterials*, edited by Sneha Mohan Bhagyaraj, Oluwatobi Samuel Oluwafemi, and Sabu Thomas, 335–64. Cambridge: Woodhead Publishing—Elsevier. https://doi.org/10.1016/B978-0-08-101973-3.00012-2.

Mehradabi, Bahareh A.T., Sonia Eskandari, Umema Khan, Rembert D. White, and John R. Regalbuto. 2017. "A Review of Preparation Methods for Supported Metal Catalysts." *Advances in Catalysis* 61: 1–35. https://doi.org/10.1016/bs.acat.2017.10.001.

Mirzaei, A., and G. Neri. 2016. "Microwave-Assisted Synthesis of Metal Oxide Nanostructures for Gas Sensing Application: A Review." *Sensors & Actuators: B: Chemical* 237: 749–75. https://doi.org/10.1016/j.snb.2016.06.114.

Mohan Bhagyaraj, Sneha, and Oluwatobi Samuel Oluwafemi. 2018. "Nanotechnology: The Science of the Invisible." In *Synthesis of Inorganic Nanomaterials*, edited by Sneha Mohan Bhagyaraj, Oluwatobi Samuel Oluwafemi, and Nandakumar Kalarikkal, 1–18. Cambridge: Woodhead Publishing—Elsevier. https://doi.org/10.1016/B978-0-08-101975-7.00001-4

Mohebbi, Saeed, Mohammad Rostamizadeh, and Davood Kahforoushan. 2020. "Effect of Molybdenum Promoter on Performance of High Silica MoO_3/B-ZSM- 5 Nanocatalyst in Biodiesel Production." *Fuel* 266: 117063. https://doi.org/10.1016/j.fuel.2020.117063.

Mourdikoudis, Stefanos, Roger M. Pallares, and Nguyen T. K. Thanh. 2018. "Characterization Techniques for Nanoparticles: Comparison and Complementarity upon Studying Nanoparticle Properties." *Nanoscale* 10: 12871–934. https://doi.org/10.1039/C8NR02278J.

Munnik, Peter, Petra E. De Jongh, and Krijn P. De Jong. 2015. "Recent Developments in the Synthesis of Supported Catalysts." *Chemical Reviews* 115 (14): 6687–6718. https://doi.org/10.1021/cr500486u.

Na, Kyungsu, Qiao Zhang, and Gabor A. Somorjai. 2014. "Colloidal Metal Nanocatalysts: Synthesis, Characterization, and Catalytic Applications." *Journal of Cluster Science* 25: 83–114. https://doi.org/10.1007/s10876-013-0636-6.

Nawaz, Muhammad, Yassine Sliman, Ismail Ercan, Michele K. Lima-Tenório, Ernandes T. Tenório-Neto, Chariya Kaewsaneha, and Abdelhamid Elaissari. 2019. "Magnetic and pH-Responsive Magnetic Nanocarriers." In *Stimuli Responsive Polymeric Nanocarriers for Drug Delivery Applications*, edited by Abdel Salam Hamdy Makhlouf and Nedal Y. Abu-Thabit, 37–85. Woodhead Publishing—Elsevier. https://doi.org/10.1016/B978-0-08-101995-5.00002-7.

Nayebzadeh, Hamed, Mohammad Haghighi, and Naser Saghatoleslami. 2019. "Texture/Phase Evolution during Plasma Treatment of Microwave-Combustion Synthesized KOH/ Ca12 Al14 O33-C Nanocatalyst for Reusability Enhancement in Conversion of Canola Oil to Biodiesel." *Renewable Energy* 139: 28–39. https://doi.org/10.1016/j.renene.2019.01.122.

Nunes, Daniela, Ana Pimentel, and Lídia Santos. 2019. "Synthesis, Design, and Morphology of Metal Oxide Nanostructures." In *Metal Oxide Nanostructures*, edited by Daniela Nunes, Lidia Santos, Luis Pereira, Rodrigo Martins, Ana Pimentel, Pedro Barquinha, and Elvira Fortunato, 21–57. Elsevier. https://doi.org/10.1016/B978-0-12-811512-1.00002-3.

Ogunkunle, Oyetola, and Noor A Ahmed. 2019. "A Review of Global Current Scenario of Biodiesel Adoption and Combustion in Vehicular Diesel Engines." *Energy Reports* 5: 1560–79. https://doi.org/10.1016/j.egyr.2019.10.028.

Otor, Hope O, Joshua B Steiner, García-Sancho C., and Ana Carolina Alba-rubio. 2020. "Encapsulation Methods for Control of Catalyst Deactivation: A Review." *ACS Catalysis* 10 (14): 7630–56. https://doi.org/10.1021/acscatal.0c01569.

Polshettiwar, Vivek, Rafael Luque, Aziz Fihri, Haibo Zhu, Mohamed Bouhrara, and Jean-marie Basset. 2011. "Magnetically Recoverable Nanocatalysts." *Chemical Reviews* 111: 3036–75.

Polshettiwar, Vivek, and Tewodros Asefa. 2013. "Introduction to Nanocatalysis." In *Nanocatalysis Synthesis and Applications*, edited by Vivek Polshettiwar and Tewodros Asefa, 1–9. Wiley. https://doi.org/10.1002/9781118609811.ch1.

Rane, Ajay Vasudeo, Krishnan Kanny, V. K. Abitha, and Sabu Thomas. 2018. "Methods for Synthesis of Nanoparticles and Fabrication of Nanocomposites." In *Synthesis of Inorganic Nanomaterials*, edited by Sneha Mohan Bhagyaraj, Nandakumar Kalarikkal, and Sabu Thomas, 121–39. Woodhead Publishing - Elsevier. https://doi.org/10.1016/B978-0-08-101975-7.00005-1.

Rashid, Umer, Soroush Soltani, Saud Ibrahim Al-resayes, and Imededdine Arbi Nehdi. 2018. "Metal Oxide Catalysts for Biodiesel Production." In *Metal Oxides in Energy Technologies*, edited by Yuping Wu, 303–19. Elsevier. https://doi.org/10.1016/B978-0-12-811167-3.00011-0

Ravi, Aiswarya, Baskar Gurunathan, Naveenkumar Rajendiran, Sunita Varjani, Edgard Gnansounou, Ashok Pandey, Simming You, Jegannathan Kenthorai Raman, and Praveenkumar Ramanujam. 2020. "Contemporary Approaches towards Augmentation of Distinctive Heterogeneous Catalyst for Sustainable Biodiesel Production." *Environmental Technology & Innovation* 19: 100906. https://doi.org/10.1016/j.eti.2020.100906.

Sádaba, Irantzu, Manuel López Granados, Anders Riisager, and Esben Taarning. 2015. "Deactivation of Solid Catalysts in Liquid Media: The Case of Leaching of Active Sites in Biomass Conversion Reactions." *Green Chemistry* 17: 4133–45. https://doi.org/10.1039/C5GC00804B.

Salinas, Daniela, Catherine Sepúlveda, Néstor Escalona, J. L. Gfierro, and Gina Pecchi. 2017. "Sol-Gel La2O3-ZrO2 Mixed Oxide Catalysts for Biodiesel Production." *Journal of Energy Chemistry* 27 (2): 565–72. https://doi.org/10.1016/j.jechem.2017.11.003.

Sánchez, Raquel, Carlos Sánchez, Charles-Philippe Lienemann, and José-Luis Todoli. 2015. "Metal and Metalloid Determination in Biodiesel and Bioethanol." *Journal of Analytical Atomic Spectroscopy* 30: 64–101. https://doi.org/10.1039/C4JA00202D.

Santos, Evelyn C.S., Thiago C. Dos Santos, Renato B. Guimarães, Lina Ishida, Rafael S. Freitas, and Célia M. Ronconi. 2015. "Guanidine-Functionalized Fe_3O_4 Magnetic Nanoparticles as Basic Recyclable Catalysts for Biodiesel Production." *RSC Advances*. https://doi.org/10.1039/c5ra07331f.

Seffati, Kambiz, Hossein Esmaeili, Bizhan Honarvar, and Nadia Esfandiari. 2020. "AC/ $CuFe_2O_4$ @ CaO as a Novel Nanocatalyst to Produce Biodiesel from Chicken Fat." *Renewable Energy* 147: 25–34. https://doi.org/10.1016/j.renene.2019.08.105.

Serio, Martino Di, Riccardo Tesser, Antonio D Angelo, and Marco Trifuoggi. 2010. "Heterogeneous Catalysis in Biodiesel Production: The Influence of Leaching." *Topics in Catalysis* 53 (November 2015): 811–19. https://doi.org/10.1007/s11 244-010-9467-y.

Shard, Alexander G. 2020. "X-Ray Photoelectron Spectroscopy." In *Characterization of Nanoparticles*, edited by Vasile-Dan Hodoroaba, Wolfgang E.S. Unger and Alexander G. Shard 349–71. Elsevier. https://doi.org/10.1016/B978-0-12-814 182-3.00019-5.

Sharma, Swati, Varun Saxena, Anupriya Baranwal, Pranjal Chandra, and Lalit Mohan Pandey. 2018. "Engineered Nanoporous Materials Mediated Heterogeneous Catalysts and Their Implications in Biodiesel Production." *Materials Science for Energy Technologies* 1 (1): 11–21. https://doi.org/10.1016/j.mset.2018.05.002.

Sietsma, Jelle R. A., A. Jos Van Dillen, Petra E. De Jongh, and Krijn P. De Jong. 2006. "Application of Ordered Mesoporous Materials as Model Supports to Study Catalyst Preparation by Impregnation and Drying." *Studies in Surface Science and Catalysis* 162: 95–102. https://doi.org/10.1016/S0167-2991(06)80895-5.

Singh, Ashok K. 2016. "Structure, Synthesis, and Application of Nanoparticles." In *Engineered Nanoparticles*, edited by Ashok K. Singh, 19–76. Elsevier. https://doi.org/10.1016/B978-0-12-801406-6.00002-9.

Singh Chouhan, Ashish Pratap, and Anil K. Sarma. 2011. "Modern Heterogeneous Catalyst for Biodiesel Production: A Comprehensive Review." *Renewable & Sustainable Energy Reviews* 15 (July 2014): 4378–99.

Singh, Subhash C., Dinesh Pratap Singh, Jai Singh, Pawan K. Dubey, Radhey Shyam Tiwari, and Onkar Nath Srivastava. 2010. "Metal Oxide Nanostructures; Synthesis, Characterizations and Applications." In *Encyclopedia of Semiconductor Nanotechnology*, edited by Admad Umar, 1000. American Scientific.

Soytaş, Serap Hayat, Oğuzhan Oğuz, and Yusuf Ziya Menceloğlu. 2019. "Polymer Nanocomposites With Decorated Metal Oxides." In *Polymer Composites with Functionalized Nanoparticles: Synthesis, Properties and Applications*, edited by Krzystof Pielichowski and Tomasz M. Majka, 287–323. Elsevier. https://doi.org/10.1016/B978-0-12-814064-2.00009-3.

Tang, Ying, Jingfang Xu, Jie Zhang, and Yong Lu. 2013. "Biodiesel Production from Vegetable Oil by Using Modified CaO as Solid Basic Catalysts." *Journal of Cleaner Production* 42: 198–203. https://doi.org/10.1016/j.jclepro.2012.11.001.

Tapia, Juan, Nancy Y. Acelas, Diana López, and Andrés Moreno. 2017. "NiMo-Sulfide Supported on Activated Carbon to Produce Renewable Diesel." *Journal of the Faculty of Science Univ. Sci.* 22 (1): 71–85. https://doi.org/10.11144/Javeriana.SC22-1.nsoa.

Toba, Makoto, Yohko Abe, Hidetoshi Kuramochi, Masahiro Osako, T. Mochizuki, and Yuji Yoshimura. 2011. "Hydrodeoxygenation of Waste Vegetable Oil over Sulfide Catalysts." *Catalysis Today* 164 (1): 533–37. https://doi.org/10.1016/j.cattod.2010.11.049.

Vahid, Behgam Rahmani, Mohammad Haghighi, Javad Toghiani, and Shervin Alaei. 2018. "Hybrid-Coprecipitation vs. Combustion Synthesis of Mg-Al Spinel Based Nanocatalyst for Efficient Biodiesel Production." *Energy Conversion and Management* 160: 220–29. https://doi.org/10.1016/j.enconman.2018.01.030.

Varma, Arvind, Alexander S. Mukasyan, Alexander S. Rogachev, and Khachatur V. Manukyan. 2016. "Solution Combustion Synthesis of Nanoscale Materials." *Chemical Reviews* 116 (23): 14493–586. https://doi.org/10.1021/acs.chemrev.6b00279.

Verma, Alkadevi, Madhulata Shukla, and Indrajit Sinha. 2019. "Introductory Chapter: Salient Features of Nanocatalysts." In *Nanocatalysts*, edited by Indrajit Sinha and Madhulata Shukla, 1–8. IntechOpen. https://doi.org/10.5772/intechopen.86209.

Wabeke, Jared T., Hazim Al-Zubaidi, Clara P. Adams, Liyana A Wajira Ariyadasa, Setare Tahmasebi Nick, Ali Bolandi, Robert Y. Ofoli, and Sherine O. Obare. 2014. "Synthesis of Nanoparticles for Biomass Conversion Processes." In *Green Technologies for the Environment*, edited by Sherine O. Obare and Rafael Luque, 219–46. American Chemical Society. https://doi.org/10.1021/bk-2014-1186.ch012.

Wang, Anping, Hu Li, Heng Zhang, Hu Pan, and Song Yang. 2018. "Efficient Catalytic Production of Biodiesel with Acid-Base Bifunctional Rod-Like Ca-B Oxides by the Sol-Gel Approach." *Materials* 12: 83. https://doi.org/10.3390/ma12010083.

Wang, Anping, Putla Sudarsanam, Yufei Xu, Heng Zhang, Hu Li, and Zong Yang. 2020. "Functionalized Magnetic Nano-Sized Materials for Efficient Biodiesel Synthesis via Acid-Base/Enzyme Catalysis." *Green Chemistry* 22: 2977–3012. https://doi.org/10.1039/D0GC00924E.

Yoosuk, Boonyawan, Paphawee Sanggam, Sakdipat Wiengket, and Pattarapan Prasassarakich. 2019. "Hydrodeoxygenation of Oleic Acid and Palmitic Acid to Hydrocarbon-Like Biofuel over Unsupported Ni-Mo and Co-Mo Sulfide Catalysts." *Renewable Energy* 139: 1391–99. https://doi.org/10.1016/j.renene.2019.03.030

Zhang, Ni, Huiyuan Xue, and Rongrong Hu. 2018. "The Activity and Stability of CeO_2@CaO Catalysts for the Production of Biodiesel." *RSC Advances* 8: 32922–29. https://doi.org/10.1039/c8ra06884d.

Zhang, Pingbo, Qiuju Han, Mingming Fan, and Pingping Jiang. 2014. "Magnetic Solid Base Catalyst CaO / CoFe2O4 for Biodiesel Production: Influence of Basicity and Wettability of the Catalyst in Catalytic Performance." *Applied Surface Science* 317: 1125–30. https://doi.org/10.1016/j.apsusc.2014.09.043.

6

Magnetic Nanomaterials and Their Relevance in Transesterification Reactions

António B. Mapossa and Michael Daramola

CONTENTS

DOI: 10.1201/9781003120858-6

6.1 Introduction

Catalysis is becoming a strategic field of science because it represents a new way to meet the challenges of energy and sustainability. These challenges are becoming major concerns of the global vision of societal challenges and the global economy. Societal pressure has been at the origin of the concept of green chemistry, which is becoming a leitmotiv in any important projects dealing with this strategic science domain. The concept of green chemistry, which makes catalysis science even more creative, has become an integral part of sustainability (Polshettiwar, Luque, Fihri, Zhu, Bouhrara & Basset, 2011).

Nanotechnology is the understanding and control of matter at dimensions of roughly 1–100 nm, where unique phenomena enable novel applications. The physical and chemical properties of nanomaterials tend to be exceptionally closely dependent on their size and shape or morphology. As a result, materials scientists are focusing their efforts on developing simple and effective methods for fabricating nanomaterials with controlled size and morphology and hence tailoring their properties. An important aspect of the nanoscale is that the smaller a nanoparticle gets, the larger its relative surface area becomes (Mathew & Juang, 2007).

Nowadays, nanomagnetic nanoparticles (MNPs) are very attractive materials due to their outstanding physical properties and high applicability in nanotechnology (Andjelković, Šuljagić, Lakić, Jeremić, Vulić, & Nikolić, 2018). These materials are widely used as heterogeneous catalysts due to their magnetic properties (Sundararajan & Srinivasan, 1991) and considerable chemical and thermal stability (Maposssa, Dantas, Silva, Kiminami, Costa, & Daramola, 2020b; Hajalilou & Mazlan, 2016). The methods of preparation of magnetic nanoparticles play a principal role in the physicochemical properties of the nanomagnetic catalyst obtained. For biodiesel production by transesterification reactions, the pioneering research in this field using magnetic nanoparticles was reported by Dantas et al. (2012). Having reported promising results (Dantas et al., 2012; Dantas, Santos, Cunha, Kiminami, & Costa, 2013), other researchers have been exploring the use of magnetic nanoparticles for biodiesel production. For instance, Zhang, Han, Fan, & Jiang (2014) synthesized $CaO/CoFe_2O_4$ and evaluated it for soybean oil transesterification into biodiesel. These authors reported enhanced catalytic performance for the catalyst when compared to $CaO/ZnFe_2O_4$ and $CaO/MnFe_2O_4$, with a yield of 87% from $CaO/CoFe_2O_4$. In the same manner, Seo et al. (2014) studied the effect of different sizes of functionalized particles of barium ferrite ($BaFe_{12}O_{19}$) in microalgae biofuel production with promising results reported.

This chapter, therefore, gives an overview on the magnetic nanoparticles (MNPs) that are used in transesterification reactions, different methods of synthesis of magnetic nanoparticles, reusability of MNPs in transesterification reactions for biodiesel production, as well as the factors or reactional conditions that affect the biodiesel yield. Furthermore, characterization techniques that describe the morphology, structural, surface area, chemistry composition and magnetic properties of nanoparticles are addressed in depth.

6.2 Fundamental Aspects of Magnetic Nanoparticles (MNPs)

To understand the behaviour of magnetic nanoparticles it is necessary to determine their physicochemical parameters such as their size, shape and magnetic properties. The different types of magnetic nanoparticles and the oxides known as ferrites are promising candidates for catalytic applications. As defined, spinel ferrite is composed of iron oxides, which may be altered using other transition metal oxides with a general formula of AB_2O_4, where A represents a divalent metal ion (Melo, Silva, Moura, De Menezes, & Sinfrônio, 2015). The magnetic properties of ferrites are tightly bound to the position of the divalent cations in the crystal structure. Ferrites crystallize in a spinel structure (cubic space group Fm3d), where divalent and trivalent cations are arranged among tetrahedral and octahedral sites (Deepty et al., 2019). For example, magnetic divalent cations (Ni^{2+}) have a strong preference for the octahedral sites, and thus, $NiFe_2O_4$ is an inverse spinel. In contrast, diamagnetic divalent cations, such as Zn^{2+}, occupy tetrahedral sites. Therefore, the structure of $ZnFe_2O_4$ is a normal spinel. Due to the opposite path of crystallization, the properties of $NiFe_2O_4$ and $ZnFe_2O_4$ are diametrically different, i.e., $NiFe_2O_4$ is ferrimagnetic with a curie temperature ≈ 858 K, while $ZnFe_2O_4$ shows antiferromagnetic ordering below 9 K. The composition of a ferrite can be modified without compromising its basic crystalline structure, indicating that the properties of the materials can be easily tuned just by varying the ratio of the divalent cations (Andjelković, Šuljagić, Lakić, Jeremić, Vulić, & Nikolić, 2018). As another example, the mixed nickel–zinc ferrites have general site occupancy $(Zn_xFe_{1-x})_{tetrahedral}[Ni_{1-x}Fe_{1+x}]_{octahedral}O_4$, where the composition varies from $NiFe_2O_4$ ($x = 0$) to $ZnFe_2O_4$ ($x = 1$), resulting in the redistribution of metal ions over the tetrahedral and octahedral sites and modification of the properties (Andjelković, Šuljagić, Lakić, Jeremić, Vulić, & Nikolić, 2018; Costa, Morelli, & Kiminami, 2001). Other magnetic nanoparticles that can be employed in transesterification reactions are iron oxides such as hematite (Fe_2O_3) and magnetite (Fe_3O_4).

6.3 Methods of Synthesis of MNPs

Several popular methods of synthesis to achieve the required shape, particle size, crystallite, dispersity, textural and magnetic properties of MNPs have been reported in various papers (Gopalan Sibi, Verma, & Kim, 2020; Majidi, Zeinali-Sehrig, Farkhani, Soleymani-Goloujeh, & Akbarzadeh, 2016; Ali, Hira-Zafar, Ul-Haq, Phull, Ali, & Hussain, 2016; Akbarzadeh, Samiei, & Davaran, 2012; Sun et al., 2007). These methods of synthesis of MNPs include co-precipitation, sol–gel, thermal decomposition, solvothermal, sonochemical, hydrothermal and combustion. Among these methods, we focus on co-precipitation, sol–gel, hydrothermal and combustion. These chemical methods are highly appreciable because they are simple, tractable and efficient, in which the size, composition and even the shape of the NPs can be managed (Ali, Hira-Zafar, Ul-Haq, Phull, Ali, & Hussain, 2016; Wu, He, & Jiang, 2008). The high quality of magnetic nanoparticles (MNPs) is directly associated with the method of synthesis. During the synthesis of magnetic nanoparticles (MNPs), several conditions such as atmosphere, reducing agents, surfactants, precursors, solvents, and other process parameters such as pressure, reaction temperature and time should carefully be adjusted to control the properties of the MNPs. More details of the methods of synthesis of MNPs and examples are presented below.

6.3.1 Co-Precipitation Method

Co-precipitation is one of the most widely used methods for the preparation MNPs (i.e. Fe_3O_4, γFe_2O_3) in aqueous solutions. The typical co-precipitation methods involve mixing ferric (Fe^{3+}) and ferrous (Fe^{2+}) ions in a 1:2 molar ratio in a highly basic aqueous solution at room temperature or at elevated temperatures (Konecny, Covarrubias, & Wang, 2017). For example, the magnetic Fe_3O_4 nanoparticles (magnetite) are obtained in an inert atmosphere (Reaction 6.1):

$$Fe^{2+} + 2Fe^{3+} + 8OH^- \rightarrow Fe_3O_4 + 4H_2O \quad (6.1)$$

Since magnetite (Fe_3O_4) nanoparticles are not stable, they are easily oxidized to hematite (Fe_2O_3) or maghemite based on Reaction (6.2). Furthermore, a scheme of preparation of magnetic nanoparticles (Fe_3O_4) by the co-precipitation method is shown in Figure 6.1.

$$Fe_3O_4 + 2H^+ \rightarrow Fe_2O_3 + Fe^{2+} + H_2O \quad (6.2)$$

Co-precipitation is the most popular method because of the mild reaction conditions, short reaction time and easy scalability. However, this method has very limited control over particle size and shape, and usually produces magnetic nanoparticles with a broad size distribution (Konecny, Covarrubias, & Wang, 2017). In addition, the saturation magnetization values of the MNPs prepared through co-precipitation are usually much lower than the bulk value. Maaz, Karim, Mumtaz, Hasanain, Liu, & Duan, (2009) prepared magnetic nanoparticles of $NiFe_2O_4$ by the co-precipitation method. X-ray diffraction (XRD) confirmed the formation of single-phase nickel ferrite nanoparticles in the range 8–28 nm. The saturation moment of all the samples was found to be much below the bulk value of nickel ferrite, which was attributed to the disordered surface spins or inert layer in these nanoparticles (Maaz, Karim, Mumtaz, Hasanain, Liu, & Duan, 2009). Additional studies into the successful of the co-precipitation method in the preparation of MPNs have been reported by researchers (Vijayaprasath et al., 2016; El Moussaoui et al., 2016; Aphesteguy, Kurlyandskaya, De Celis, Safronov, & Schegoleva, 2015; Liu, Jia, Wu, Ran, Zhang, & Wu, 2011b; Jadhav, Shirsath, Toksha, Patange, Shengule, & Jadhav, 2010; Rashad, Elsayed, Moharam, Abou-Shahba, & Saba, 2009). The common microstructure of MNPs obtained by this method and confirmed by transmission electron microscopy (TEM) technique is shown in Figure 6.3a.

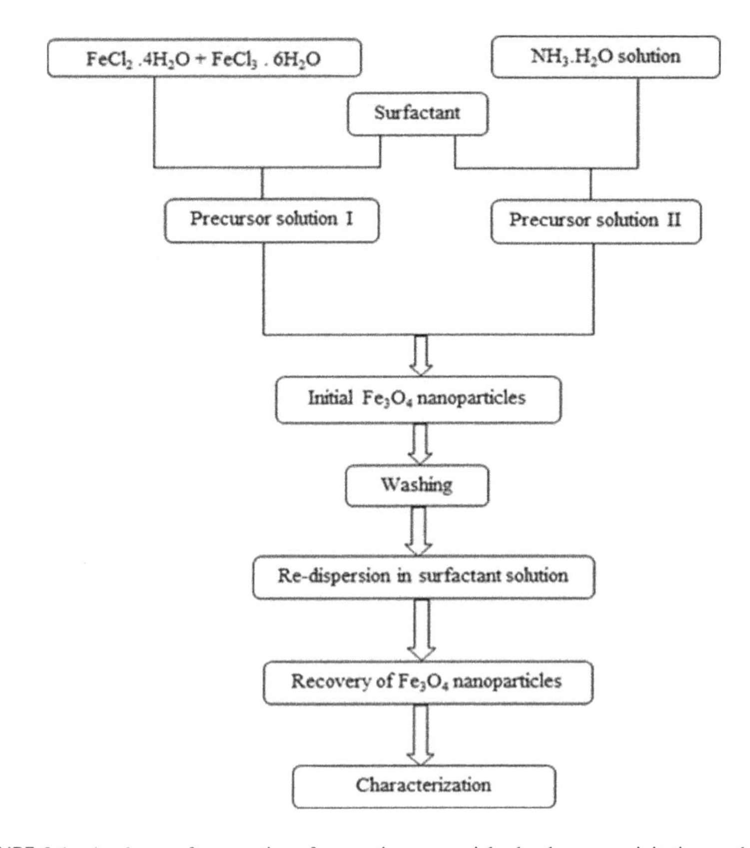

FIGURE 6.1 A scheme of preparation of magnetic nanoparticles by the co-precipitation method.

6.3.2 Hydrothermal Method

The hydrothermal method is considered highly effective for the preparation of magnetic nanoparticles due to its ability to control the particle size and offering flexibility in surface modification. The hydrothermal synthesis of nanoparticles is based on a wet-chemical synthesis in sealed reactors or autoclaves at high vapour pressure (from 0.3 to 4 MPa) and elevated temperature (130 to 250 °C) (Konecny, Covarrubias, & Wang, 2017; Ali, Hira-Zafar, Ul-Haq, Phull, Ali, & Hussain, 2016). The reaction conditions including reaction time, temperature, types of oxidants and solvents play a crucial role in determining the characteristics of the synthesized MNPs. The hydrothermal method has some advantages such as low temperature for synthesis, high purity, simple reactions, cost effectiveness and good dispersibility of the MNPs (Gopalan Sibi, Verma, & Kim, 2020). In addition to these advantages, this method has some disadvantages, such as high pressure during the synthesis of MNPs. Despite the disadvantages, the success of the hydrothermal method has been reported by many researchers. For example, work carried out by Nejati and Zabihi (2012) evaluated the magnetic properties of $NiFe_2O_4$ nanoparticles prepared by the hydrothermal method. The results showed that the nanoparticles exhibited superparamagnetic behaviour with saturation magnetization of 39.60 emu/g. Furthermore, XRD showed a single phase of $NiFe_2O_4$ with a particle size of 12 nm and high crystallinity. In addition, $CoFe_2O_4$ nanoparticles were synthesized by the hydrothermal method. The superparamagnetic behaviour was observed by the VSM technique where the saturation magnetization was 113 emu/g (Zhao, Zhang, & Xing, 2008). Additional studies have reported the successful use of the hydrothermal method in the preparation of magnetic nanoparticles (Wu et al., 2014; Zhao, Wu, Guan, & Han, 2007; Ji, Tang, Ren, Zhang, Gu, & Du, 2004; Rozman, & Drofenik, 1995). The common microstructure of MNPs obtained by this method and confirmed by the transmission electron microscopy (TEM) technique is shown in Figure 6.3b.

6.3.3 Sol–Gel Method

The sol–gel method is one of the chemical methods for the preparation of magnetic nanoparticles (MNPs) where the term 'sol' is characterized by the initial chemical solution and the 'gel' is the polymeric chain that immobilizes the liquid phase in its interstices. Traditionally, the formation of the 'sol' occurs through hydrolysis and condensation of metal alkoxide precursors. Therefore, this phase is usually defined as a colloidal solution, as it covers a high number of systems (Danks, Hall, & Schnepp, 2016). In this method, the transition of the species from the 'sol' to 'gel' state occurs, where the colloidal particles of the 'sol' come together to form fibres that intertwine, forming a porous solid. After this stage, there is a consolidation phase of the materials, depending on the drying and calcination conditions (Brinker & Scherer, 2013). Factors such as solvent type, temperature, precursors, catalysts, pH, additives and mechanical agitation need to be considered in the sol–gel method because these parameters can influence the kinetics, growth and hydrolysis as well as condensation reactions (Majidi, Zeinali-Sehrig, Farkhani, Soleymani-Goloujeh, & Akbarzadeh, 2016). For magnetic nanoparticles (MNPs), the sol–gel method offers advantages such as good homogeneity, low cost and high purity. Studies have shown use of the sol–gel method to prepare magnetic nanoparticles. For example, a study carried out by Kumar, Prameela, Rao, Kiran, and Rao (2020) investigated the structural and magnetic properties of copper-substituted nickel–zinc nanoparticles synthesized using the sol–gel method. The single-spinel phase structure of all the samples was observed by XRD technique with a crystallize size between 22 to 33 nm. An FTIR technique confirmed the tetrahedral and octahedral sites, with characteristic bands of magnetic nanoparticle-type ferrites. Superparamagnetic behaviour of MNPs was confirmed by VSM technique with saturation magnetization around 74 emu/g, suggesting that this parameter contributes for easy separation of the MNPs using a field magnet during the transesterification reaction. Additional studies have been reported (Bhagwat, Humbe, More, & Jadhav, 2019; Ramana, Rao, & Rao, 2018; Jalaiah, & Babu, 2017; Atif, Nadeem, Grössinger, & Turtelli, 2011; Srivastava, Chaubey, & Ojha, 2009; Zahi, Hashim, & Daud, 2007; Chen, & He, 2001) that demonstrated a good structural, morphology, thermal and magnetic properties of nanoparticles prepared by the sol–gel method. The common microstructure of MNPs obtained by this method and confirmed by the transmission electron microscopy (TEM) technique is shown in Figure 6.3c.

6.3.4 Combustion Synthesis Method

The combustion synthesis method involves mixtures of redox reactions containing the metal ions of interest, such as oxidizing reagents, and a reducing agent (fuel) such as maleic hydrazine, citric acid, urea, glycine, N-methyl urea, hexamethylenetetramine, L-alanine, starch and L-arginine as a reducing agent (Kombaiah, Vijaya, Kennedy, Bououdina, Al-Lohedan, & Ramalingam, 2017; Costa, Kiminami, & Morelli, 2009; Hwang, Tsai, Huang, Peng, & Chen, 2005; Costa, Morelli, & Kiminami, 2001; Segadaes, Morelli, & Kiminami, 1998; Jain, Adiga, & Verneker, 1981). This definition is based on the thermodynamic concepts of the chemistry of propellants and explosives (Jain, Adiga, & Verneker, 1981).

An example of the preparation of magnetic nanoparticle nickel ferrite ($NiFe_2O_4$) by combustion reaction using urea as fuel is presented (Reaction 6.3) (Mapossa, Dantas, Silva, Kiminami, Costa, & Daramola, 2020b). According to the principle of chemistry of propellants the various valences of elements of the oxidizing agent and reducing agent are represented as follows Ni = +2; Zn = +2; Fe = +3; C = +4; H = +1; O = –2. The total valence of urea is equal to +6 and total valences of nitrates [$Ni(NO_3)_2$: $Fe(NO_3)_3$: 1:2] for oxidizing agent pure nickel ferrite ($NiFe_2O_4$) is equal to –40. Thus, the combination of total valences of reducing agent (fuel, urea) and oxidizing agent obtained the simple expression as follows: –40+6n = 0; n = 6.67 mol of urea in the combustion reaction. Since it is a complete combustion, the stoichiometric reaction ($\Phi e = 1$) follows the definition described for the oxygen balance equals zero. To accomplish this, the content of oxygen from the nitrates is completely oxidized by the reducing agent (fuel) in the mixture (Hwang, Tsai, Huang, Peng, & Chen, 2005).

$$Ni(NO_3)_2.6H_2O_{(s)} + 2Fe(NO_3)_3.9H_2O_{(s)} + 6.67(NH_2)_2CO_{(s)} \rightarrow NiFe_2O_{4(s)} + 6.67CO_{2(g} + 37.33H_2O_{(g)} + 10.67N_{2(g)} \quad (6.3)$$

For example, from the combustion reaction represented above, some parameters such as enthalpy of reaction (ΔH^o) and the adiabatic flame temperature of MNPs can be calculated using the thermodynamic concepts (Phadatare, Salunkhe, Khot, Sathish, Dhawale, & Pawar, 2013). The formation enthalpy values of reactants and products are found in the literature, while n represents the number of mols in the combustion reaction. The enthalpy of the reaction is expressed by the formula represented by Equation (6.1) as follows:

$$\Delta H^o = \sum n(H_f^o)_{products} - \sum n(H_f^o)_{reactants} \quad (6.1)$$

In the combustion reaction, the adiabatic flame time can be expressed by the formula represented by Equation (6.2) (Phadatare, Salunkhe, Khot, Sathish, Dhawale, & Pawar, 2013).

$$Q = -\Delta H^o = \int_{298}^{T_{ad}} (\sum n.C_p)_{products} .dT \quad (6.2)$$

where Q represents the heat absorbed by the products under adiabatic condition, and C_p represents the heat capacity of the products at constant pressure.

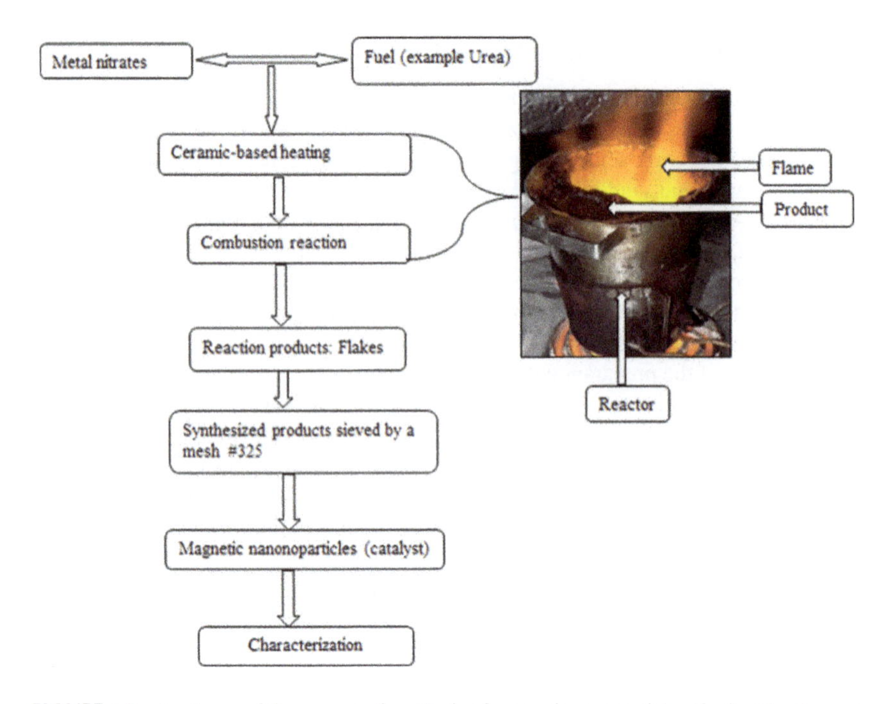

FIGURE 6.2 A scheme of the process of synthesis of magnetic nanoparticles (ferrites) by the combustion method.

Nowadays, the development of magnetic nanoparticles by the combustion reaction synthesis method (process flow chart shown in Figure 6.2) is gaining attention from researchers. In the literature, various studies have demonstrated good results for magnetic nanoparticles synthesized by combustion methods. Some studies are described as follows. For example, magnetic nanoparticles of Ni-Zn ferrites were obtained by the combustion method using urea as the reducing agent (fuel) (Costa, Tortella, Morelli, Kaufman, & Kiminami, 2002). Magnetic nanoparticles with surface areas of 63.89 m^2/g were obtained. The crystallize size of the nanoparticles calculated by the XRD was 18.6 nm. In addition, the pure nickel ferrite ($NiFe_2O_4$) with a surface area of 55.21 m^2/g and crystallite size of 18 nm was synthesized by the combustion method using urea as the reducing reagent (Costa, Lula, Kiminami, Gama, De Jesus, & Andrade, 2006). The temperature used during the combustion reaction was between 250–450 °C.

Santos, Costa, and Andrade (2012) reported that glycine as the reducing reagent or fuel was very effective in producing major phases of nickel ferrite ($NiFe_2O_4$) and zinc ferrite ($ZnFe_2O_4$) by the combustion reaction. The crystallite size and surface area results obtained by XRD and BET were 44 and 27 nm and 73 and 115 m^2/g for $NiFe_2O_4$ and $ZnFe_2O_4$, respectively. Additionally, Džunuzović et al. (2015) used hydrated acetic acid as fuel with metal nitrates to prepare magnetic $Ni_{1-x}Zn_xFe_2O_4$ ferrites with different zinc compositions represented by (x = 0.0; 0.3; 0.5; 0.7; 1.0) via the combustion method. The additional concentration of zinc increased the crystallite sizes obtained by XRD from 38 to 45 nm. The ferrimagnetic property for the system of ferrites with saturation of magnetization varying between 43.12–78.68 emu/g was obtained. The morphological, structural and magnetic properties of nanoparticles of nickel ferrite ($NiFe_2O_4$) prepared by the combustion method were affected by the ratio of glycine/nitrate (Alarifi, Deraz, & Shaban, 2009). The results showed that the combustion method was effective for obtaining nickel ferrite nanoparticles with sizes varying between 4–62 nm and inverse spinel structure identified by X-ray diffraction. Furthermore, the results of the magnetization exhibited a ferromagnetic characteristic of the nanoparticles of $NiFe_2O_4$ samples with the saturation magnetization between 2.387–57 emu/g. Priyadharsini, Pradeep, Rao and Chandrasekaran (2009) reported the effect of glycine on the structural, spectroscopic and magnetic properties of Ni–Zn ferrite nanoparticles prepared by the combustion method. The single-phase spinel structure formation with crystallite sizes between 10–20 nm was identified by XRD. The SEM morphologically exhibited agglomerated flakes of nanoparticles containing large pores on the surface. The nanoparticles showed saturation magnetization values ranging from 4 to 27 emu/g. $NiFe_2O_4$ nanoparticles were synthesized by the combustion reaction (Maensiri, Masingboon, Boonchom & Seraphin, 2007). These authors reported that the X-ray diffraction results showed that the nanoparticles have only the inverse spinel structure without the presence of any other phase impurities. The ferromagnetic property of the $NiFe_2O_4$ ferrites, with saturation magnetization values between 26.4–42.5 emu/g, was obtained. Nanoparticles of Ni–Zn ferrite were synthesized by the combustion method using citric acid as the reducing agent. The effect of the concentration of Ni^{2+} ions on the magnetic and dielectric properties was reported (Kambale, Adhate, Chougule, & Kolekar, 2010). The crystallite sizes of nanoparticles of Ni–Zn ferrites varied between 53–71 nm with an increased composition of Ni^{2+} ions. Additionally, the magnetic property increased with an increase in the concentration of Ni^{2+} ions with saturation of magnetization (M_s) values between 2.90–67.62 emu/g.

Several fuels have different properties such as heat of combustion, reducing valency and decomposition temperature (Phadatare, Salunkhe, Khot, Sathish, Dhawale, & Pawar, 2013). These properties of fuels influence the mechanism of combustion. However, the authors also reported that the fuels also influenced the structural and magnetic properties (MNPs) (Phadatare, Salunkhe, Khot, Sathish, Dhawale, & Pawar, 2013). Thus, a study of the influence of different fuels on the combustion mechanism, structural and magnetic properties of $NiFe_2O_4$ nanoparticles prepared by combustion method was done by Phadatare, Salunkhe, Khot, Sathish, Dhawale and Pawar (2013). Different fuels such as polyvinyl alcohol (PVA), glycine and urea were used as fuel. The results showed that the prepared $NiFe_2O_4$ nanoparticles using glycine showed a highly agglomerated porous foam-like structure of nanoparticles, which is characteristic of combustion synthesis. The crystallite sizes obtained by XRD were 28.35, 29.20 and 32.62 nm for $NiFe_2O_4$ nanoparticles prepared using polyvinyl alcohol (PVA), glycine and urea, respectively. Furthermore, the saturation magnetization (M_s) values were 39, 41 and 44 emu/g for $NiFe_2O_4$ nanoparticles prepared using polyvinyl alcohol (PVA), glycine and urea, respectively.

Therefore, studies have shown excellent results of crystallite size, surface area, magnetic properties and morphological properties via the combustion method, despite

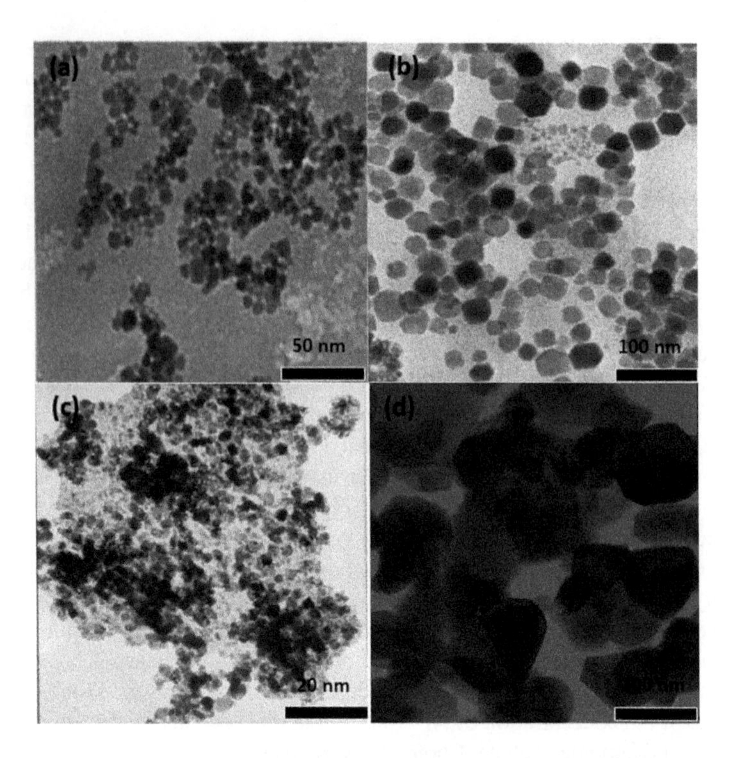

FIGURE 6.3 TEM micrographs of magnetic $NiFe_2O_4$ nanoparticles prepared by: (a) co-precipitation method (Shi, Ding, Liu, & Wang, 1999); (b) hydrothermal method (Wang, 2006); (c) sol–gel method (Chen and He, 2001) and (d) combustion method (Phadatare, Salunkhe, Khot, Sathish, Dhawale, & Pawar, 2013). The transmission electron microscopy (TEM) technique confirmed the difference in the morphology of the MNPs obtained by different method of synthesis. (Republished with permission from Elsevier.)

some differences because the authors used different reagents (for example, fuels). Therefore, different types of reagents in the synthesis and the quantity of reagents affect the properties of the final material obtained. The common microstructure of MNPs obtained by this method and confirmed by transmission electron microscopy (TEM) technique is shown in Figure 6.3d.

Table 6.1 summarizes the advantages and disadvantages of four methods (co-precipitation, hydrothermal, sol–gel and combustion) of synthesis of magnetic nanoparticles. In terms of simplicity of synthesis, co-precipitation is the preferred method. In terms of size and morphology control of the nanoparticles, the combustion method appears to be the best method developed to date. The hydrothermal method is a relatively little explored method for the preparation of magnetic nanoparticles, although it allows the synthesis of high-quality nanoparticles (Lu, Salabas, & Schüth, 2007). Among the methods developed so far, the co-precipitation method has been much more explored to synthesize magnetic nanoparticles and it can obtain MNPs on a large scale.

TABLE 6.1

Comparison of Methods of Synthesis of MNPs

Method	Synthesis	Temperature, °C	Time	Size distribution	Shape control	Yield	References
Co-precipitation	Very simple, ambient conditions	20–90	Minutes	Relatively narrow	Not good	High/scalable	(Lu, Salabas, & Schüth, 2007).
Hydrothermal	Simple, high pressure	220	Hours; days	Very narrow	Very good	Medium	(Lu, Salabas, & Schüth, 2007).
Sol–gel	Simple	≥ 70	Hours	Very narrow	Good	High/scalable	(Danks, Hall, & Schnepp, 2016).
Combustion	Very simple	≥ 600	Minutes	Very narrow	Very good	High/scalable	(Costa, Morelli, & Kiminami, 2001)

6.4 Characterization of MNPs

The characterization of magnetic nanoparticles can provide useful information about their morphology, structure, specific surface area, pore diameter, magnetic properties, and so on, which can facilitate a better understanding of the structure–activity relationships of MNPs. This information is most important to understand which factor (for example, surface area) contributes to the catalytic activity of MNPs for efficient biodiesel production. Therefore, the fundamental characterization techniques, including X-ray diffraction (XRD), can provide information about the types and crystal forms of particles, and energy-dispersive X-ray diffraction (EDXD), transmission electron microscopy (TEM) and scanning electron microscopy (SEM) can provide more information about the morphology, internal structure and particle size of the catalysts, while Fourier transform infrared spectroscopy (FTIR) can provide detailed molecular information on the nature of adsorbed species on a catalyst surface, and vibrating sample magnetometry (VSM) provides information about the magnetic properties. Other characterization techniques include zeta potential measurements and the temperature programmed desorption (TPD) technique. Details of each of these characterization techniques are described below.

X-ray diffraction (XRD). The XRD spectra are used for calculating the crystallographic identity of the produced material and phase purity, and for determining the mean particle size based on the broadening of the most prominent peak in the XRD profile (Faraji, Yamini, & Rezaee, 2010). The Debye-Scherrer formula represented by Equation (6.3) (Mapossa, Dantas, Kiminami, Silva, & Costa, 2015) is used to calculate the crystallite size (D) from the extension X-ray line (d_{311}) by the secondary diffraction line deconvolution of the polycrystalline cerium (standard), where λ is the wavelength of the X-ray beam, β is the full width at half maximum (FWHM), θ is the Bragg scattering angle and K (0.89) is the shape factor.

$$D = \frac{K\lambda}{\beta cos\theta} \tag{6.3}$$

Additionally, Faraji, Yamini and Rezaee (2010) reported that in a diffraction pattern, the intensity can be used to quantify the proportion of magnetic nanoparticles formed in a mixture by comparing experimental peaks and reference peak intensities. Maaz, Karim, Mumtaz, Hasanain, Liu and Duan (2009) reported the XRD of cubic $NiFe_2O_4$ nanoparticles with an average crystallite size of 28 nm (Figure 6.4). The XRD peaks were compared to the standard PDF card number 742081 for inverse cubic nickel ferrite. Furthermore, the XRD peaks suggested that the prepared magnetic $NiFe_2O_4$ nanoparticles have a polycrystalline nature.

6.4.1 Energy-Dispersive X-Ray Diffraction (EDXD)

This technique provides a wide advantage of being carried out on the suspension and is used to improve the knowledge of fine structural details. EDX can be

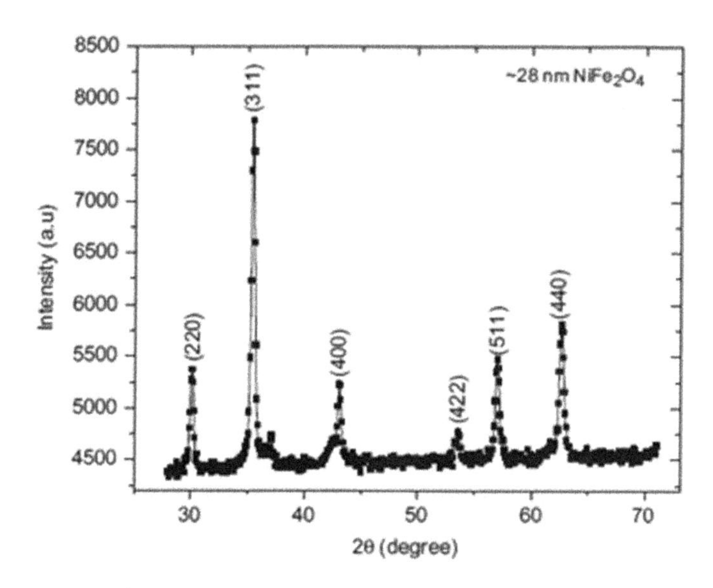

FIGURE 6.4 X-ray diffraction pattern of magnetic $NiFe_2O_4$ nanoparticles synthesized by the co-precipitation method. (Maaz, Karim, Mumtaz, Hasanain, Liu, & Duan, 2009. Republished with permission from Elsevier.)

used to provide an elemental analysis and determination of the chemical composition of prepared magnetic nanoparticles (MNPs). From the EDXD technique, the ratio of the elements in the MNPs structure can be estimated (Faraji, Yamini, & Rezaee, 2010).

6.4.2 Scanning Electron Microscopy (SEM) and Transmission Electron Microscopy (TEM)

These methods are widely used to determine the shape, size distribution and morphology of prepared particles in the micro to nano range. This information is important for the evaluation of catalyst performance. Commonly, a rough and porous surface can provide a larger specific surface area and more active sites, while a smoother surface has fewer active sites that are accessible to the reactants (Wang, Sudarsanam, Xu, Zhang, Li, & Yang, 2020). The resolution of the SEM is lower compared to TEM, and it is not effective for MNPs with particle sizes lower than 20 nm (Faraji, Yamini, & Rezaee, 2010). However, the method of synthesis of MNPs is a crucial step in SEM and TEM technique characterizations, as it can affect the aggregation or morphology of MNPs. Figure 6.5a shows the SEM micrographs of magnetic Fe_3O_4 nanoparticles impregnated in calcinated eggshell, in which the presence of a rough flake structure was confirmed. Meanwhile, the TEM micrograph confirmed the presence of spherical Fe_3O_4 nanoparticles in the structure of calcined eggshell (Figure 6.5b) (Chingakham, David, & Sajith, 2019).

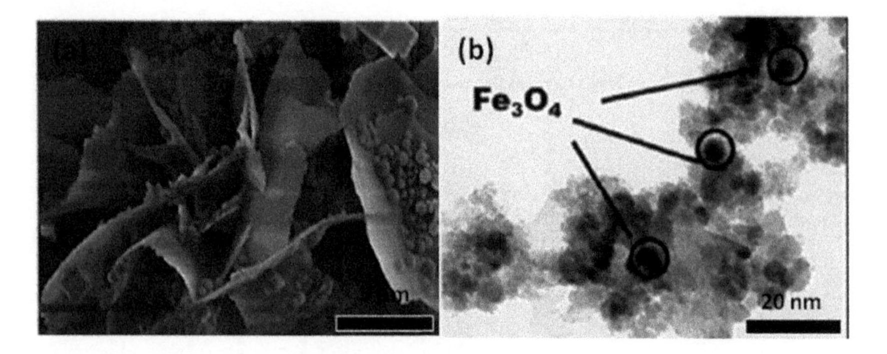

FIGURE 6.5 (a) SEM micrograph of Fe_3O_4 loaded in calcinated eggshell and (b) TEM micrograph of Fe_3O_4 loaded in calcinated eggshell. (Chingakham, David, & Sajith, 2019. Republished with permission from Elsevier.)

6.4.3 N_2 Adsorption–Desorption Technique

The determination of the specific surface area and porosity (pore volume and pore diameter) is extremely important when it is intended to use magnetic nanoparticles as catalysts, as there is a correlation between catalytic activity and textural properties. A large specific surface area can effectively load more active sites with a larger capacity. In the meantime, suitable pore size and high pore volume facilitate the catalytic reaction (Wang, Sudarsanam, Xu, Zhang, Li, & Yang, 2020). To obtain such information, the construction of isotherms is very important, as their shape can reveal many details about the characteristics of these materials. An isotherm occurs through the relationship between the amount of adsorbed and desorbed gas for a catalyst, at constant temperature, as a function of the gas pressure. Normally, they are constituted by means of graphs, in which the volume of adsorbed gas (V) depends on the relative pressure (P/P_o), where P_o is the saturation pressure. A type I isotherm is related to micropore adsorption, limited to a few molecular layers, where the pores slightly exceed the adsorbents' molecular diameter, whereas types II and IV isotherms are commonly found in adsorption measures, occurring in a non-porous system or with pores in the range of mesopores or macropores. The inflection point of the isotherm corresponds to the formation of the first adsorbed layer that covers the entire surface of the material, already in the type IV isotherm, indicating the presence of micropores associated with mesopores. Finally, types III and V isotherms are related to very weak interactions in systems containing macro- and mesopores (Ambroz, Macdonald, Martis, & Parkin, 2018; Mapossa, Dantas, Diniz, Silva, Kiminami, & Costa, 2017). Figure 6.6 shows six different types of BET isotherms. A study carried out by Dantas, Leal, Mapossa, Cornejo and Costa (2017) obtained a high biodiesel yield using magnetic $Ni_{0.1}Cu_{0.3}Zn_{0.5}Fe_2O_4$ catalyst with a high surface area of 29.89 $m^2\ g^{-1}$ compared to the 65% biodiesel yield obtained using magnetic catalyst $Ni_{0.3}Cu_{0.2}Zn_{0.5}Fe_2O_4$ with a low surface area of 23.15 $m^2\ g^{-1}$.

6.4.4 Fourier Transform Infrared Spectroscopy (FTIR)

FTIR is the most important technique for structural elucidation, particularly functional group detection of the magnetic nanoparticles. For example, iron oxide magnetic

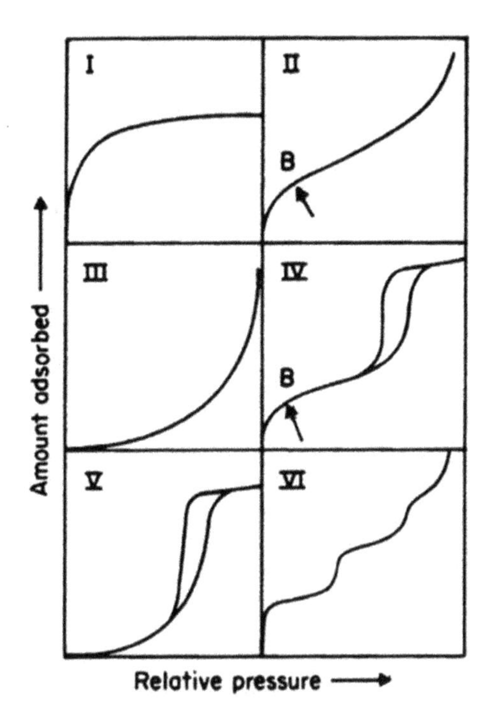

FIGURE 6.6 International Union of Pure and Applied Chemistry (IUPAC) classification of BET isotherms. (Muttakin, Mitra, Thu, Ito, & Saha, 2018. Republished with permission from Elsevier.)

nanoparticles (type spinel ferrites) are distinguished with two interstitial sites of tetrahedral (A) and octahedral (B) sites by FTIR (Mapossa, Dantas, Silva, Kiminami, Costa, & Daramola, 2020b). Waldron (1955) reported that the vibrational frequency (υ_1) around 600–500 cm^{-1} is related to the Fe^{3+}–O^{2-} complex at the tetrahedral site (A). Then the vibrational frequency (υ_2) around 450–350 cm^{-1} is related to the Fe^{3+}–O^{2-} and M^{2+}–O^{2-} complexes (where $M^{2+} = Zn^{2+}$, Mn^{2+}, Fe^{2+}, etc.) at the octahedral (B) site. The FTIR spectra of magnetic $NiFe_{2-x}Cr_xO_4$ nanoparticles (Figure 6.7), revealed the presence of two strong absorption bands, υ_1 and υ_2, which lie in the expected range of cubic spinel-type ferrites. The major bands between 560–570 cm^{-1} correspond to vibrations of the metal at the tetrahedral site, while the low-frequency bands between 396–412 cm^{-1} prove the octahedral sites (Gabal, Kosa, & El Muttairi, 2014).

6.4.5 Vibrating Sample Magnetometry (VSM)

The vibrating sample magnetometer (VSM) is a widely used technique to investigate the magnetic properties of magnetic nanoparticles (MNPs) as a function of an applied external magnetic field usually between –3 and 3 Tesla. Like most conventional magnetization probes, the technique is not element specific, but it rather measures the whole magnetization (Faraji, Yamini, & Rezaee, 2010). Based on the obtained VSM curve at low and room temperatures, the magnetic behaviour of the MNPs as well as information of the magnetic parameters such as saturation magnetization (M_s),

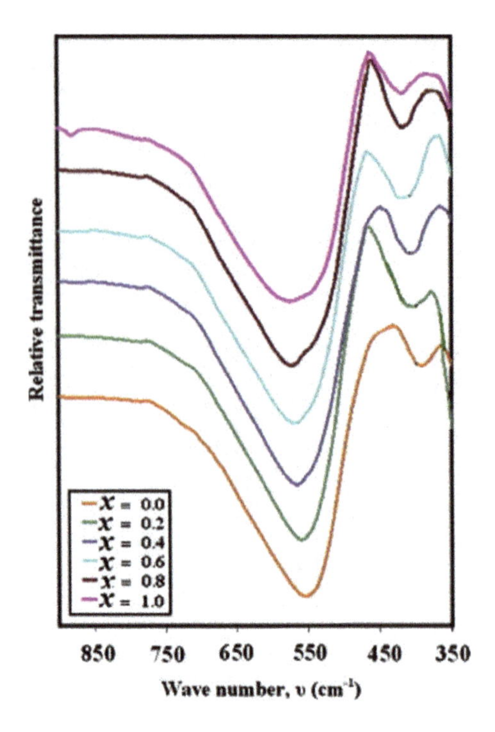

FIGURE 6.7 FTIR spectra of $NiFe_{2-x}Cr_xO_4$ nanoparticles. (Gabal, Kosa, & El Muttairi, 2014. Republished with permission from Elsevier.)

remnant magnetization (M_r) and coercive field (H_c) can be determined (Maposssa, Dantas, Diniz, Silva, Kiminami, & Costa, 2017; Dantas, Leal, Maposssa, Silva, & Costa, 2016). Ferromagnetic properties can be acceptable after supporting the active sites, such that the catalyst can be separated from the reaction system quickly and efficiently (Wang, Sudarsanam, Xu, Zhang, Li, & Yang, 2020). A study carried out by Yan et al. (2007) employed the VSM technique to evaluate the magnetic property of magnetic nanoparticles (Fe_3O_4). Therefore, the superparamagnetic behaviour of nanoparticles was observed (Figure 6.8).

6.4.6 Zeta Potential Analysis

The zeta potential (ζ) gives information about the potential stability and electric charges on the surface of the MNPs, which is a property of the change of acidity in the catalyst due to the absorption and desorption of ions and protons (Maposssa, Dantas, Silva, Kiminami, Costa, & Daramola, 2020b). The isoelectric point, also referred to as the point of zero charge, is the pH at which the particles in suspension have a net charge of zero and no mobility in the electric field (Faraji, Yamini, & Rezaee, 2010). Work carried out by Maposssa, Dantas and Costa (2020a) evaluated the electric charges of the nanomagnetic catalyst ($Ni_{0.5}Zn_{0.5}Fe_2O_4$) by zeta potential measurements. This study suggested that the MNPs had good stability and a positive surface charge affected positively during transesterification with an excellent biodiesel yield.

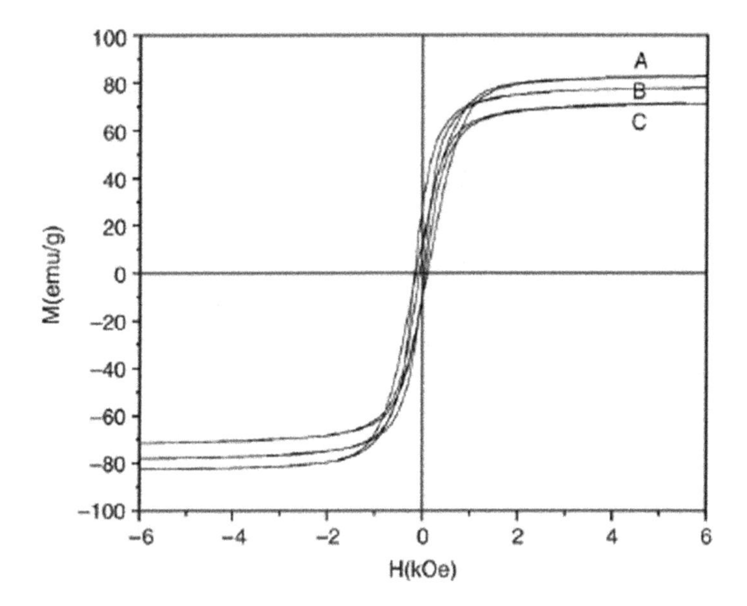

FIGURE 6.8 VSM curves for the different sizes of Fe_3O_4 nanoparticles at room temperature: (A) 130 nm, (B) 60 nm and (C) 15 nm. (Yan et al., 2007. Republished with permission from Elsevier.)

6.4.7 Temperature Programmed Desorption (TPD) Technique

Temperature programmed desorption has been developed especially for the field of catalysis, because the technique allows one to study the interaction of reaction gases with solid surfaces, thereby making it a powerful tool for both the evaluation of active sites on magnetic heterogeneous catalyst surfaces and the understanding of the mechanisms of catalytic reactions including adsorption, surface reaction and desorption. The desorption rate is measured as a function of temperature, either by detecting the evolved gas or by monitoring the retained species, for example, gravimetrically (Ishii & Kyotani, 2016). The most common molecules used for TPD are ammonia (TPD-NH_3) for acidic sites and carbon dioxide (TPD-CO_2) for basic sites. The practicality of TPD-NH_3 and its experimental details have been well reported by many researchers in the field of catalysis (Mapossa, Dantas, & Costa, 2020a; Mapossa, Dantas, Silva, Kiminami, Costa, & Daramola, 2020b; Dantas, Leal, Mapossa, Cornejo, & Costa, 2017).

6.5 Application of Magnetic Nanoparticles in Transesterification Reactions

In transesterification reactions, the magnetic nanoparticles are considered as valued substitutes for conventional heterogeneous materials (Shylesh, Schweizer, Demeshko, Schünemann, Ernst, & Thiel, 2009) due to their excellent magnetic behaviour. These materials act as promising heterogeneous nanocatalysts due to the presence of high thermal stability, higher number of active sites and a larger surface area that improve the catalytic performance of nanoparticles in catalytic reactions. Furthermore, the vast advantage of nanomagnetic particles as catalysts is because, from reactants, this material is easily separated by an external magnetic field without loss of the nanocatalyst and so it can be reused for up to several runs. This reduces waste generation and simplifies the other processes involved in the chemical reaction. For example, a study conducted by Maposssa, Dantas and Costa (2020a) showed a catalytic performance of $Ni_{0.5}Zn_{0.5}Fe_2O_4$ as magnetic nanocatalyst in the transesterification reaction where it exhibited an excellent yield of biodiesel of 92% under the following reaction conditions: 9:1 of methanol/soybean oil molar ratio and, 2% of catalyst loading at 180 °C in 3 h. The high yield of biodiesel was attributed to the strong acid sites on the surface of the catalyst, suitable morphology, reasonable surface area and zeta potential of the catalyst. Furthermore, the presence of Cu^{2+} ions in $Ni_{0.5}Zn_{0.5}Fe_2O_4$ magnetic nanoparticles facilitated an increase of 5.5–85% in the conversion values in methyl esters obtained from soybean oil transesterification (Dantas, Leal, Maposssa, Cornejo, & Costa, 2017). The authors also reported that the strong active sites of nanoparticles provided an increase in the surface area/volume ratio, culminating in high catalytic activities. Therefore, findings from these studies suggest that the development of magnetic nanocatalysts could provide an environmentally friendly platform for biodiesel production. In addition, more studies showing the catalytic activity of MNPs in transesterification reactions for biodiesel production as well as their reaction parameters are presented in Table 6.2.

According to Table 6.2, these studies showed that the catalytic performance of the magnetic nanoparticle-based catalysts applied in transesterification reactions provided a biodiesel production range from 64£ to 98%. However, a study carried out by Xie, Han and Wang (2018) reported the catalytic performance of sodium silicate/Fe_3O_4–MCM-41 in the transesterification reaction where the magnetic nanocatalyst provided a high yield of biodiesel of up to 99%. Furthermore, soybean oil remains the most studied source of biodiesel production, and is efficient and effective. Methanol is the preferred alcohol as it is inexpensive and has excellent physical and chemical properties, such as a higher reactivity due to it being a polar molecule (Farooq et al., 2019; Musa, 2016; Dai, Chen, Wang, & Chen, 2016).

6.6 Influence of Parameters in Transesterification Reactions for Biodiesel Production

Biodiesel is considered as a substitute for fossil-based diesel fuels because of its renewability and environmental friendliness. It is generally produced by transesterification of oils of free fatty acids with alcohol (the preference is methanol) as shown in Figure 6.9. Therefore, a high biodiesel yield is achieved by optimizing the reaction conditions (parameters) that affect the transesterification reactions. These parameters include catalyst (MNPs) loading, molar ratio of oil/alcohol, reaction temperature and reaction time. Several studies have reported the influence of such parameters in transesterification reactions (Maposssa, Dantas, & Costa, 2020a; Baskar, Selvakumari, & Aiswarya, 2018; Feyzi, Hassankhani, & Rafiee, 2013). Details of the effects of these parameters in transesterification reactions are described below.

TABLE 6.2
Previous Studies Show the Biodiesel Yield via Transesterification Reactions Using Different Parameters and Different MNP Catalysts

Magnetic nanoparticles (MNPs)	Chemical synthesis method	Reaction time (h)	Reaction temperature (°C)	Molar ratio (oil: alcohol)	Alcohol name	Feed	Catalyst amount (wt%)	Biodiesel yield (%)	References
$MgO/MgFe_2O_4$	Microwave-assisted combustion	3	110	1:12	Methanol	Sunflower oil	3	92.5	(Amani, Haghighi, & Rahmanivahid, 2019)
$Ni_{0.5}Zn_{0.5}Fe_2O_4$	Combustion	3	180	1:9	Methanol	Soybean oil	2	92.1	(Mapossa, Dantas, & Costa, 2020a).
$Ni_{0.5}Zn_{0.5}Fe_2O_4$	Combustion	1	180	1:5	Ethanol	Different oils	3	84–95	(Dantas et al., 2021)
MgO/Fe_2O_3-SiO_2		4.1	70	1:12	Methanol	*Camelina sativa* seed oil	4.9	99	(Rahimi, Kahrizi, Feyzi, Ahmadvandi, & Mostafaei, 2021)
Citrus sinensis peel ash (CSPA)@Fe_3O_4	Co-precipitation	3	65	1:6	Methanol	Waste cooking oil	6	98	(Changmai., Rano, Vanlalveni, & Rokhum, 2021)
$H_6PV_3MoW_8O_{40}/Fe_3O_4$/ZIF-8	Co-precipitation	10	160	1:30	Methanol	Soybean oil	6	92.6	(Xie, Gao, & Li, 2021)
Fe_3O_4@ZIF-8/TiO_2	Co-precipitation	1.2	50	1:30	Ethanol and methanol	Oleic acid	6	80 and 93	(Sabzevar, Ghahramaninezhad, & Shahrak, 2021)
Fe_3O_4-CeO_2-25K	Co-precipitation	2	65	1:7	Methanol	Rapeseed oil	4.5	96.13	(Ambat, Srivastava, Haapaniemi, & Sillanpää, 2019)
$Zn_{0.5}Mn_{0.5}Fe_2O_4$	Sonication-assisted microwave	0.8	65	1:21	Methanol	Cooking oil	4	98.3	(Ashok, Kennedy, & Vijaya, 2019)

(*continued*)

TABLE 6.2 (Continued)

Previous Studies Show the Biodiesel Yield via Transesterification Reactions Using Different Parameters and Different MNP Catalysts

Magnetic nanoparticles (MNPs)	Chemical synthesis method	Reaction time (h)	Reaction temperature (°C)	Molar ratio (oil: alcohol)	Alcohol name	Feed	Catalyst amount (wt%)	Biodiesel yield (%)	References
$Zn_{0.5}Cu_{0.5}Fe_2O_4$	Sonication assisted microwave	0.7	65	1:21	Methanol	Waste cooking oil	4	98.9	(Ashok, Ratnaji, Kennedy, & Vijaya, 2020)
Ni-doped ZnO	Co-precipitation	1	55	1:8	Methanol	Castor oil	11	95.2	(Baskar, Selvakumari, & Aiswarya, 2018)
$Ni_{0.2}Cu_{0.3}Zn_{0.5}Fe_2O_4$	Combustion	1	180	1:20	Methanol	Soybean oil	4	85	(Dantas, Leal, Mapossa, Cornejo, & Costa, 2017)
Fe_3O_4-eggshell	Co-precipitation	2	65	1:12	Methanol	Pongamia oil	2	98	(Chingakham, David, & Sajith, 2019)
$LiFe_5O_8$-$LiFeO_2$	Solid-state reaction	2	65	1:36	Methanol	Soybean oil	8	96.5	(Dai, Y.-M., Wang, Y.-F. & Chen, 2018)
$Ni_{0.5}Zn_{0.5}Fe_2O_4$	Combustion	1	180	1:12	Methanol	vfSoybean oil	2	91.4	(Dantas, Leal, Cornejo, Kiminami, & Costa, 2020)
$Ni_{0.5}Zn_{0.5}Fe_2O_4$	Combustion	1	180	1:15	Methanol	Soybean oil	3	94	(Farias et al., 2020)
$Cs/Al/Fe_3O_4$	Precipitation	2	58	1:14	Methanol	Sunflower oil	4	94.8	(Feyzi, Hassankhani, & Rafiee, 2013)
$Ca/Fe_3O_4@SiO_2$	Sol–gel	5	65	1:15	Methanol	Sunflower oil	8	97	(Feyzi and Norouzi, 2016)
$KF/CaO–Fe_3O_4$	Co-precipitation	3	65	1:12	Methanol	Stillingia oil	4	95	(Hu, Guan, Wang, & Han, 2011)
$K/Fe_2O_3–Al_2O_3$	Co-precipitation and impregnation	6	65	1:12	Methanol	Microalgae	4	95.6	(Kazemifard, Nayebzadeh, Saghatoleslami, & Safakish, 2019)

Mn-ZnO/PEG	Precipitation	4	60	1:15	Methanol	Microalgae	3.5	87.5	(Raj, Bharathiraja, Vijayakumar, Arokiyaraj, Iyyappan, & Kumar, 2019)
$K_2CO_3/\gamma Al_2O_3/$ sepiolite/(0.3) Fe_3O_4	Co-precipitation	4	78	1:12	Ethanol	Sunflower oil	5	88	(Silveira Junior et al., 2018)
$Ca(OH)_2/Fe_3O_4$	Not reported	0.6	65	1:12	Methanol	Mixed jatropha–castor oil	7	95	(Chang, Lin, Jhang, Cheng, Chen, & Mao, 2017)
CaO/Fe_3O_4	Precipitation	5	65	1:20	Methanol	Palm seed oil	10	69.7	Ali, Al-Hydary, & Al-Hattab, 2017)
$MgFe_2O_4$@CaO	Precipitation	3	70	1:12	Methanol	Soybean oil	1	98.3	(Liu, Zhang, Fan, & Jiang, 2016)
$Ni_{0.3}Zn_{0.7}Fe_2O_4$	Combustion	1	180	1:12	Methanol	Soybean oil	2	94	(Mapossa, Dantas, Silva, Kiminami, Costa, & Daramola, 2020b)
$ZnFe_2O_4$	Combustion	1	180	1:12	Methanol	Soybean oil	2	71	(Mapossa, Dantas, Kiminami, Silva, & Costa, 2015)
$SO_4^{2-}/Fe_3O_4-Al_2O_3$	Co-precipitation and wet impregnation	6	120	1:9	Methanol	Microalgae	8	87.6	(Safakish, Nayebzadeh, Saghatoleslami, & Kazemifard, 2020)
$KOH/Fe_3O_4@\gamma-Al_2O_3$	Not reported	5.5	65	1:16	Methanol	Canola oil	6.5	97.4	(Ghalandari, Taghizadeh, & Rahmani, 2019)
Sodium silicate/ Fe_3O_4–MCM-41	Co-precipitation	8	65	1:25	Methanol	Soybean oil	3	99.2	(Xie, Han, & Wang, 2018)
Fe_3O_4–GO/lipase	Co-precipitation	60	40	1:4	Methanol	Soybean oil	25	92.8	(Xie and Huang, 2018)
$ZnO/BiFeO_3$	Co-precipitation	6	65	1:15	Methanol	Canola oil	4	95.43	(Salimi and Hosseini, 2019)

(continued)

TABLE 6.2 (Continued)

Previous Studies Show the Biodiesel Yield via Transesterification Reactions Using Different Parameters and Different MNP Catalysts

Magnetic nanoparticles (MNPs)	Chemical synthesis method	Reaction time (h)	Reaction temperature (ºC)	Molar ratio (oil: alcohol)	Alcohol name	Feed	Catalyst amount (wt%)	Biodiesel yield (%)	References
Guanidine-functionalized Fe_3O_4	Co-precipitation and sonication	24	120	1:30	Methanol	Soybean oil	10	96	(Santos, dos Santos, Guimarães, Ishida, Freitas, & Ronconi, 2015)
$CaO/CuFe_2O_4$	Precipitation	4	70	1:15	Methanol	Chicken fat	3	94.52	(Seffati, Honarvar, Esmaeili, & Esfandiari, 2019)
Activated Carbon/ $CuFe_2O_4$@CaO	Not reported	4	65	1:12	Methanol	Chicken fat	3	95.6	(Seffati, Esmaeili, Honarvar, & Esfandiari, 2020)
$Ca/Al/Fe_3O_4$	Not reported	3	65	1:15	Methanol	Rapeseed oil	6	98.71	(Tang, Wang, Zhang, Li, S., Tian, & Wang, 2012)
Fe_3O_4@HKUST-1 composites	Solvothermal	3	65	1:30	Methanol	Soybean oil	1.2	92.3	(Xie and Wan, 2018)
$CaFe_2O_4$-$Ca_2Fe_2O_5$ based catalysts	Co-precipitation and calcination	0.5	99.8	1:15	Methanol	Soybean oil	4	85.4	(Xue, Luo, Zhang, & Fang, 2014)
$CaFe_2O_4$-$Ca_2Fe_2O_5$-Fe_3O_4-Fe	Co-precipitation and calcination	0.5	99.8	1:15	Methanol	Jatropha oil	4	78.2	(Xue, Luo, Zhang, & Fang, 2014)
$CaO/CoFe_2O_4$	Hydrothermal	5	70	1:15	Methanol	Soybean oil	1	87.4	(Zhang, Han, Fan & Jiang, 2014)
Na_2SiO_3@Fe_3O_4/C	Not reported	1.3	55	1:7	Methanol	Soybean oil	7	97.9	(Zhang, Fang, & Wang, 2015)
Na_2SiO_3@Fe_3O_4/C	Not reported	1.3	55	1:7	Methanol	Jatropha oil	7	94.7	(Zhang, Fang, & Wang, 2015)

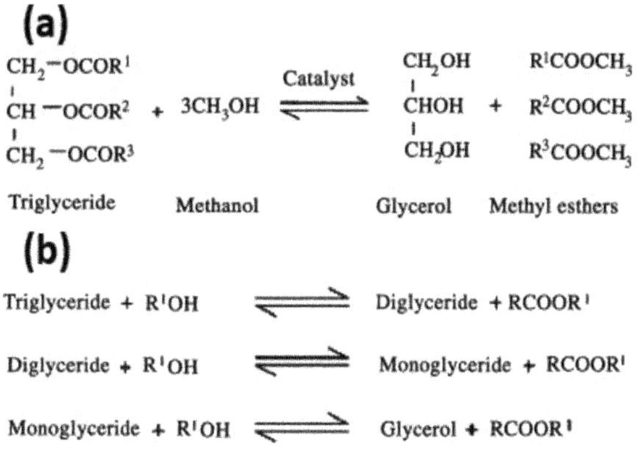

(a)

$$\begin{array}{l} CH_2-OCOR^1 \\ CH-OCOR^2 \\ CH_2-OCOR^3 \end{array} + 3CH_3OH \underset{}{\overset{\text{Catalyst}}{\rightleftharpoons}} \begin{array}{l} CH_2OH \\ CHOH \\ CH_2OH \end{array} + \begin{array}{l} R^1COOCH_3 \\ R^2COOCH_3 \\ R^3COOCH_3 \end{array}$$

Triglyceride Methanol Glycerol Methyl esthers

(b)

$$\text{Triglyceride} + R^1OH \rightleftharpoons \text{Diglyceride} + RCOOR^1$$

$$\text{Diglyceride} + R^1OH \rightleftharpoons \text{Monoglyceride} + RCOOR^1$$

$$\text{Monoglyceride} + R^1OH \rightleftharpoons \text{Glycerol} + RCOOR^1$$

FIGURE 6.9 Transesterification reaction. (a) General equation for transesterification of triglycerides. (b) Stages of transesterification of triglycerides. (Tapanes, Aranda, de Mesquita Carneiro, & Antunes, 2008. Republished with permission from Elsevier.)

6.6.1 Effect of Catalyst (MNPs) Amount

The catalytic reaction is influenced by the amount of MPNs. Various studies have reported that the conversion of oil to biodiesel increases with an increase in the amount of catalyst and an additional increase in the catalyst amount beyond the optimum value shows a reduction in biodiesel yield due to a decrease in the availability of active sites (Mapossa, Dantas, & Costa, 2020a; Ambat, Srivastava, Haapaniemi, & Sillanpää, 2019; Baskar, Selvakumari, & Aiswarya, 2018). Therefore, the high additional amount of catalyst aids in saponification of oil, which finally inhibits the reaction. Specifically, work done by Ashok, Kennedy and Vijaya (2019) reported the biodiesel yield when varying the pure and manganese-doped zinc ferrite (ZMF) nanocatalyst amount from 1 wt% to 7 wt%. The results showed that transesterification strongly depends upon the catalyst amount with the maximum biodiesel yield (Figure 6.10a). While raising the catalyst amount from 1 wt% to 4 wt%, the initial reagents were adsorbed on magnetic nanocatalyst active centres (Omar and Amin, 2011), leading to the formation of reaction products. Also, an increase in the concentration of the catalyst will lead to greater exposure of the catalyst surface area, which in turn would increase the product yield in the reaction mixture. However, catalyst loading beyond 4 wt% had a negative impact, resulting in a low biodiesel yield (Mazaheri et al., 2018) due to the rise in viscosity of the reactants followed by subsequent mixing and mass transfer resistance problems (Liu, Su, Liu, Li, & Solomon, 2011a).

6.6.2 Effect of Alcohol/Oil Molar Ratio

The biodiesel conversion rate significantly increases as the oil to alcohol molar ratio is raised. However, the biodiesel yield is adversely affected on having an alcohol content above the optimum amount due to the higher solubility of glycerol to ester phase, resulting in difficulty in separation of the biodiesel (Ambat, Srivastava, Haapaniemi, & Sillanpää, 2019). Banihani (2016) and Ayetor, Sunnu Parbey (2015) also reported

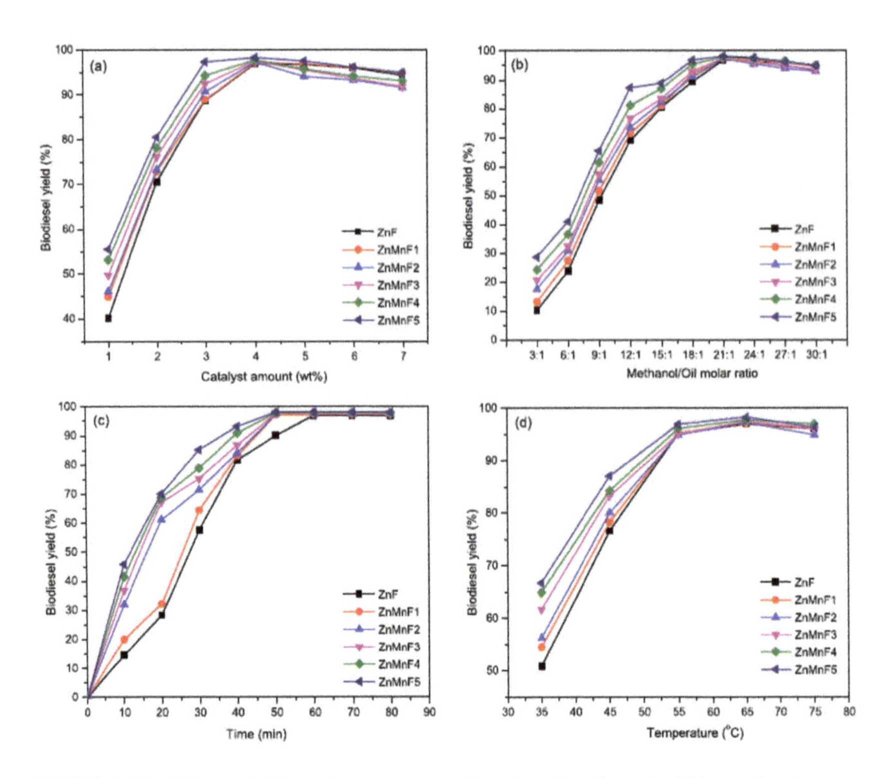

FIGURE 6.10 Effects of: (a) catalyst amount (methanol to oil molar ratio 21:1, reaction temperature 65 °C, 50 min reaction time); (b) methanol/oil molar ratio (catalyst amount 4 wt%, reaction temperature 65 °C, 50 min reaction time); (c) reaction time (catalyst amount 4 wt%, methanol/oil molar ratio 21:1, reaction temperature 65 °C); (d) reaction temperature (methanol/oil molar ratio 21:1, catalyst amount 4 wt%, 50 min reaction time). The magnetic nanoparticles $ZnFe_2O_4$, $Zn_{0.9}Mn_{0.1}Fe_2O_4$, $Zn_{0.8}Mn_{0.2}Fe_2O_4$, $Zn_{0.7}Mn_{0.3}Fe_2O_4$, $Zn_{0.6}Mn_{0.4}Fe_2O_4$, and $Zn_{0.5}Mn_{0.5}Fe_2O_4$ were labelled as ZnF, ZnMnF1, ZnMnF2, ZnMnF3, ZnMnF4 and ZnMnF5. (Ashok, Kennedy, & Vijaya, 2019. Republished with permission from Elsevier.)

that an excess amount of alcohol above the optimum limit leads to increased solubility of glycerol into the ester phase, thereby encouraging the reverse reaction between glycerol and ester, which reduces the biodiesel yield. Figure 6.10b shows the effect of the methanol/oil molar ratio on the biodiesel yield using a nanocatalyst. It was found that the maximum biodiesel yield was achieved with a 21:1 methanol/oil molar ratio (Ashok, Kennedy, & Vijaya, 2019).

6.6.3 Reaction Time

The reaction time also plays a major role in the transesterification reaction using nanomagnetic catalysts. In general, increasing the time increases the biodiesel yield until it reaches its maximum. After that, the percentage of fatty acid methyl esters (FAMEs) remains almost constant, without much of a reduction in the ester content (Ambat, Srivastava, Haapaniemi, & Sillanpää, 2019). This is due to the transesterification reaction which is a reversible process which is conducted for a reduction of the higher yield of fatty acid methyl esters and the production of soap (Mapossa, Dantas, & Costa, 2020a). Figure 6.10c shows the effect of reaction time in transesterification reactions using nanomagnetic particles (Ashok, Kennedy, & Vijaya, 2019). In addition, Xie, Gao and Li (2021) stated that from 4 to 12 h, the oil conversion ascended monotonously along with an increment in the reaction duration, until the maximum conversion rate of 92.6% was achieved at a reaction time of 10 h.

6.6.4 Reaction Temperature

The catalytic performance of heterogeneous catalyst in transesterification reactions is very temperature-dependent. The biodiesel yield increases gradually with an enhancement in the reaction temperature and results in maximum yield of fatty acid methyl esters. This is attributed to the decrease in oil viscosity and an increase in miscibility with alcohol caused by an increase in temperature (Mapossa, Dantas, & Costa, 2020a). However, the biodiesel yield can be reduced with an excessive rise in temperature because high temperature favours methanol vaporization as well as saponification reaction (Singh, Bux, & Sharma, 2016; Abbah, Nwandikom, Egwuonwu, & Nwakuba, 2016; Eevera, T., Rajendran, K. & Saradha, 2009). Figure 6.10d shows the effect of reaction temperature evaluated in transesterification reactions using nanomagnetic particles of a pure and manganese-doped zinc ferrite system.

6.7 Recovery and Recycling of MNPs during Transesterification

As a very promising strategy, magnetic nanoparticles employed as heterogeneous catalysts in transesterification reactions exhibit striking characteristics, such as large surface area, mobility, actives sites and high mass transference. In addition, they can easily be recovered with consequent recyclability by applying an external magnetic field (Netto, Toma, & Andrade, 2013). Due to their excellent environmental compatibility, the use of such magnetic nanoparticles represents an effective green chemistry approach, since it prolongs, through the successive recovery cycles, the useful lifetime of the catalyst. The recovery of magnetic nanoparticles by an external magnetic field has more advantages than a conventional process in that recovery with the heterogeneous catalyst via filtration or centrifugation is time consuming and challenging, especially when the catalyst particles are smaller than the micrometre size range. Several studies have now reported the implication of reusability of nanomagnetic nanoparticles by a magnetic field (Figure 6.11). Work done Chingakham, David and Sajith (2019) reported the reusability of the Fe_3O_4 nanoparticle-impregnated eggshell catalyst in transesterification reactions where the catalytic activity was found to be retained for up to 10 cycles and without considerable loss of stability (Figure 6.12). Furthermore, the reusability of KF/CaO–Fe_3O_4 was studied by carrying out reaction cycles (Hu, Guan, Wang, & Han, 2011). The catalyst, after 3 h of transesterification reaction, was

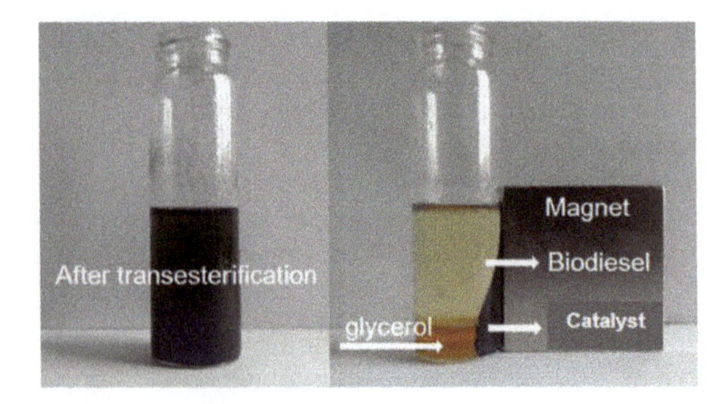

FIGURE 6.11 Image of the nanomagnetic catalyst recovered by magnet after transesterification reaction (Li et al., 2020). Republished with permission from Elsevier.

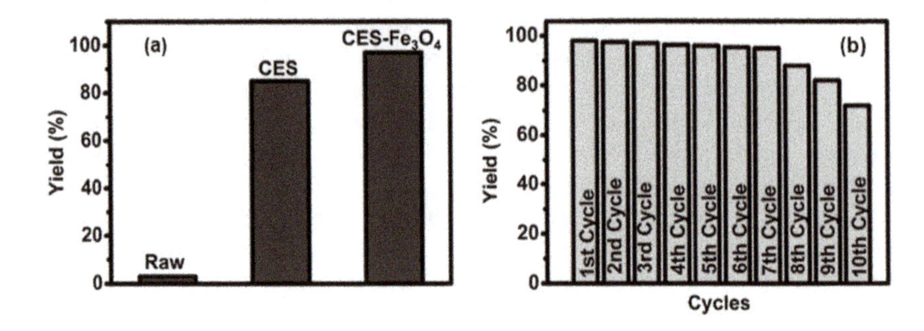

FIGURE 6.12 (a) Biodiesel yield comparison. (b) Reusability of the CES–Fe_3O_4. (Chingakham, David, & Sajith, 2019. Republished with permission from Elsevier.)

separated using a permanent magnet. The results showed that the catalyst maintained sustained catalytic activity even after being reused for 14 runs with no apparent loss of catalytic activity and stability. Table 6.3 shows additional results of the activity of reused magnetic nanoparticles during transesterification reactions.

The results reported by researchers in Table 6.3 demonstrates that the activity and stability of the catalyst is not greatly affected. Therefore, the reduction in the yield could be due to the mass loss of the nanocatalysts by repeated washing after magnetic separation. Furthermore, Ashok, Kennedy and Vijaya (2019) reported that the slightly decreased biodiesel yield after the fourth cycles may be due to factors such as deactivation of active sites and a change in the specific basicity of the catalyst. The same idea was corroborated by Baskar, Selvakumari and Aiswarya (2018) when they attributed the decreased biodiesel yield after fifth reuses of catalyst to the deposition of organic materials on the surface of the catalyst and a reduction in active sites. Therefore, due to the high catalytic efficiency, stability and magnetic recoverability of magnetic nanoparticles (MNPs), these catalysts have a potential application in transesterification reactions for biodiesel production. However, it must be noted that, due to the very presence of multiple and expensive metal constituents, its large-scale application and economic viability is questionable.

TABLE 6.3

Reusability of Magnetic Nanoparticles (MNPs) during Transesterification Reactions for Biodiesel Production

Magnetic nanoparticles	Cycles of reusability of MNPs	Biodiesel yield (%)	References
$MgO/MgFe_2O_4$	5	78.8	(Amani, Haghighi, & Rahmanivahid, 2019)
Fe_3O_4-CeO_2-25K	5	80.94	(Ambat, Srivastava, Haapaniemi, & Sillanpää, 2019)
$Zn_{0.5}Mn_{0.5}Fe_2O_4$	10	93.8	(Ashok, Kennedy, & Vijaya, 2019)
$Zn_{0.5}Cu_{0.5}Fe_2O_4$	4	96	(Ashok, Ratnaji, Kennedy, & Vijaya, 2020)
Ni/ZnO	5	85	(Baskar, Selvakumari, & Aiswarya, 2018)
Mn-ZnO/PEG	6	73.5	(Raj, Bharathiraja, Vijayakumar, Arokiyaraj, Iyyappan, & Kumar, 2019)
$MgFe_2O_4$@CaO	5	89	(Liu, Zhang, Fan, & Jiang, 2016)
$CaO/CoFe_2O_4$	5	80	(Zhang, Han, Fan, & Jiang, 2014)
$Cs/Al/Fe_3O_4$	4	88	(Feyzi, Hassankhani, & Rafiee, 2013)
K/Fe_2O_3–Al_2O_3	6	85.6	(Kazemifard, Nayebzadeh, Saghatoleslami, & Safakish, 2019)

6.8 Challenges of MNPs in Transesterification

Biodiesel has become more attractive recently because of its environmental benefits and the fact that it is made from renewable resources. The current challenges include its cost and the limited availability of fat and oil resources. There are two aspects to the cost of biodiesel: the costs of raw material (fats and oils) and the cost of processing. The cost of raw materials accounts for 60–75% of the total cost of biodiesel fuel (Bisen, Sanodiya, Thakur, Baghel, & Prasad, 2010). Genetic engineering can also be used to engineer new crops and improve the oil levels in existing crops, to provide sufficient renewable raw materials for biodiesel production. Since direct usage of vegetable oils as biodiesel is impractical, many processes have been developed to convert them into a suitable form. Furthermore, kinetics studies are important to identify optimal conditions for MPN-catalysed transesterification and this needs much research focus.

The magnetic properties of solid acid catalysts are excellent for separating the catalyst from the reaction medium. However, since the surface of magnetic oxide materials is unstable under acidic conditions and very sensitive, the particles can easily be aggregated into large clusters. Such large clusters cause anisotropic dipolar interactions, which result in loss of their catalytic activity (Vasić, Hojnik-Podrepšek, Knez, & Leitgeb, 2020; Xin et al., 2014; Moghanian, Mobinikhaledi, Blackman, & Sarough-Farahani, 2014).

6.9 Conclusions and Future Outlook

In this chapter, the application of magnetic nanoparticles (MNPs) as heterogeneous catalysts in transesterification reactions for biodiesel production has been addressed. Magnetic nanoparticles can be prepared by different methods such as sol–gel, co-precipitation, hydrothermal, microwave and combustion. MNPs have several advantages including that they are easily separated by an external magnetic field without loss of the nanocatalyst and they can be reused for up to several runs in experiments. In most studies, magnetic nanoparticles presented excellent catalytic activity during transesterification reactions for biodiesel production. Furthermore, some studies have shown that MNPs were still effective after 14 successive cycling runs. It is important to note that magnetic nanoparticles as catalysts have a large surface area as well as high mass transfer area, which are ideal criteria in catalytic applications. The catalytic activity of MNPs as catalysts is a direct result of their intrinsic characteristics as well as their synthesis method, nevertheless, catalytic performance can be influenced by conditions that are imposed on these materials to prepare them for a given application. Thus, better accuracy of the surface reactivity can confirm the materials as catalytic active, and this has been done in some specific studies, such as to measure the surface area, to study the shape and distribution of the pores, and to study the acid and basic sites present on the surface of the nanoparticles. It can be said that the structure, morphology, chemical and thermal state, and surface reactivity display a strong interaction with the nature, quantity and strength of the active sites. Studies have demonstrated that the immobilization of lipase on magnetic nanoparticles increases the stability of the enzyme in the reaction mixture and the enzyme can be separated from the reaction mixture and reused. However, the effect of magnetic nanoparticles on enzyme functioning needs to be studied. Additionally, we know that the method of synthesis plays a principal role in the physicochemical properties of the nanomagnetic catalyst obtained. However, the synthesis of magnetic nanoparticles and their relevance in transesterification still require much development in order to achieve the greatest optimization for large-scale biodiesel production.

Finally, studies of the application of MNPs in transesterification reactions for biodiesel production remain few in number, and so more studies are still required.

Conflicts of Interest

The authors declare no conflicts of interest.

Authors' Contributions

All authors contributed to defining the scope and content of the chapter, and assembled, analysed and reviewed the data. All authors read and approved the final chapter.

REFERENCES

Abbah, E., Nwandikom, G., Egwuonwu, C. & Nwakuba, N. (2016). Effect of reaction temperature on the yield of biodiesel from neem seed oil. *American Journal of Energy Science,* 3, 16–20.

Akbarzadeh, A., Samiei, M. & Davaran, S. (2012). Magnetic nanoparticles: preparation, physical properties, and applications in biomedicine. *Nanoscale Research Letters,* 7, 144.

Alarifi, A., Deraz, N. & Shaban, S. (2009). Structural, morphological and magnetic properties of $NiFe_2O_4$ nano-particles. *Journal of Alloys and Compounds,* 486, 501–506.

Ali, A., Hira-Zafar, M. Z., Ul-Haq, I., Phull, A. R., Ali, J. S. & Hussain, A. (2016). Synthesis, characterization, applications, and challenges of iron oxide nanoparticles. *Nanotechnology, Science And applications,* 9, 49.

Ali, M. A., Al-Hydary, I. A. & Al-Hattab, T. A. (2017). Nano-magnetic catalyst $CaO\text{-}Fe_3O_4$ for biodiesel production from date palm seed oil. *Bulletin of Chemical Reaction Engineering & Catalysis,* 12, 460–468.

Amani, T., Haghighi, M. & Rahmanivahid, B. (2019). Microwave-assisted combustion design of magnetic Mg–Fe spinel for MgO-based nanocatalyst used in biodiesel production: influence of heating-approach and fuel ratio. *Journal of Industrial and Engineering Chemistry,* 80, 43–52.

Ambat, I., Srivastava, V., Haapaniemi, E. & Sillanpää, M. (2019). Nano-magnetic potassium impregnated ceria as catalyst for the biodiesel production. *Renewable Energy,* 139, 1428–1436.

Ambroz, F., Macdonald, T. J., Martis, V., & Parkin, I. P. (2018). Evaluation of the BET theory for the characterization of meso and microporous MOFs. *Small Methods,* 2, 1800173.

Andjelković, L., Šuljagić, M., Lakić, M., Jeremić, D., Vulić, P. & Nikolić, A. S. (2018). A study of the structural and morphological properties of Ni–ferrite, Zn–ferrite and Ni–Zn–ferrites functionalized with starch. *Ceramics International,* 44, 14163–14168.

Aphesteguy, J., Kurlyandskaya, G., De Celis, J., Safronov, A. & Schegoleva, N. (2015). Magnetite nanoparticles prepared by co-precipitation method in different conditions. *Materials Chemistry and Physics,* 161, 243–249.

Ashok, A., Kennedy, L. J. & Vijaya, J. J. (2019). Structural, optical and magnetic properties of $Zn1\text{-}xMnxFe_2O_4$ ($0 \leq x \leq 0.5$) spinel nano particles for transesterification of used cooking oil. *Journal of Alloys and Compounds,* 780, 816–828.

Ashok, A., Ratnaji, T., Kennedy, L. J., & Vijaya, J. J. (2020). Magnetically separable $Zn1\text{-}xCuxF_2O_4$ ($0 \leq x \leq 0.5$) nanocatalysts for the transesterification of waste cooking oil. *Advanced Powder Technology*, 31(6), 2573–2585.

Atif, M., Nadeem, M., Grössinger, R. & Turtelli, R. S. (2011). Studies on the magnetic, magnetostrictive and electrical properties of sol–gel synthesized Zn doped nickel ferrite. *Journal of Alloys and Compounds,* 509, 5720–5724.

Ayetor, G. K., Sunnu, A. & Parbey, J. (2015). Effect of biodiesel production parameters on viscosity and yield of methyl esters: *Jatropha curcas*, *Elaeis guineensis* and *Cocos nucifera*. *Alexandria Engineering Journal,* 54, 1285–1290.

Banihani, F. F. (2016). Transesterification and production of biodiesel from waste cooking oil: effect of operation variables on fuel properties. *American Journal of Chemical Engineering,* 4, 154–160.

Baskar, G., Selvakumari, I. A. E. & Aiswarya, R. (2018). Biodiesel production from castor oil using heterogeneous Ni doped ZnO nanocatalyst. *Bioresource Technology,* 250, 793–798.

Bisen, P.S., Sanodiya, B.S., Thakur, G.S., Baghel, R.K., & Prasad, G.B.K.S., (2010). Biodiesel production with special emphasis on lipase-catalyzed transesterification. *Biotechnology Letters,* 32, 1019–1030.

Bhagwat, V., Humbe, A. V., More, S. & Jadhav, K. (2019). Sol-gel auto combustion synthesis and characterizations of cobalt ferrite nanoparticles: different fuels approach. *Materials Science and Engineering: B*, 248, 114388.

Brinker, C. J. & Scherer, G. W. (2013). *Sol-gel science: the physics and chemistry of sol-gel processing*, Academic Press.

Chang, K.-L., Lin, Y.-C., Jhang, S.-R., Cheng, W. L., Chen, S.-C. & Mao, S.-Y. (2017). Rapid Jatropha-Castor biodiesel production with microwave heating and a heterogeneous base catalyst nano-Ca (OH) 2/Fe_3O_4. *Catalysts,* 7, 203.

Changmai, B., Rano, R., Vanlalveni, C. & Rokhum, L. (2021). A novel *Citrus sinensis* peel ash coated magnetic nanoparticles as an easily recoverable solid catalyst for biodiesel production. *Fuel,* 286, 119447.

Chen, D.-H. & He, X.-R. (2001). Synthesis of nickel ferrite nanoparticles by sol-gel method. *Materials Research Bulletin,* 36, 1369–1377.

Chingakham, C., David, A. & Sajith, V. (2019). Fe_3O_4 nanoparticles impregnated eggshell as a novel catalyst for enhanced biodiesel production. *Chinese Journal of Chemical Engineering,* 27, 2835–2843.

Costa, A., Kiminami, R. & Morelli, M. (2009). *Combustion synthesis processing of nanoceramics. Handbook of nanoceramics and their based nanodevices (Synthesis and Processing).* Ed. American Scientific Publishers, 1, 375–392.

Costa, A., Lula, R., Kiminami, R., Gama, L., De Jesus, A. & Andrade, H. (2006). Preparation of nanostructured $NiFe_2O_4$ catalysts by combustion reaction. *Journal of Materials Science,* 41, 4871–4875.

Costa, A. C. F., Morelli, M. R. & Kiminami, R. H. (2001). Combustion synthesis: effect of urea on the reaction and characteristics of Ni-Zn ferrite powders. *Journal of Materials Synthesis and Processing,* 9, 347–352.

Costa, A., Tortella, E., Morelli, M., Kaufman, M. & Kiminami, R. (2002). Effect of heating conditions during combustion synthesis on the characteristics of Ni 0.5 Zn 0.5 Fe 2 O 4 nanopowders. *Journal of Materials Science,* 37, 3569–3572.

Dai, Y.-M., Chen, K.-T., Wang, P.-H. & Chen, C.-C. (2016). Solid-base catalysts for biodiesel production by using silica in agricultural wastes and lithium carbonate. *Advanced Powder Technology,* 27, 2432–2438.

Dai, Y.-M., Wang, Y.-F. & Chen, C.-C. (2018). Synthesis and characterization of magnetic $LiFe_5O_8$-$LiFeO_2$ as a solid basic catalyst for biodiesel production. *Catalysis Communications,* 106, 20–24.

Danks, A. E., Hall, S. R. & Schnepp, Z. (2016). The evolution of 'sol–gel'chemistry as a technique for materials synthesis. *Materials Horizons,* 3, 91–112.

Dantas, J., Leal, E., Cornejo, D., Kiminami, R. & Costa, A. (2020). Biodiesel production evaluating the use and reuse of magnetic nanocatalysts Ni0. 5Zn0. 5Fe2O4 synthesized in pilot-scale. *Arabian Journal of Chemistry,* 13, 3026–3042.

Dantas, J., Leal, E., Maposса, A. B., Cornejo, D. R., & Costa, A. C. F. M. (2017). Magnetic nanocatalysts of Ni0. 5Zn0. 5Fe2O4 doped with Cu and performance evaluation in transesterification reaction for biodiesel production. *Fuel,* 191, 463–471.

Dantas, J., Leal, E., Mapossa, A. B., Silva, A. S. A. & Costa, A. C. F. D. M. (2016). Síntese, caracterização e performance catalítica de nanoferritas mistas submetidas a reação de transesterificação e esterificação via rota metílica e etílica para biodiesel. *Matéria (Rio de Janeiro),* 21, 1080–1093.

Dantas, J., Leal, E., Mapossa, A.B., et al. (2021). Biodiesel production on bench scale from different sources of waste oils by using NiZn magnetic heterogeneous nanocatalyst. *International Journal of Energy Research.* https://doi.org/10.1002/er.6577.

Dantas, J., Santos, J. R. D., Cunha, R. B. L., Kiminami, R. H. G., & Costa, A. C. F. (2013). Use of Ni-Zn ferrites doped with Cu as catalyst in the transesterification of soybean oil to methyl esters. *Materials Research,* 16, 625–627.

Dantas, J., Silva, A. S., Santos, P. T. A, et al. (2012). Evaluation of catalyst Ni0. 4Cu0. 1Zn0. 5Fe2O4 on methyl esterification of free fatty acid present in cottonseed oil. *Materials Science Forum,* 727, 1302–1307.

Deepty, M., Srinivas, C., Kumar, E. R., et al. (2019). XRD, EDX, FTIR and ESR spectroscopic studies of co-precipitated Mn–substituted Zn–ferrite nanoparticles. *Ceramics International,* 45, 8037–8044.

Džunuzović, A., Ilić, N., Petrović, M. V., et al. (2015). Structure and properties of Ni–Zn ferrite obtained by auto-combustion method. *Journal of Magnetism and Magnetic Materials,* 374, 245–251.

Eevera, T., Rajendran, K. & Saradha, S. (2009). Biodiesel production process optimization and characterization to assess the suitability of the product for varied environmental conditions. *Renewable Energy,* 34, 762–765.

El Moussaoui, H., Mahfoud, T., Habouti, S., et al. (2016). Synthesis and magnetic properties of tin spinel ferrites doped manganese. *Journal of Magnetism and Magnetic Materials,* 405, 181–186.

Faraji, M., Yamini, Y. & Rezaee, M. (2010). Magnetic nanoparticles: synthesis, stabilization, functionalization, characterization, and applications. *Journal of the Iranian Chemical Society,* 7, 1–37.

Farias, A. F., de Araújo, D. T., da Silva, A. L., et al. (2020). Evaluation of the catalytic effect of ZnO as a secondary phase in the Ni0. 5Zn0. 5Fe2O4 system and of the stirring mechanism on biodiesel production reaction. *Arabian Journal of Chemistry*, 13(6), 5788–5799.

Farooq, M., Ramli, A., Naeem, A., et al. (2019). A green route for biodiesel production from waste cooking oil over base heterogeneous catalyst. *International Journal of Energy Research,* 43, 5438–5446.

Feyzi, M., Hassankhani, A. & Rafiee, H. R. (2013). Preparation and characterization of $Cs/Al/Fe_3O_4$ nanocatalysts for biodiesel production. *Energy Conversion and Management,* 71, 62–68.

Feyzi, M. & Norouzi, L. (2016). Preparation and kinetic study of magnetic $Ca/Fe_3O_4@SiO_2$ nanocatalysts for biodiesel production. *Renewable Energy,* 94, 579–586.

Gabal, M. A., Kosa, S., & El Muttairi, T. S. (2014). Magnetic dilution effect of nanocrystalline $NiFe_2O_4$ synthesized via sucrose-assisted combustion route. *Ceramics International,* 40, 675–681.

Ghalandari, A., Taghizadeh, M. & Rahmani, M. (2019). Statistical optimization of the biodiesel production process using a magnetic core-mesoporous shell KOH/Fe3O4@ γ-Al2O3 nanocatalyst. *Chemical Engineering & Technology,* 42, 89–99.

Gopalan Sibi, M., Verma, D. & Kim, J. (2020). Magnetic core–shell nanocatalysts: promising versatile catalysts for organic and photocatalytic reactions. *Catalysis Reviews,* 62, 163–311.

Hajalilou, A. & Mazlan, S. A. (2016). A review on preparation techniques for synthesis of nanocrystalline soft magnetic ferrites and investigation on the effects of microstructure features on magnetic properties. *Applied Physics A,* 122, 680.

Hu, S., Guan, Y., Wang, Y. & Han, H. (2011). Nano-magnetic catalyst KF/CaO–Fe_3O_4 for biodiesel production. *Applied Energy,* 88, 2685–2690.

Hwang, C.-C., Tsai, J.-S., Huang, T.-H., Peng, C.-H. & Chen, S.-Y. (2005). Combustion synthesis of Ni–Zn ferrite powder—influence of oxygen balance value. *Journal of Solid State Chemistry,* 178, 382–389.

Ishii, T., & Kyotani, T. (2016). Temperature programmed desorption. In *Materials science and engineering of carbon*, ed. M. Inagaki, and F. Kang, 287–305. Tsinghua University Press.

Jadhav, S. S., Shirsath, S. E., Toksha, B., Patange, S., Shengule, D. & Jadhav, K. (2010). Structural and electric properties of zinc substituted NiFe2O4 nanoparticles prepared by co-precipitation method. *Physica B: Condensed Matter,* 405**,** 2610–2614.

Jain, S., Adiga, K. & Verneker, V. P. (1981). A new approach to thermochemical calculations of condensed fuel-oxidizer mixtures. *Combustion and Flame,* 40, 71–79.

Jalaiah, K. & Babu, K. V. (2017). Structural, magnetic and electrical properties of nickel doped Mn-Zn spinel ferrite synthesized by sol-gel method. *Journal of Magnetism and Magnetic Materials,* 423, 275–280.

Ji, G., Tang, S., Ren, S., Zhang, F., Gu, B. & Du, Y. (2004). Simplified synthesis of single-crystalline magnetic CoFe2O4 nanorods by a surfactant-assisted hydrothermal process. *Journal of Crystal Growth,* 270, 156–161.

Kambale, R., Adhate, N., Chougule, B. & Kolekar, Y. (2010). Magnetic and dielectric properties of mixed spinel Ni–Zn ferrites synthesized by citrate–nitrate combustion method. *Journal of Alloys and Compounds,* 491, 372–377.

Kazemifard, S., Nayebzadeh, H., Saghatoleslami, N. & Safakish, E. (2019). Application of magnetic alumina-ferric oxide nanocatalyst supported by KOH for in-situ transesterification of microalgae cultivated in wastewater medium. *Biomass and Bioenergy,* 129, 105338.

Kombaiah, K., Vijaya, J. J., Kennedy, L. J., Bououdina, M., Al-Lohedan, H. A. & Ramalingam, R. J. (2017). Studies on *Opuntia dilenii* haw mediated multifunctional ZnFe2O4 nanoparticles: optical, magnetic and catalytic applications. *Materials Chemistry and Physics,* 194, 153–164.

Konecny, A.P.; Covarrubias, J.; Wang, H. Magnetic nanoparticle design and application in magnetic hyperthermia. In *Magnetic Nanomaterials: Applications in Catalysis and Life Sciences*; Bossmann, S.H., Wang, H., Eds.; RSC: London, 2017.

Kumar, S. J., Prameela, P., Rao, K. S., Kiran, J. & Rao, K. (2020). Structural and magnetic properties of copper-substituted nickel–zinc nanoparticles prepared by sol-gel method. *Journal of Superconductivity and Novel Magnetism,* 33, 693–705.

Li, H., Wang, Y., Ma, X., et al. (2020). A novel magnetic CaO-based catalyst synthesis and characterization: enhancing the catalytic activity and stability of CaO for biodiesel production. *Chemical Engineering Journal,* 391, 123549.

Liu, H., Su, L., Liu, F., Li, C. & Solomon, U. U. (2011a). Cinder supported K2CO3 as catalyst for biodiesel production. *Applied Catalysis B: Environmental,* 106, 550–558.

Liu, Y., Jia, S., Wu, Q., Ran, J., Zhang, W. & Wu, S. (2011b). Studies of Fe3O4-chitosan nanoparticles prepared by co-precipitation under the magnetic field for lipase immobilization. *Catalysis Communications,* 12, 717–720.

Liu, Y., Zhang, P., Fan, M. & Jiang, P. (2016). Biodiesel production from soybean oil catalyzed by magnetic nanoparticle MgFe2O4@ CaO. *Fuel,* 164, 314–321.

Lu, A. H., Salabas, E. E. L. & Schüth, F. (2007). Magnetic nanoparticles: synthesis, protection, functionalization, and application. *Angewandte Chemie International Edition,* 46, 1222–1244.

Maaz, K., Karim, S., Mumtaz, A., Hasanain, S., Liu, J. & Duan, J. (2009). Synthesis and magnetic characterization of nickel ferrite nanoparticles prepared by co-precipitation route. *Journal of Magnetism and Magnetic Materials,* 321, 1838–1842.

Maensiri, S., Masingboon, C., Boonchom, B. & Seraphin, S. (2007). A simple route to synthesize nickel ferrite (NiFe2O4) nanoparticles using egg white. *Scripta Materialia,* 56, 797–800.

Majidi, S., Zeinali Sehrig, F., Farkhani, S. M., Soleymani Goloujeh, M. & Akbarzadeh, A. (2016). Current methods for synthesis of magnetic nanoparticles. *Artificial Cells, Nanomedicine, and Biotechnology,* 44, 722–734.

Mapossa, A. B., Dantas, J., Diniz, V. C. S., Silva, M. R., Kiminami, R. H. G. A., & Costa, A. C. F. M. (2017). Síntese e caracterização do ferroespinélio Ni0,7Zn0,3Fe2O4: avaliação de desempenho na esterificação metílica e etílica. *Cerâmica,* 63, 223–232.

Mapossa, A. B., Dantas, J., Kiminami, A., Silva, M. R. & Costa, A. (2015). Síntese do ferroespinélio ZnFe2O4 e avaliação do seu desempenho em reações de esterificação e transesterificação via rota metílica. *Revista Eletrônica de Materiais e Processos,* 10, 137–143.

Mapossa, A. B., Dantas, J., & Costa, A. C. (2020a). Transesterification reaction for biodiesel production from soybean oil using Ni0. 5Zn0. 5Fe2O4 nanomagnetic catalyst: kinetic study. *International Journal of Energy Research,* 44, 6674–6684.

Mapossa, A. B., Dantas, J., Silva, M. R., Kiminami, R. H., Costa, A. C. F. & Daramola, M. O. (2020b). Catalytic performance of NiFe2O4 and Ni0. 3Zn0. 7Fe2O4 magnetic nanoparticles during biodiesel production. *Arabian Journal of Chemistry,* 13, 4462–4476.

Mathew, D. S. & Juang, R.-S. (2007). An overview of the structure and magnetism of spinel ferrite nanoparticles and their synthesis in microemulsions. *Chemical Engineering Journal,* 129, 51–65.

Mazaheri, H., Ong, H. C., Masjuki, H. H., et al. (2018). Rice bran oil based biodiesel production using calcium oxide catalyst derived from *Chicoreus brunneus* shell. *Energy,* 144, 10–19.

Melo, R., Silva, F., Moura, K., De Menezes, A. & Sinfrônio, F. (2015). Magnetic ferrites synthesised using the microwave-hydrothermal method. *Journal of Magnetism and Magnetic Materials,* 381, 109–115.

Moghanian, H., Mobinikhaledi, A., Blackman, A. G., & Sarough-Farahani, E. (2014). Sulfanilic acid-functionalized silica-coated magnetite nanoparticles as an efficient, reusable and magnetically separable catalyst for the solvent-free synthesis of 1-amido-and 1-aminoalkyl-2-naphthols. *RSC Advances,* 4, 28176–28185.

Musa, I. A. (2016). The effects of alcohol to oil molar ratios and the type of alcohol on biodiesel production using transesterification process. *Egyptian Journal of Petroleum,* 25, 21–31.

Muttakin, M., Mitra, S., Thu, K., Ito, K., & Saha, B. B. (2018). Theoretical framework to evaluate minimum desorption temperature for IUPAC classified adsorption isotherms. *International Journal of Heat and Mass Transfer,* 122, 795–805.

Nejati, K. & Zabihi, R. (2012). Preparation and magnetic properties of nano size nickel ferrite particles using hydrothermal method. *Chemistry Central Journal,* 6, 23.

Netto, C. G., Toma, H. E. & Andrade, L. H. (2013). Superparamagnetic nanoparticles as versatile carriers and supporting materials for enzymes. *Journal of Molecular Catalysis B: Enzymatic,* 85, 71–92.

Omar, W. N. N. W. & Amin, N. A. S. (2011). Biodiesel production from waste cooking oil over alkaline modified zirconia catalyst. *Fuel Processing Technology,* 92, 2397–2405.

Phadatare, M., Salunkhe, A., Khot, V., Sathish, C., Dhawale, D. & Pawar, S. (2013). Thermodynamic, structural and magnetic studies of NiFe2O4 nanoparticles prepared by combustion method: effect of fuel. *Journal of Alloys and Compounds,* 546, 314–319.

Polshettiwar, V., Luque, R., Fihri, A., Zhu, H., Bouhrara, M. & Basset, J.-M. (2011). Magnetically recoverable nanocatalysts. *Chemical Reviews,* 111, 3036–3075.

Priyadharsini, P., Pradeep, A., Rao, P. S. & Chandrasekaran, G. (2009). Structural, spectroscopic and magnetic study of nanocrystalline Ni–Zn ferrites. *Materials Chemistry and Physics,* 116, 207–213.

Rahimi, T., Kahrizi, D., Feyzi, M., Ahmadvandi, H.R. & Mostafaei, M. (2021). Catalytic performance of MgO/Fe2O3-SiO2 core-shell magnetic nanocatalyst for biodiesel production of *Camelina sativa* seed oil: optimization by RSM-CCD method. *Industrial Crops and Products,* 159, 113065.

Raj, J. V. A., Bharathiraja, B., Vijayakumar, B., Arokiyaraj, S., Iyyappan, J. & Kumar, R. P. (2019). Biodiesel production from microalgae *Nannochloropsis oculata* using heterogeneous poly ethylene glycol (PEG) encapsulated ZnOMn2+ nanocatalyst. *Bioresource Technology,* 282, 348–352.

Ramana, P., Rao, K. S. & Rao, K. (2018). Influence of iron content on the structural and magnetic properties of Ni-Zn ferrite nanoparticles synthesized by PEG assisted sol-gel method. *Journal of Magnetism and Magnetic Materials,* 465, 747–755.

Rashad, M., Elsayed, E., Moharam, M., Abou-Shahba, R. & Saba, A. (2009). Structure and magnetic properties of NixZn1 – xFe2O4 nanoparticles prepared through co-precipitation method. *Journal of Alloys and Compounds,* 486, 759–767.

Rozman, M. & Drofenik, M. (1995). Hydrothermal synthesis of manganese zinc ferrites. *Journal of the American Ceramic Society,* 78, 2449–2455.

Sabzevar, A.M., Ghahramaninezhad, M. & Shahrak, M.N. (2021). Enhanced biodiesel production from oleic acid using TiO2-decorated magnetic ZIF-8 nanocomposite catalyst and its utilization for used frying oil conversion to valuable product. *Fuel,* 288, 119586.

Safakish, E., Nayebzadeh, H., Saghatoleslami, N., & Kazemifard, S. (2020). Comprehensive assessment of the preparation conditions of a separable magnetic nanocatalyst for biodiesel production from algae. *Algal Research,* 49, 101949.

Salimi, Z. & Hosseini, S. A. (2019). Study and optimization of conditions of biodiesel production from edible oils using ZnO/BiFeO3 nano magnetic catalyst. *Fuel,* 239, 1204–1212.

Santos, E. C., dos Santos, T. C., Guimarães, R. B., Ishida, L., Freitas, R. S., & Ronconi, C. M. (2015). Guanidine-functionalized Fe3O4 magnetic nanoparticles as basic recyclable catalysts for biodiesel production. *RSC Advances*, 5(59), 48031–48038.

Santos, P. T. A., De Melo Costa, A. C. F., & Andrade, H. M. C. (2012). Preparation of NiFe2O4 and ZnFe2O4 samples by combustion reaction and evaluation of performance in reaction water gas shift reaction—WGSR. *Materials Science Forum,* 727, 1290–1295.

Seffati, K., Esmaeili, H., Honarvar, B. & Esfandiari, N. (2020). AC/CuFe2O4@ CaO as a novel nanocatalyst to produce biodiesel from chicken fat. *Renewable Energy,* 147, 25–34.

Seffati, K., Honarvar, B., Esmaeili, H. & Esfandiari, N. (2019). Enhanced biodiesel production from chicken fat using CaO/CuFe2O4 nanocatalyst and its combination with diesel to improve fuel properties. *Fuel,* 235, 1238–1244.

Segadaes, A. M., Morelli, M. R. & Kiminami, R. G. (1998). Combustion synthesis of aluminium titanate. *Journal of the European Ceramic Society,* 18, 771–781.

Seo, J. Y., Lee, K., Lee, S. Y., et al. (2014). Effect of barium ferrite particle size on detachment efficiency in magnetophoretic harvesting of oleaginous *Chlorella* sp. *Bioresource Technology,* 152, 562–566.

Shi, Y., Ding, J., Liu, X., & Wang, J. (1999). $NiFe_2O_4$ ultrafine particles prepared by co-precipitation/mechanical alloying. *Journal of Magnetism and Magnetic Materials,* 205, 249–254.

Shylesh, S., Schweizer, J., Demeshko, S., Schünemann, V., Ernst, S. & Thiel, W. R. (2009). Nanoparticle supported, magnetically recoverable oxodiperoxo molybdenum complexes: efficient catalysts for selective epoxidation reactions. *Advanced Synthesis & Catalysis,* 351, 1789–1795.

Silveira Junior, E. G., Justo, O. R., Perez, V. H., et al. (2018). Extruded catalysts with magnetic properties for biodiesel production. *Advances in Materials Science and Engineering*, 2018. https://doi.org/10.1155/2018/3980967

Singh, V., Bux, F. & Sharma, Y. C. (2016). A low cost one pot synthesis of biodiesel from waste frying oil (WFO) using a novel material, β-potassium dizirconate (β-K2Zr2O5). *Applied Energy,* 172, 23–33.

Srivastava, M., Chaubey, S. & Ojha, A. K. (2009). Investigation on size dependent structural and magnetic behavior of nickel ferrite nanoparticles prepared by sol–gel and hydrothermal methods. *Materials Chemistry and Physics,* 118, 174–180.

Sun, J., Zhou, S., Hou, P., et al. (2007). Synthesis and characterization of biocompatible Fe3O4 nanoparticles. *Journal of Biomedical Materials Research Part A,* 80, 333–341.

Sundararajan, R. & Srinivasan, V. (1991). Catalytic decomposition of nitrous oxide on CuxCo3—xO4 spinels. *Applied Catalysis,* 73, 165–171.

Tang, S., Wang, L., Zhang, Y., Li, S., Tian, S. & Wang, B. (2012). Study on preparation of Ca/Al/Fe3O4 magnetic composite solid catalyst and its application in biodiesel transesterification. *Fuel Processing Technology,* 95, 84–89.

Tapanes, N. C. O., Aranda, D. A. G., de Mesquita Carneiro, J. W., & Antunes, O. A. C. (2008). Transesterification of *Jatropha curcas* oil glycerides: theoretical and experimental studies of biodiesel reaction. *Fuel,* 87, 2286–2295.

Vasić, K., Hojnik Podrepšek, G., Knez, Ž., & Leitgeb, M. (2020). Biodiesel production using solid acid catalysts based on metal oxides. *Catalysts,* 10, 237.

Vijayaprasath, G., Murugan, R., Asaithambi, S., et al. (2016). Structural and magnetic behavior of Ni/Mn co-doped ZnO nanoparticles prepared by co-precipitation method. *Ceramics International,* 42, 2836–2845.

Waldron, R. (1955). Infrared spectra of ferrites. *Physical Review,* 99, 1727.

Wang, A., Sudarsanam, P., Xu, Y., Zhang, H., Li, H., & Yang, S. (2020). Functionalized magnetic nanosized materials for efficient biodiesel synthesis via acid–base/enzyme catalysis. *Green Chemistry*, 22(10), 2977–3012.

Wang, J. (2006). Prepare highly crystalline $NiFe_2O_4$ nanoparticles with improved magnetic properties. *Materials Science and Engineering: B*, 127(1), 81–84.

Wu, R., Liu, J.-H., Zhao, L., et al. (2014). Hydrothermal preparation of magnetic Fe3O4@C nanoparticles for dye adsorption. *Journal of Environmental Chemical Engineering,* 2, 907–913.

Wu, W., He, Q. & Jiang, C. (2008). Magnetic iron oxide nanoparticles: synthesis and surface functionalization strategies. *Nanoscale Research Letters,* 3, 397.

Xie, W., Gao, C. & Li, J. (2021). Sustainable biodiesel production from low-quantity oils utilizing H6PV3MoW8O40 supported on magnetic Fe3O4/ZIF-8 composites. *Renewable Energy, 168*, 927–937.

Xie, W., Han, Y. & Wang, H. (2018). Magnetic Fe3O4/MCM-41 composite-supported sodium silicate as heterogeneous catalysts for biodiesel production. *Renewable Energy,* 125, 675–681.

Xie, W. & Huang, M. (2018). Immobilization of Candida rugosa lipase onto graphene oxide Fe3O4 nanocomposite: characterization and application for biodiesel production. *Energy Conversion and Management,* 159, 42–53.

Xie, W. & Wan, F. (2018). Basic ionic liquid functionalized magnetically responsive Fe3o4@ Hkust-1 composites used for biodiesel production. *Fuel,* 220, 248–256.

Xin, T., Ma, M., Zhang, H., et al. (2014). A facile approach for the synthesis of magnetic separable Fe3O4@ TiO2, core–shell nanocomposites as highly recyclable photocatalysts. *Applied Surface Science,* 288, 51–59.

Xue, B.-J., Luo, J., Zhang, F. & Fang, Z. (2014). Biodiesel production from soybean and Jatropha oils by magnetic CaFe2O4–Ca2Fe2O5-based catalyst. *Energy,* 68, 584–591.

Yan, A., Liu, X., Qiu, G., et al. (2007). A simple solvothermal synthesis and characterization of round-biscuit-like Fe3O4 nanoparticles with adjustable sizes. *Solid State Communications,* 144, 315–318.

Zahi, S., Hashim, M. & Daud, A. (2007). Synthesis, magnetic properties and microstructure of Ni–Zn ferrite by sol–gel technique. *Journal of Magnetism and Magnetic Materials,* 308, 177–182.

Zhang, F., Fang, Z. & Wang, Y.-T. (2015). Biodiesel production directly from oils with high acid value by magnetic Na2SiO3@ Fe3O4/C catalyst and ultrasound. *Fuel,* 150, 370–377.

Zhang, P., Han, Q., Fan, M. & Jiang, P. (2014). Magnetic solid base catalyst CaO/CoFe2O4 for biodiesel production: influence of basicity and wettability of the catalyst in catalytic performance. *Applied Surface Science,* 317, 1125–1130.

Zhao, D., Wu, X., Guan, H. & Han, E. (2007). Study on supercritical hydrothermal synthesis of CoFe2O4 nanoparticles. *Journal of Supercritical Fluids,* 42, 226–233.

Zhao, L., Zhang, H., Xing, Y., et al. (2008). Studies on the magnetism of cobalt ferrite nanocrystals synthesized by hydrothermal method. *Journal of Solid State Chemistry,* 181, 245–252.

7

Biomaterial-Based Nanocatalysts in Biodiesel Synthesis

Wanison A.G. Pessoa Jr., Ingrity Suelen Costa Sá, Mitsuo L. Takeno, Silma de S. Barros, Marcia S.F. Lira, Ana E.M. de Freitas, Edson Pablo Silva, and Flávio A. de Freitas*

CONTENTS

7.1 Introduction

Petroleum derivatives, a non-renewable source, are commonly used in the production of energy, fuels and chemicals.[1] However, due to its excessive consumption, the scarcity of these fossil resources has been increased.[2] In addition, fossil fuels emit various toxic gases that contribute to exacerbation of the greenhouse effect, such as SO_2, CO_2, etc., which are influencers in global warming, threatening the existence of all living beings.[3] The use of renewable energy sources, because of the great damage caused by fossil fuels, has been the subject of several studies as alternatives for the generation of cleaner energy.[4]

Biodiesel is a biofuel without sulphur and polycyclic aromatic compounds, with low HC and CO emissions as compared to traditional fuels, which makes it less polluting.[5]

* *Corresponding Author* freitas.flavio@yahoo.com.br

DOI: 10.1201/9781003120858-7

It has been shown to be an excellent alternative to clean energy due to its low toxicity, and ecological and renewable nature.[6] It has use as a promising alternative in diesel engines, due to its characteristics that give it a significant performance improvement, in addition to not having to modify the engine.[7]

Between 2000 and 2017, there was a large increase in biodiesel production, from 16 billion to 143 billion litres, and it has been estimated that between 2017 and 2024, about 12 billion gallons will be produced.[8] The world's largest producers of biodiesel are Brazil, the United States, Malaysia, Argentina, Spain, Belgium and Germany, serving about 80% of the global demand.[9] However, the production of biodiesel presents a major obstacle in commercialization, due to the high cost of the raw material (oil), in addition to the product purification process using homogeneous catalysts, which, consequently, impacts production.[10] In the quest to reduce costs, several edible and non-edible oils have been used, such as soybean oil, rapeseed, jojoba, peanuts, palm, sunflower, castor and jatropha.[11] The use of microbial oil, residues from oil refining and waste cooking oil have also been reported.[12] However, the use of edible oils in the production of biodiesel makes it economically ineffective, due to its competition with foods available for human consumption.[13] Thus, the use of residues of edible oils, animal fats and inedible biomass has been an alternative in the production of biodiesel.[14] Another extremely important factor is related to the acidity of the used oil and the type of catalyst applied in these reactions. If the oil has high acidity (usually above 2 mg KOH/g of oil) the catalyst will be deactivated or it will lead to the formation of a large amount of soap, which will increase production costs considerably.[15]

Biodiesel is obtained through the transesterification of these oils (triglycerides), as well as the esterification of fatty acids, as shown in Figure 7.1. These reactions between raw materials and alcohols usually need a high activation energy. Thus, they are catalysed using homogeneous catalysts (soluble in the reaction medium) or heterogeneous (solid insoluble in the reaction medium).[16]

The most commonly used homogeneous alkaline catalysts are NaOH, KOH and methoxides, and the acids are H_2SO_4 and H_3PO_4.[17] These, by their nature, considerably accelerate the rate of reaction and production of biodiesel. However, they have numerous disadvantages, such as high catalyst consumption, long reaction times, high amount of solvent for oil extraction, equipment corrosion, high toxicity, and a large amount of water required to purify biodiesel to eliminate the catalyst, among others.[18] Therefore, these disadvantages increase the cost of biodiesel production, resulting in the search for new heterogeneous catalysts from renewable sources.[19]

Several studies have focused on the development of heterogeneous catalysts for transesterification and esterification reactions. These are divided into basic, acidic or bifunctional types, where bio-based catalysts are usually basic or used as a support for the production of higher alkalinity or bifunctional catalysts (acid and base sites).[20] These can be easily obtained from various agro-industrial residues, such as brewery waste,[21] cupuaçu seeds,[22] tucumã fruit peels,[23] pineapple leaves,[24] palm kernels,[25] bamboo,[26] etc. These biomasses have received great interest for presenting sustainable and promising potential in the synthesis of solid catalysts (heterogeneous), presenting oil conversion rates in biodiesel greater than 98%, in addition to other qualities, such as low-cost, reuse and reduction of the environmental impact when they are mistakenly discarded.[27]

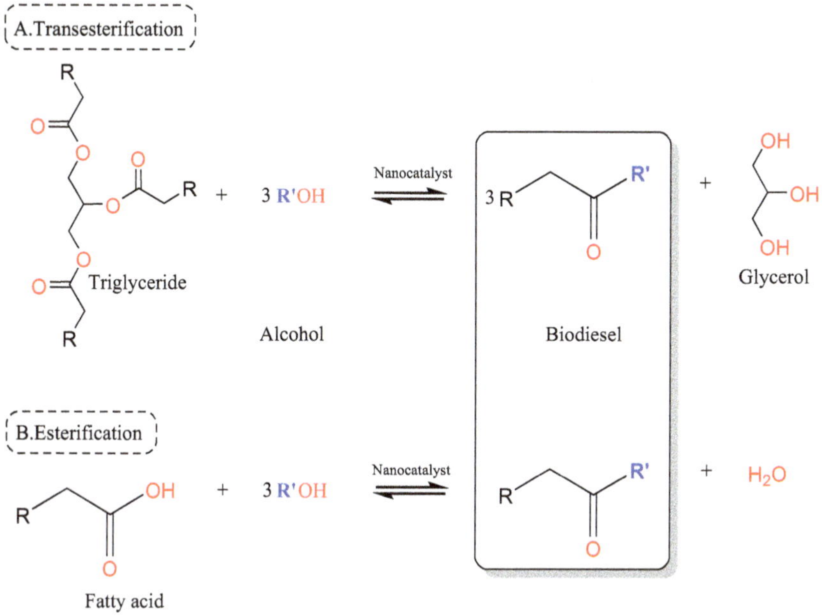

FIGURE 7.1 Reaction schemes for the transesterification (a) and esterification reactions (b).

A class of catalysts that has been attracting a lot of attention is the so-called nanocatalysts. Several studies have shown that reducing the particle size of the catalyst facilitates access to active sites by increasing the surface area. Thus, the combination of obtaining a green catalyst, using renewable sources and efficiency, because of their size, makes these materials highly attractive given the growing economic and environmental concerns of industries in general.

7.2 Biodiesel Synthesis

Biodiesel is produced mainly through transesterification reactions (Figure 7.1a), a reversible reaction where triglycerides react with short-chain alcohols, such as methanol and ethanol, in the presence of an acidic or basic nanocatalyst, forming fatty acid methyl esters (FAME) or fatty acid ethyl esters (FAEE), with glycerol as a by-product.[28]

In the transesterification reaction, the sources of lipids may be of animal origin, such as pork and chicken fats,[29] and different vegetable oils such as *Pongamia pinnata* oil,[30] *Hydnocarpus wightianus* oil,[31] *Elaeis guineensis* oil,[32] *Helianthus annuus* oil,[33] *Ricinus communis* oil,[34] and *Glycine max* oil.[35] However, the use of edible oils in the production of biodiesel makes it economically unattractive, due to the fact that in addition to competing with human consumption, the cost of production is also higher.[36]

Operating costs for biodiesel production include expenses associated with raw materials: oil, catalyst, alcohol, washing water, and others, where around 70–80% of these costs are related to the lipid source used.[37] Thus, the use of residues such as frying oils, animal fats, and inedible biomass has become an alternative in the production of biodiesel. Edible oil residues produced in the food industry, such as soapstocks,[38] and also the waste produced in restaurants, homes and shopping malls can be converted into biodiesel through esterification and/or transesterification. An example of the use of these fatty residues is the study by Foroutan et al.,[39] where the heterogeneous CaO/MgO nanocatalyst was synthesized from waste chicken eggshells and applied to the biodiesel synthesis using fatty residues, obtaining a 98.4% yield under the following reaction conditions: time 7.08 h, temperature 69.37 °C, methanol:oil molar ratio 16.7:1 and 4.6 wt% of catalyst. Krishnamurthy et al.[40] achieved a 96.9% yield on the conversion of dairy waste scum into FAME when catalysed by nano-CaO produced from snail shells. Another raw material widely used as a source of lipids is algal biomass, such as *Chlorella pyrenoidosa*[41] and microalgae.[42]

Nanocatalysts from biomaterials can also be applied in another method of biodiesel synthesis: esterification (Figure 7.1b), where free fatty acids, such as oleic acid and acids present in industrial sludge and oil residues, react with alcohol forming the fatty acid alkyl esters and water.[43] For example, Suresh et al.[44] successfully used *Cinnamomum tamala* extract (rich in eugenol) as a reducing agent in the synthesis of CuO nanoparticles and applied it in the synthesis of biodiesel by esterification of fatty residues, obtaining a 97.8% conversion rate.

Since the transesterification reaction has slow kinetics, the use of catalysts is necessary to accelerate the process. Currently, the main catalysts used in the industry are homogeneous, such as KOH and NaOH (basic) and H_2SO_4 and HCl (acids).[45] When comparing the catalytic activity of these catalysts according to their acidity and basicity, the base catalysts show high catalytic activity in shorter reaction times, lower temperatures and greater conversion efficiency when compared to acid catalysts in the transesterification process.[46] However, the use of homogeneous catalyst has major disadvantages such as the high cost of production, the formation of soap (mainly for base catalysts), the high amount of effluents produced due to the washing process for the purification of biodiesel, in addition to the difficult separation of the medium and the fact that this catalyst cannot be reused.[47] Acids still have the disadvantages of being highly toxic, releasing gas and causing corrosion.[48] By applying heterogeneous catalysts, it is possible to minimize the above-mentioned disadvantages, since the catalyst is unlikely to form soap, and drastically reduces purification costs and wastewater generation.[49] In addition, they present a great advantage, which is the possibility of recycling, reducing significantly the production costs.[50] Therefore, heterogeneous catalysts, and especially those obtained from biomaterials (renewable sources), show promise as a green alternative for several processes since they are easy to recover, reusable and considered sustainable.[51] Also, the use of wastes to obtain a catalyst can bring economic and environmental benefits through waste reduction, in addition to producing a low-cost catalyst.

Some heterogeneous solid base catalysts obtained from biomaterials have been reported in the literature, such as CaO. This catalyst has been obtained from the calcination of various materials based on calcium carbonate ($CaCO_3$). Another example of catalysts is activated carbon, such as that produced by Araujo et al.[52] These authors used açaí (*Euterpe oleracea*) seeds, waste produced in large quantities in Amazonas (northern Brazil), to produce the catalyst from partial carbonization. Then, they increased their acidity through the sulphonation process and applied them in esterification and transesterification reactions, where a soybean oil was modified with 20% oleic acid and catalysed by the sulphonated carbon, yielding up to 80% of methyl esters (biodiesel).

Several studies are being carried out in order to increase the catalytic performance of these materials, one of these methods is to reduce the particle size of the material down to the nano scale.[53] Since nanocatalysts have a larger surface area and pore volume size than conventional catalysts, they have greater catalytic efficiency in the synthesis of biodiesel. Also, due to their high catalytic activity, they exhibit less saponification, good rigidity and greater capacity for reuse.[54]

7.3 Biomaterial-Based Nanocatalysts

Nanocatalysts (catalysts in nanometric dimensions: 10^{-9} m) have stood out when applied in the synthesis of biodiesel, where the size of these particles is usually confirmed by microscopy and laser diffraction techniques. These catalysts have several advantages over catalysts with larger particle sizes, such as resistance to saponification, high surface area and good catalytic activity due to the increased surface area/volume ratio of the material.[55] It has been reported that the smaller the size of the catalyst, the higher the reaction rates due to diffusion forces.[56] Thus, the development of nanomaterials has been the focus of several studies for application in biofuels and, with the growing concerns about sustainability and preservation of the environment, in addition to the reduction of natural energy resources and greater demand for fuels, researchers have developed nanocatalysts derived from renewable sources. Among the biomaterials studied, those from agro-industrial and animal by-products have been evident.[57]

These biomaterial-based nanocatalysts are concentrated in wastes rich in alkaline elements, such as potassium (K), sodium (Na), magnesium (Mg) and calcium (Ca),[58] where, calcined at low temperatures, the activated carbon is obtained (retaining a part of the organic matter) and at high temperatures, mainly alkaline oxides are obtained.[59] Table 7.1 shows several nanocatalysts based on biomaterials applied in the synthesis of biodiesel.

It is possible to observe that all biomaterials have some alkali elements in their composition, where several studies show the effectiveness of the catalysts in converting oils into biodiesel in the function of these elements.[60] In addition, to improve the catalytic properties, other compounds are added to the biomaterial, such as Zn,[61] Fe_3O_4,[62] MgO,[63] and Au,[64] among others.

7.3.1 Biomass-Based Nanocatalysts from Agro-Industrial Wastes

The increase in demand for food and other basic resources extracted from nature has intensified agricultural and agro-industrial activities, generating tons of waste. The synthesis of nanocatalysts from these residues has become a promising area, which removes a potentially toxic agro-industrial by-product from the environment, adding value to it through the production of heterogeneous catalysts. These by-products are usually discarded in the environment for natural decomposition, or they are incinerated,

TABLE 7.1

Biomaterial-Based Nanocatalysts: Source, Processes of Obtaining (Calcination), and Performance in the Biodiesel Synthesis

		Calcination			Biodiesel		
Biomaterial source	**Nanocatalyst**	**Temp. (°C)**	**Time (h)**	**Particle size (nm)**	**Oil applied**	**Conversion (%)**	**Yield (%)**
Chicken eggshell	Zn/CaO[1]	900	3	39.1	Cooking oil waste	96.74	–
	Fe_3O_4/CaO[2]	900	4	1629	*Pongamia pinnata* raw oil	–	98.0
	CaO/MgO[3]	900	4	50	Waste edible oils	–	98.4
	CaO[4]	900	4	46.1	Microalgae biomass	92.30	–
	$CuFe_2O_4/CaO$[5]	800	4	50	Chicken fat oil	–	94.5
	CaO[6]	900	3	25–100	*Chlorella pyrenoidosa*	–	93.4
Leaves of the lotus tree	Activated carbon/$CuFe_2O_4/CaO$[7]	400	2	<100	Chicken fat oil	–	95.6
Elephant-ear pod	K, Mg, Ca and Fe[8]	700	4	–	$Fe_2(SO_4)_3$ pretreated rubber seed oil blend	–	98.8
Moringa leaves	$CaCO_3$ $CaMg(CO_3)_2$, $(K_2Ca(CO_3)_2)$[9]	500	2	36.5	Soybean oil	–	86.7
Raw sugar beet agro-industrial waste	CaO[10]	800	2	25–49	Sunflower oil	–	93.0
Snail shell	CaO[11]	900	4	40	*Hydnocarpus wightiana*	–	98.9
					Dairy waste scum		96.9
Açai seeds	C/SO_3H*[12]	400 °C and sulphonated at 80 °C for 3 h.	3	–	Oleic acid	–	88
					Acidified soybean oil		80
Silver Croaker stones	CaO[13]	800 °C, hydrated and re-calcined at 600 °C	3	–	Soybean oil	98.9	97.4
Cinnamomum tamala	CuO*[14]	–	–	18.5–20	Pig tallow oil	–	97.8

* Nanocatalyst applied in the esterification reaction.

– Not determined or evaluated.

[1] Borah et al., "Transesterification of Waste Cooking Oil for Biodiesel Production Catalyzed by Zn Substituted Waste Egg Shell Derived CaO Nanocatalyst."

[2] Chingakham, David, and Sajith, "Fe3O4 Nanoparticles Impregnated Eggshell as a Novel Catalyst for Enhanced Biodiesel Production."

[3] Foroutan et al., "Transesterification of Waste Edible Oils to Biodiesel Using Calcium Oxide@magnesium Oxide Nanocatalyst."

[4] Pandit and Fulekar, "Biodiesel Production from Scenedesmus Armatus Using Egg Shell Waste as Nanocatalyst."

[5] Seffati et al., "Enhanced Biodiesel Production from Chicken Fat Using CaO/CuFe 2 O 4 Nanocatalyst and Its Combination with Diesel to Improve Fuel Properties."

[6] Ahmad et al., "Optimization of Direct Transesterification of Chlorella Pyrenoidosa Catalyzed by Waste Egg Shell Based Heterogenous Nano – CaO Catalyst."

[7] Seffati et al., "AC/CuFe2O4@CaO as a Novel Nanocatalyst to Produce Biodiesel from Chicken Fat."

[8] Falowo, Oloko-Oba, and Betiku, "Biodiesel Production Intensification via Microwave Irradiation-Assisted Transesterification of Oil Blend Using Nanoparticles from Elephant-Ear Tree Pod Husk as a Base Heterogeneous Catalyst."

[9] Aleman-Ramirez et al., "Preparation of a Heterogeneous Catalyst from Moringa Leaves as a Sustainable Precursor for Biodiesel Production."

[10] Abdelhady et al., "Efficient Catalytic Production of Biodiesel Using Nano-Sized Sugar Beet Agro-Industrial Waste."

[11] Krishnamurthy, Sridhara, and Ananda Kumar, "Optimization and Kinetic Study of Biodiesel Production from Hydnocarpus Wightiana Oil and Dairy Waste Scum Using Snail Shell CaO Nano Catalyst."

[12] Araujo et al., "Magnetic Acid Catalyst Produced from Acai Seeds and Red Mud for Biofuel Production."

[13] Takeno et al., "A Novel CaO-Based Catalyst Obtained from Silver Croaker (Plagioscion Squamosissimus) Stone for Biodiesel Synthesis: Waste Valorization and Process Optimization."

[14] Suresh, Sivarajasekar, and Balasubramani, "Enhanced Ultrasonic Assisted Biodiesel Production from Meat Industry Waste (Pig Tallow) Using Green Copper Oxide Nanocatalyst: Comparison of Response Surface and Neural Network Modelling."

releasing a high amount of CO_2 into the atmosphere and contributing to the increase of the greenhouse effect.[65] The result of this inefficient management results in an increase in adverse environmental impacts caused by human activity.[66] Subsequently, it is necessary to develop methods to recycle these residues, aiming to remove the environmental impacts, in addition to adding value by developing a new product. In this context, agro-industries would benefit from the implementation of greener processes that are environmentally friendly, mainly around solid waste management.[67] The use of biomaterials (agro-industrial waste) to obtain nanocatalysts has been frequently reported, where the general synthesis process can be seen in Figure 7.2.

An example is the work of Falowo et al.,[68] where a nanocatalyst rich in K, Mg, Ca and Fe has been synthesized from the calcination of elephant-ear tree pod husk at 700 ° C/4 h with a pore size 61.81 nm and a high surface area (1.2960 m^2/g) with a yield of 98.8% of biodiesel from pretreated oil via transesterification using microwave heating. They also showed the possibility of recycling the catalyst with a 74.68% yield after the fourth cycle. Another good example comes from rice production, which generates about 5.6 × 10^9 tons of rice husks annually, making it one of the main agro-industrial wastes.[69] The high silica content in this husk makes it a promising material to produce catalysts and also catalytic supports. Hence, several researchers have applied this waste as a basis for the synthesis of nanocatalysts. Recently, Hazmi et al.[70] synthesized a bifunctional nanocatalyst from rice husk doped with K_2O and Fe by the wet impregnation method and applied it in the simultaneous transesterification/esterification of waste cooking oil bearing high acidity. The nanocatalyst showed a 98.6% conversion rate of the oil into biodiesel, which was reused up to three times with conversions greater than 90% and up to five times with conversion rates close to 80%.

Another well-studied source of biomass, which generates large amounts of waste, is plant leaves. Barros et al.[71] obtained a catalyst derived from pineapple leaves (residue from fruit production). This heterogeneous biocatalyst showed good catalytic activity in the transesterification of soybean oil (98% in 30 min), probably due to the high content of alkaline elements, including calcium, potassium, sodium and magnesium. In a similar work, Aleman-Ramirez et al.[72] calcined *Moringa oleifera* leaves to obtain the ashes, where the energy-dispersive X-ray spectroscopy (EDS) analysis showed that the majority composition of elements were potassium, calcium, magnesium, phosphorus and sulphur. The obtained nanocatalyst showed a yield of 86.7% biodiesel in 2 h. In addition, the catalyst was reused three times, decreasing its activity.

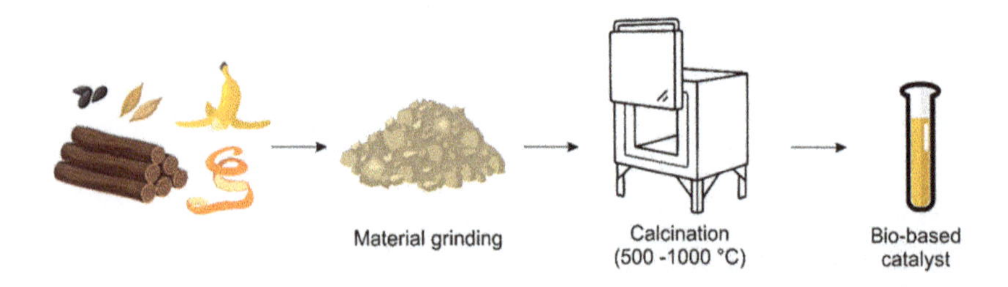

FIGURE 7.2 Obtaining low-cost catalysts from different renewable agro-industrial wastes.

Seffati et al.[73] synthesized activated carbon from lotus leaves and then impregnated $CuFe_2O_4$ nanoparticles. The material obtained was then encapsulated with CaO. Thus, the synthesized AC/$CuFe_2O_4$-CaO proved to be an excellent nanocatalyst, resulting in a chicken fat to biodiesel conversion rate of 95.63%.

Fruit peels and seeds have been used to obtain catalysts applied in the production of biodiesel.[74] Although many studies do not confirm the particle size, it has been shown to be very efficient.

7.3.2 CaO-Based Nanocatalysts

Currently, materials developed from bioderivatives have gained great interest in the scientific community, mainly due to the increased demand for solid waste treatment. Among these, there are several sources of calcium that can be used in the preparation of nanocatalysts for the synthesis of biodiesel, such as bones,[75] crustacean shells,[76] molluscs,[77] eggshells,[78] snail shells,[79] mussel shells,[80] seashells,[81] clamshells[82] and agro-industrial residues such as sugar beet.[83] The synthesis process is very similar to that presented for the synthesis of catalysts obtained from agro-industrial wastes by calcination, as schematized in Figure 7.3. For example, Takeno et al.,[84] looking for a new source of CaO, calcined the stone of the silver croaker, a residue found on the head of this fish that is widely consumed in Brazil. They showed that calcination generated a CaO-based catalyst with high activity, resulting in a conversion rate of approximately 99% of soybean oil into biodiesel and the ability to maintain its efficiency for six consecutive cycles without having to re-calcinate before each reuse.

The most used animal source biomaterial to produce low-cost and high efficiency nanocatalysts is eggshell. The use of eggshell to obtain a solid catalyst was first indicated in 2009.[85] Since then, several works have been published aiming at improving the methods of synthesis and optimization of reaction parameters (temperature, reaction time, catalyst load and alcohol:oil molar ratio). For example, Abdelhady et al.,[86] used beet residue to synthesize nanometric calcium oxide (CaO). This nanocatalyst was used in the transesterification of sunflower oil, showing a conversion rate of 93%. Teo et al.[87] synthesized a super-base nanocatalyst for biodiesel production through the processes of decomposition, adsorption and precipitation assisted by an amphoteric surfactant (betaine). The nanocatalyst exhibited a high activity to produce biodiesel, reaching a conversion rate of 97% in 300 min at 60 °C. In addition, it also revealed that it is possible to reduce the reaction time by almost half when increasing temperature.

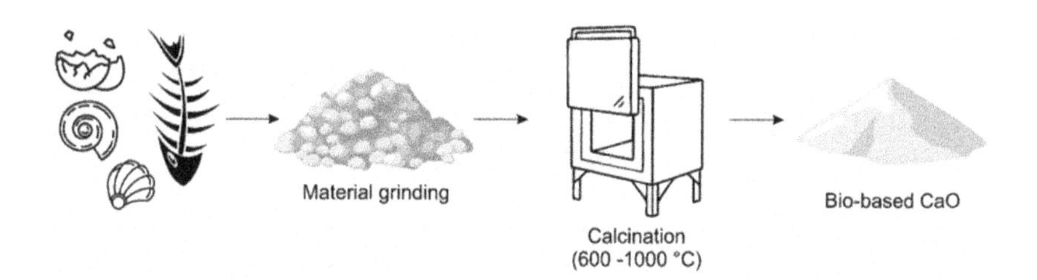

FIGURE 7.3 Obtaining CaO catalysts from different $CaCO_3$-based materials.

In the same vein, Foroutan et al.[88] prepared a CaO-MgO nanocatalyst derived from chicken eggshell residues for biodiesel production using edible oil residues. The results showed a high conversion rate that reached 98.37%. Chingakham et al.[89] synthesized CaO from chicken eggshells followed by the impregnation of magnetic nanoparticles of Fe_3O_4 to produce a new catalyst. This study showed that application of the material obtained in the synthesis of biodiesel gave a yield of 98%, and it could be reused up to seven times. Bet-Moushoul et al.[90] obtained CaO/Au nanoparticles and applied them as a catalyst in the transesterification of sunflower oil, where the sources of CaO used were eggshells, mussel shells, calcite and dolomite. The catalytic activity of these nanomaterials was compared with commercial CaO, showing the high reuse capacity of the prepared nanocatalysts (10 times without loss of activity).

7.3.3 Disadvantages of Bio-Based Nanocatalysts

Although the use of wastes in obtaining catalysts is a promising and advantageous activity when compared to other catalysts already in the market, they usually have some disadvantages in relation to their use, such as a decrease in catalytic activity due to the leaching process. This inconvenience can be eased when applied in reactions with high temperature.[91]

There remain challenges to be overcome for the better use of green catalysts. Firstly, many catalysts obtained from renewable resources are composed mainly of CaO, K_2O, Ca or K, and they often suffer from deactivation, poisoning and leaching problems in the reaction medium, where these are related to the formation of calcium diglyceride on the catalyst surface during the reaction, as well as the leaching of Ca^{2+} and K^+ ions from the catalyst surface.[92] Some studies have also shown that the calcination temperature can greatly influence these effects.[93]

7.4 Optimization of Biodiesel Synthesis

In the synthesis of biodiesel using nanocatalysts obtained from biomaterials, normally, the catalytic activity is evaluated through transesterification or esterification reactions, which will depend on its acidic or basic nature. Thus, in this process the parameters, namely temperature, reaction time, molar ratio alcohol:oil or fatty acid and catalyst load, are varied, where the optimization can be OPAT (one parameter at a time) or by a formal design of the experiment.

7.4.1 Temperature

Temperature is an extremely important factor in the production of biodiesel. It is known that its increase favours the reaction, since it reduces the viscosity of the reaction medium, increasing the mass transfer rate from the medium to the catalyst surface,[94] in addition to increasing entropy and, consequently, the number of collisions between molecules,[95] resulting in high oil–biodiesel conversion rates.

The temperature used in catalysis studies to obtain a greater conversion rate depends basically on the alcohol used. When methanol is used, most catalytic studies are done at temperatures close to the boiling point of alcohol: 64.7 and 78.4 °C for methanol and ethanol, respectively.[96] However, some studies use temperature ranging from room temperature[97] to higher temperatures (190 °C).[98] Some authors have also reported the use of butanol to produce butyl biodiesel, and, for that, they have used temperatures above its boiling point (117.7 °C).[99]

The reaction temperature above the boiling point is explained because the transesterification reaction occurs at the interface created between the alcohol and the oil or fatty acid, since they are immiscible. Thus, temperatures close to the boiling point of alcohol improve the miscibility of alcohol in the oil and, consequently, increase the reaction rate.[100]

The temperature variation in the optimization process also allows the activation energy of the reaction to be obtained through the Arrhenius equation, showing how the synthesized nanocatalysts reduce the reaction energy barrier.[101]

7.4.2 Reaction Time

Time is another important reaction parameter in the synthesis of biodiesel, especially in regard to industrial production. Several studies have varied this parameter in order to achieve high performance in the shortest time.[102]

The reaction time range found in several studies on transesterification applying bio-based catalysts varies from 0.5 to 8 h of reaction.[103] Usually, in order to reduce the reaction time, other parameters – temperature, alcohol content, and/or catalyst load – are increased. Teo et al.[104] showed that it is possible to reduce the reaction time in half by increasing the temperature of the medium.

7.4.3 Alcohol: Oil Molar Ratio

The variation in the volume of alcohol with respect to the oil amount follows the principle of Le Chatelier, since an increase in the concentration of a reagent benefits the synthesis of the product,[105] in addition to decreasing the medium viscosity[106] which, as was shown earlier, favours the transfer of mass in the medium (increased reaction rate). In this sense, alcohol appears in excess, usually presenting molar ratios from 5[107] to 20 mol/mol,[108] although some studies have much higher molar ratios, from 40 to 150 mol/mol.[109]

For this purpose, methanol is usually used because it requires lower temperatures and is more reactive than ethanol and butanol.[110] However, other alcohols (e.g. ethanol) are less toxic, in addition to being obtained from renewable sources.[111]

7.4.4 Catalyst Load

The catalyst load in a reaction is of paramount importance, since, in lower amounts it can result in lower conversions and longer reaction times; and high concentrations can increase viscosity and, consequently, reduce conversion,[112] as already described. The catalyst load is always related to the limiting reagent (triglyceride or fatty acid).

To be considered a catalyst, minimal amounts of the catalytic agent are suitable for the transformation of large quantities of reagents.[113] However, there is a very extensive catalyst load range (from 1 to 20 wt%),[114] and it is possible to find studies with up to 100 wt% of the catalyst.[115]

7.5 Conclusion

Nanocatalysts are usually more efficient than catalysts of larger sizes, mainly because they have a larger surface area and, consequently, more active sites available. They can be obtained from the calcination of various residues, such as bones, eggshells, peels and seeds fruits, oyster shells, etc., where the catalytic activity of these bio-based materials in transesterification reactions is related to the concentration of the prominent elements [usually potassium (K), sodium (Na), magnesium (Mg) and calcium (Ca)]. In addition to the high catalytic activity in the synthesis of biodiesel, these catalysts are low-cost and reduce the possible environmental impact caused by agro-industrial wastes and, due to their heterogeneous nature, they can be recycled, being reused several times. These biomaterials can also be used as catalytic support, increasing the basicity of the catalyst, or giving it a bifunctional character (acidic and basic) when supporting another active phase. Some works have presented large amounts of catalysts to obtain conversion or high yields, not meeting the requirement to be classified as such. Although efficient and low-cost, these catalysts still need to be extensively studied to improve their stability, avoiding their deactivation and contamination of the biofuel, since most present leaching, and the presence of metals in biodiesel increases the speed of degradation by oxidation.

Notes

1 Sudarsanam et al., "Advances in Porous and Nanoscale Catalysts for Viable Biomass Conversion."
2 Caliskan, "Environmental and Enviroeconomic Researches on Diesel Engines with Diesel and Biodiesel Fuels."
3 Sharma and Singh, "Microalgal Biodiesel: A Possible Solution for India's Energy Security"; Changmai, Sudarsanam, and Rokhum, "Biodiesel Production Using a Renewable Mesoporous Solid Catalyst"; Caliskan, "Environmental and Enviroeconomic Researches on Diesel Engines with Diesel and Biodiesel Fuels."
4 Gupta and Rathod, "Waste Cooking Oil and Waste Chicken Eggshells Derived Solid Base Catalyst for the Biodiesel Production: Optimization and Kinetics"; Athar and Zaidi, "A Review of the Feedstocks, Catalysts, and Intensification Techniques for Sustainable Biodiesel Production."
5 Mendonça et al., "Application of Calcined Waste Cupuaçu (*Theobroma grandiflorum*) Seeds as a Low-Cost Solid Catalyst in Soybean Oil Ethanolysis: Statistical Optimization"; Mendonça et al., "New Heterogeneous Catalyst for Biodiesel Production from Waste Tucumã Peels (*Astrocaryum aculeatum Meyer*): Parameters Optimization Study."
6 Saravanan Arumugamurthy et al., "Conversion of a Low Value Industrial Waste into Biodiesel Using a Catalyst Derived from Brewery Waste: An Activation and Deactivation Kinetic Study"; Rudreshwar Balinge and Balakrishnan, "Mini Review on Recent Progress toward Sustainable Production of Biodiesel from Biomass."

7 Singh et al., "A Comprehensive Review of Physicochemical Properties, Production Process, Performance and Emissions Characteristics of 2nd Generation Biodiesel Feedstock: *Jatropha curcas*"; de Sá Barros et al., "Pineapple (*Ananás comosus*) Leaves Ash as a Solid Base Catalyst for Biodiesel Synthesis."
8 Rezania et al., "Review on Transesterification of Non-Edible Sources for Biodiesel Production with a Focus on Economic Aspects, Fuel Properties and by-Product Applications."
9 Kant Bhatia et al., "An Overview on Advancements in Biobased Transesterification Methods for Biodiesel Production: Oil Resources, Extraction, Biocatalysts, and Process Intensification Technologies."
10 Muhammad et al., "An Overview of the Role of Ionic Liquids in Biodiesel Reactions"; Saravanan Arumugamurthy et al., "Conversion of a Low Value Industrial Waste into Biodiesel Using a Catalyst Derived from Brewery Waste: An Activation and Deactivation Kinetic Study"; Gupta and Rathod, "Waste Cooking Oil and Waste Chicken Eggshells Derived Solid Base Catalyst for the Biodiesel Production: Optimization and Kinetics."
11 Manojkumar, Muthukumaran, and Sharmila, "A Comprehensive Review on the Application of Response Surface Methodology for Optimization of Biodiesel Production Using Different Oil Sources"; Tang and Niu, "Preparation of Carbon-Based Solid Acid with Large Surface Area to Catalyze Esterification for Biodiesel Production"; Singh et al., "A Comprehensive Review of Physicochemical Properties, Production Process, Performance and Emissions Characteristics of 2nd Generation Biodiesel Feedstock: *Jatropha curcas*"; Liu et al., "Recent Progress on Biodiesel Production from Municipal Sewage Sludge."
12 Gupta and Rathod, "Waste Cooking Oil and Waste Chicken Eggshells Derived Solid Base Catalyst for the Biodiesel Production: Optimization and Kinetics"; Athar and Zaidi, "A Review of the Feedstocks, Catalysts, and Intensification Techniques for Sustainable Biodiesel Production"; Liu et al., "Recent Progress on Biodiesel Production from Municipal Sewage Sludge."
13 Ashok et al., "Magnetically Recoverable Mg Substituted Zinc Ferrite Nanocatalyst for Biodiesel Production: Process Optimization, Kinetic and Thermodynamic Analysis."
14 Krishnamurthy et al. (2020)
15 Mendonça et al., "Application of Calcined Waste Cupuaçu (*Theobroma grandiflorum*) Seeds as a Low-Cost Solid Catalyst in Soybean Oil Ethanolysis: Statistical Optimization."
16 Liu et al., "Recent Progress on Biodiesel Production from Municipal Sewage Sludge."
17 Shan et al., "Catalysts from Renewable Resources for Biodiesel Production," 2018; Abdullah et al., "A Review of Biomass-Derived Heterogeneous Catalyst for a Sustainable Biodiesel Production."
18 Mendonça et al., "New Heterogeneous Catalyst for Biodiesel Production from Waste Tucumã Peels (*Astrocaryum aculeatum* Meyer): Parameters Optimization Study"; Mendonça et al., "Application of Calcined Waste Cupuaçu (*Theobroma grandiflorum*) Seeds as a Low-Cost Solid Catalyst in Soybean Oil Ethanolysis: Statistical Optimization"; de Sá Barros et al., "Pineapple (*Ananás comosus*) Leaves Ash as a Solid Base Catalyst for Biodiesel Synthesis."
19 Balajii and Niju, "Banana Peduncle – A Green and Renewable Heterogeneous Base Catalyst for Biodiesel Production from Ceiba Pentandra Oil"; Leung, Wu, and Leung, "A Review on Biodiesel Production Using Catalyzed Transesterification"; Martínez et al., "In-Situ Transesterification of Jatropha Curcas L. Seeds Using Homogeneous and Heterogeneous Basic Catalysts."

20 Balajii and Niju, "Banana Peduncle – A Green and Renewable Heterogeneous Base Catalyst for Biodiesel Production from Ceiba Pentandra Oil"; Chingakham, David, and Sajith, "Fe3O4 Nanoparticles Impregnated Eggshell as a Novel Catalyst for Enhanced Biodiesel Production"; Seffati et al., "AC/$CuFe_2O_4$@CaO as a Novel Nanocatalyst to Produce Biodiesel from Chicken Fat."
21 Saravanan Arumugamurthy et al., "Conversion of a Low Value Industrial Waste into Biodiesel Using a Catalyst Derived from Brewery Waste: An Activation and Deactivation Kinetic Study."
22 Mendonça et al., "Application of Calcined Waste Cupuaçu (*Theobroma Grandiflorum*) Seeds as a Low-Cost Solid Catalyst in Soybean Oil Ethanolysis: Statistical Optimization."
23 Mendonça et al., "New Heterogeneous Catalyst for Biodiesel Production from Waste Tucumã Peels (*Astrocaryum aculeatum* Meyer): Parameters Optimization Study."
24 de Sá Barros et al., "Pineapple (*Ananás Comosus*) Leaves Ash as a Solid Base Catalyst for Biodiesel Synthesis."
25 Quah et al., "Magnetic Biochar Derived from Waste Palm Kernel Shell for Biodiesel Production via Sulfonation."
26 Zhou, Niu, and Li, "Activity of the Carbon-Based Heterogeneous Acid Catalyst Derived from Bamboo in Esterification of Oleic Acid with Ethanol."
27 Quah et al., "Magnetic Biochar Derived from Waste Palm Kernel Shell for Biodiesel Production via Sulfonation."
28 de Sá Barros et al., "Pineapple (*Ananás comosus*) Leaves Ash as a Solid Base Catalyst for Biodiesel Synthesis"; Asikin-Mijan, Lee, and Taufiq-Yap, "Synthesis and Catalytic Activity of Hydration-Dehydration Treated Clamshell Derived CaO for Biodiesel Production."
29 Adepoju, "Optimization Processes of Biodiesel Production from Pig and Neem (*Azadirachta indica a.Juss*) Seeds Blend Oil Using Alternative Catalysts from Waste Biomass"; Seffati et al., "AC/$CuFe_2O_4$@CaO as a Novel Nanocatalyst to Produce Biodiesel from Chicken Fat."
30 Chingakham, David, and Sajith, "Fe_3O_4 Nanoparticles Impregnated Eggshell as a Novel Catalyst for Enhanced Biodiesel Production."
31 Krishnamurthy, Sridhara, and Ananda Kumar, "Optimization and Kinetic Study of Biodiesel Production from Hydnocarpus Wightiana Oil and Dairy Waste Scum Using Snail Shell CaO Nano Catalyst."
32 Qu et al., "Preparation of Calcium Modified Zn-Ce/Al_2O_3 Heterogeneous Catalyst for Biodiesel Production through Transesterification of Palm Oil with Methanol Optimized by Response Surface Methodology."
33 Abdelhady et al., "Efficient Catalytic Production of Biodiesel Using Nano-Sized Sugar Beet Agro-Industrial Waste."
34 Abukhadra et al., "Sonication Induced Transesterification of Castor Oil into Biodiesel in the Presence of MgO/CaO Nanorods as a Novel Basic Catalyst: Characterization and Optimization."
35 Aleman-Ramirez et al., "Preparation of a Heterogeneous Catalyst from Moringa Leaves as a Sustainable Precursor for Biodiesel Production."
36 Ashok et al., "Magnetically Recoverable Mg Substituted Zinc Ferrite Nanocatalyst for Biodiesel Production: Process Optimization, Kinetic and Thermodynamic Analysis."
37 Gebremariam and Marchetti, "Economics of Biodiesel Production: Review."

38 Reis et al., "Biodiesel Production from Fatty Acids of Refined Vegetable Oils by Heterogeneous Acid Catalysis and Microwave Irradiation."
39 Foroutan et al., "Transesterification of Waste Edible Oils to Biodiesel Using Calcium Oxide@Magnesium Oxide Nanocatalyst."
40 Krishnamurthy et al. (2020)
41 Ahmad et al., "Optimization of Direct Transesterification of Chlorella Pyrenoidosa Catalyzed by Waste Egg Shell Based Heterogenous Nano – CaO Catalyst."
42 Pandit and Fulekar, "Biodiesel Production from Microalgal Biomass Using CaO Catalyst Synthesized from Natural Waste Material"; Sakthi Vignesh et al., "Sustainable Biofuel from Microalgae: Application of Lignocellulosic Wastes and Bio-Iron Nanoparticle for Biodiesel Production"; Krishnamurthy, Sridhara, and Ananda Kumar, "Optimization and Kinetic Study of Biodiesel Production from Hydnocarpus Wightiana Oil and Dairy Waste Scum Using Snail Shell CaO Nano Catalyst."
43 Araujo et al., "Low Temperature Sulfonation of Acai Stone Biomass Derived Carbons as Acid Catalysts for Esterification Reactions"; Lacerda et al., "Esterification of Oleic Acid Using 12-Tungstophosphoric Supported in Flint Kaolin of the Amazonia."
44 Suresh et al. (2021)
45 Ravi et al., "Contemporary Approaches towards Augmentation of Distinctive Heterogeneous Catalyst for Sustainable Biodiesel Production."
46 Madhuvilakku and Piraman, "Biodiesel Synthesis by TiO_2-ZnO Mixed Oxide Nanocatalyst Catalyzed Palm Oil Transesterification Process."
47 Pessoa Junior et al., "Application of Water Treatment Sludge as a Low-Cost and Eco-Friendly Catalyst in the Biodiesel Production via Fatty Acids Esterification: Process Optimization."
48 Reis et al., "Biodiesel Production from Fatty Acids of Refined Vegetable Oils by Heterogeneous Acid Catalysis and Microwave Irradiation."
49 Mendonça et al., "Application of Calcined Waste Cupuaçu (*Theobroma grandiflorum*) Seeds as a Low-Cost Solid Catalyst in Soybean Oil Ethanolysis: Statistical Optimization"; Mendonça et al., "New Heterogeneous Catalyst for Biodiesel Production from Waste Tucumã Peels (*Astrocaryum aculeatum* Meyer): Parameters Optimization Study"; Rezania et al., "Review on Transesterification of Non-Edible Sources for Biodiesel Production with a Focus on Economic Aspects, Fuel Properties and By-Product Applications."
50 Atadashi et al., "The Effects of Catalysts in Biodiesel Production: A Review."
51 Das et al., "Cobalt-Doped CaO Catalyst Synthesized and Applied for Algal Biodiesel Production."
52 Araujo et al., "Low Temperature Sulfonation of Acai Stone Biomass Derived Carbons as Acid Catalysts for Esterification Reactions."
53 Nayebzadeh et al., "Fabrication of Carbonated Alumina Doped by Calcium Oxide via Microwave Combustion Method Used as Nanocatalyst in Biodiesel Production: Influence of Carbon Source Type"; Soltani et al., "Recent Progress in Synthesis and Surface Functionalization of Mesoporous Acidic Heterogeneous Catalysts for Esterification of Free Fatty Acid Feedstocks: A Review."
54 Teo et al., "Effective Synthesis of Biodiesel from Jatropha Curcas Oil Using Betaine Assisted Nanoparticle Heterogeneous Catalyst from Eggshell of Gallus Domesticus"; Feyzi et al., "Preparation, Characterization, Kinetic and Thermodynamic Studies of MgO-La_2O_3 Nanocatalysts for Biodiesel Production from Sunflower Oil."
55 Pandit and Fulekar, "Biodiesel Production from Scenedesmus Armatus Using Egg Shell Waste as Nanocatalyst"; Nayebzadeh et al., "Fabrication of Carbonated

Alumina Doped by Calcium Oxide via Microwave Combustion Method Used as Nanocatalyst in Biodiesel Production: Influence of Carbon Source Type"; Soltani et al., "Recent Progress in Synthesis and Surface Functionalization of Mesoporous Acidic Heterogeneous Catalysts for Esterification of Free Fatty Acid Feedstocks: A Review."

56 Thiele, "Relation between Catalytic Activity and Size of Particle"; Moshfegh, "Nanoparticle Catalysts."

57 Abdullah et al., "A Review of Biomass-Derived Heterogeneous Catalyst for a Sustainable Biodiesel Production"; Shan et al., "Catalysts from Renewable Resources for Biodiesel Production," 2018.

58 Abdullah et al., "A Review of Biomass-Derived Heterogeneous Catalyst for a Sustainable Biodiesel Production"; Shan et al., "Catalysts from Renewable Resources for Biodiesel Production," 2018.

59 Abdullah et al., "A Review of Biomass-Derived Heterogeneous Catalyst for a Sustainable Biodiesel Production."

60 de Sá Barros et al., "Pineapple (*Ananás comosus*) Leaves Ash as a Solid Base Catalyst for Biodiesel Synthesis"; Balajii and Niju, "Banana Peduncle – A Green and Renewable Heterogeneous Base Catalyst for Biodiesel Production from Ceiba Pentandra Oil."

61 Borah et al., "Transesterification of Waste Cooking Oil for Biodiesel Production Catalyzed by Zn Substituted Waste Egg Shell Derived CaO Nanocatalyst."

62 Chingakham, David, and Sajith, "Fe3O4 Nanoparticles Impregnated Eggshell as a Novel Catalyst for Enhanced Biodiesel Production."

63 Foroutan et al., "Transesterification of Waste Edible Oils to Biodiesel Using Calcium Oxide@magnesium Oxide Nanocatalyst."

64 Bet-Moushoul et al., "Application of CaO-Based/Au Nanoparticles as Heterogeneous Nanocatalysts in Biodiesel Production."

65 de Sá Barros et al., "Pineapple (*Ananás comosus*) Leaves Ash as a Solid Base Catalyst for Biodiesel Synthesis."

66 Tripathi et al., "Biomass Waste Utilisation in Low-Carbon Products: Harnessing a Major Potential Resource."

67 Abol-Fotouh et al., "Bacterial Nanocellulose from Agro-Industrial Wastes: Low-Cost and Enhanced Production by Komagataeibacter Saccharivorans MD1."

68 FALOWO et. al., "Biodiesel production intensification via microwave irradiation-assisted transesterification of oil blend using nanoparticles from elephant-ear tree pod husk as a base heterogeneous catalyst."

69 Singh, "Rice Husk Ash."

70 Hazmi et al., Supermagnetic Nano-Bifunctional Catalyst from Rice Husk: Synthesis, Characterization and Application for Conversion of Used Cooking Oil to Biodiesel

71 de Sá Barros et al., "Pineapple (*Ananás comosus*) Leaves Ash as a Solid Base Catalyst for Biodiesel Synthesis."

72 Aleman-Ramirez et al., "Preparation of a Heterogeneous Catalyst from Moringa Leaves as a Sustainable Precursor for Biodiesel Production."

73 Seffati et al., "AC/$CuFe_2O_4$@CaO as a Novel Nanocatalyst to Produce Biodiesel from Chicken Fat."

74 Mendonça et al., "New Heterogeneous Catalyst for Biodiesel Production from Waste Tucumã Peels (*Astrocaryum aculeatum* Meyer): Parameters Optimization Study"; Mendonça et al., "Application of Calcined Waste Cupuaçu (*Theobroma grandiflorum*) Seeds as a Low-Cost Solid Catalyst in Soybean Oil Ethanolysis: Statistical Optimization."

75 Nisar et al., "Enhanced Biodiesel Production from Jatropha Oil Using Calcined Waste Animal Bones as Catalyst."
76 Correia et al., "Calcium/Chitosan Spheres as Catalyst for Biodiesel Production."
77 Shan et al., "Catalysts from Renewable Resources for Biodiesel Production,".
78 Borah et al., "Transesterification of Waste Cooking Oil for Biodiesel Production Catalyzed by Zn Substituted Waste Egg Shell Derived CaO Nanocatalyst"; Pandit and Fulekar, "Biodiesel Production from Scenedesmus Armatus Using Egg Shell Waste as Nanocatalyst"; Chingakham, David, and Sajith, "Fe_3O_4 Nanoparticles Impregnated Eggshell as a Novel Catalyst for Enhanced Biodiesel Production"; Foroutan et al., "Transesterification of Waste Edible Oils to Biodiesel Using Calcium Oxide@magnesium Oxide Nanocatalyst"; Ahmad et al., "Optimization of Direct Transesterification of Chlorella Pyrenoidosa Catalyzed by Waste Egg Shell Based Heterogenous Nano – CaO Catalyst."
79 Krishnamurthy, Sridhara, and Ananda Kumar, "Optimization and Kinetic Study of Biodiesel Production from Hydnocarpus Wightiana Oil and Dairy Waste Scum Using Snail Shell CaO Nano Catalyst."
80 Hadiyanto et al., "The Development of Heterogeneous Catalyst C/CaO/NaOH from Waste of Green Mussel Shell (*Perna varidis*) for Biodiesel Synthesis."
81 Jindapon, Jaiyen, and Ngamcharussrivichai, "Seashell-Derived Mixed Compounds of Ca, Zn and Al as Active and Stable Catalysts for the Transesterification of Palm Oil with Methanol to Biodiesel."
82 Asikin-Mijan, Lee, and Taufiq-Yap, "Synthesis and Catalytic Activity of Hydration-Dehydration Treated Clamshell Derived CaO for Biodiesel Production."
83 Abdelhady et al., "Efficient Catalytic Production of Biodiesel Using Nano-Sized Sugar Beet Agro-Industrial Waste."
84 Takeno et al., "A Novel CaO-Based Catalyst Obtained from Silver Croaker (*Plagioscion squamosissimus*) Stone for Biodiesel Synthesis: Waste Valorization and Process Optimization."
85 Laca, Laca, and Díaz, "Eggshell Waste as Catalyst: A Review."
86 Abdelhady et al. 2020
87 Teo et al., "Effective Synthesis of Biodiesel from Jatropha Curcas Oil Using Betaine Assisted Nanoparticle Heterogeneous Catalyst from Eggshell of Gallus Domesticus."
88 Foroutan et al., "Transesterification of Waste Edible Oils to Biodiesel Using Calcium Oxide@magnesium Oxide Nanocatalyst."
89 Chingakham, David, and Sajith, "Fe_3O_4 Nanoparticles Impregnated Eggshell as a Novel Catalyst for Enhanced Biodiesel Production."
90 Bet-Moushoul et al., "Application of CaO-Based/Au Nanoparticles as Heterogeneous Nanocatalysts in Biodiesel Production."
91 Vadery et al., "Room Temperature Production of Jatropha Biodiesel over Coconut Husk Ash."
92 Alsharifi et al., "Biodiesel Production from Canola Oil Using Novel Li/TiO_2 as a Heterogeneous Catalyst Prepared via Impregnation Method."
93 Mendonça et al., "Application of Calcined Waste Cupuaçu (*Theobroma grandiflorum*) Seeds as a Low-Cost Solid Catalyst in Soybean Oil Ethanolysis: Statistical Optimization"; Piker et al., "A Green and Low-Cost Room Temperature Biodiesel Production Method from Waste Oil Using Egg Shells as Catalyst"; Takeno et al., "A Novel CaO-Based Catalyst Obtained from Silver Croaker (*Plagioscion squamosissimus*) Stone for Biodiesel Synthesis: Waste Valorization and Process Optimization."

94 Mansir et al., "Efficient Reaction for Biodiesel Manufacturing Using Bi-Functional Oxide Catalyst"; Mendonça et al., "New Heterogeneous Catalyst for Biodiesel Production from Waste Tucumã Peels (*Astrocaryum aculeatum* Meyer): Parameters Optimization Study."
95 Machado et al., "Study of Pressure and Temperature Influence on Rapeseed Biodiesel Oxidation Kinetics Using PetroOXY Method."
96 Liu et al., "Mixed and Ground KBr-Impregnated Calcined Snail Shell and Kaolin as Solid Base Catalysts for Biodiesel Production"; Mendonça et al., "New Heterogeneous Catalyst for Biodiesel Production from Waste Tucumã Peels (*Astrocaryum aculeatum* Meyer): Parameters Optimization Study."
97 de Moura et al., "Heterogeneous Catalysis of Babassu Oil Monitored by Thermogravimetric Analysis"; Piker et al., "A Green and Low-Cost Room Temperature Biodiesel Production Method from Waste Oil Using Egg Shells as Catalyst."
98 Wang et al., "Sustainable Biodiesel Production via Transesterification by Using Recyclable $Ca_2MgSi_2O_7$ Catalyst."
99 Faraguna et al., "Correlation Method for Conversion Determination of Biodiesel Obtained from Different Alcohols by 1H NMR Spectroscopy"; Reis et al., "Biodiesel Production from Fatty Acids of Refined Vegetable Oils by Heterogeneous Acid Catalysis and Microwave Irradiation."
100 Guan et al., "Transesterification of Vegetable Oil to Biodiesel Fuel Using Acid Catalysts in the Presence of Dimethyl Ether"; Laskar et al., "Transesterification of Soybean Oil at Room Temperature Using Biowaste as Catalyst; an Experimental Investigation on the Effect of Co-Solvent on Biodiesel Yield."
101 Mendonça et al., "New Heterogeneous Catalyst for Biodiesel Production from Waste Tucumã Peels (*Astrocaryum aculeatum* Meyer): Parameters Optimization Study"; de Sá Barros et al., "Pineapple (*Ananás comosus*) Leaves Ash as a Solid Base Catalyst for Biodiesel Synthesis"; Mendonça et al., "Application of Calcined Waste Cupuaçu (*Theobroma grandiflorum*) Seeds as a Low-Cost Solid Catalyst in Soybean Oil Ethanolysis: Statistical Optimization."
102 Mendonça et al., "New Heterogeneous Catalyst for Biodiesel Production from Waste Tucumã Peels (*Astrocaryum aculeatum* Meyer): Parameters Optimization Study"; de Sá Barros et al., "Pineapple (*Ananás comosus*) Leaves Ash as a Solid Base Catalyst for Biodiesel Synthesis"; Mendonça et al., "Application of Calcined Waste Cupuaçu (*Theobroma grandiflorum*) Seeds as a Low-Cost Solid Catalyst in Soybean Oil Ethanolysis: Statistical Optimization"; Pessoa Junior et al., "Application of Water Treatment Sludge as a Low-Cost and Eco-Friendly Catalyst in the Biodiesel Production via Fatty Acids Esterification: Process Optimization."
103 Abdullah et al., "A Review of Biomass-Derived Heterogeneous Catalyst for a Sustainable Biodiesel Production"; Shan et al., "Catalysts from Renewable Resources for Biodiesel Production," 2018.
104 Teo et al., "Effective Synthesis of Biodiesel from *Jatropha curcas* Oil Using Betaine Assisted Nanoparticle Heterogeneous Catalyst from Eggshell of Gallus Domesticus."
105 Sahani and Sharma, "Economically Viable Production of Biodiesel Using a Novel Heterogeneous Catalyst: Kinetic and Thermodynamic Investigations."
106 Abukhadra and Sayed, "K+ Trapped Kaolinite (Kaol/K+) as Low Cost and Eco-Friendly Basic Heterogeneous Catalyst in the Transesterification of Commercial Waste Cooking Oil into Biodiesel."

107 Anr et al., "Biodiesel Production from Crude Jatropha Oil Using a Highly Active Heterogeneous Nanocatalyst by Optimizing Transesterification Reaction Parameters."
108 Borah et al., "Transesterification of Waste Cooking Oil for Biodiesel Production Catalyzed by Zn Substituted Waste Egg Shell Derived CaO Nanocatalyst."
109 de Sá Barros et al., "Pineapple (*Ananás comosus*) Leaves Ash as a Solid Base Catalyst for Biodiesel Synthesis"; De Oliveira et al., "Microwave-Assisted Preparation of a New Esterification Catalyst from Wasted Flint Kaolin"; Nur Syazwani, Rashid, and Taufiq Yap, "Low-Cost Solid Catalyst Derived from Waste *Cyrtopleura costata* (Angel Wing Shell) for Biodiesel Production Using Microalgae Oil."
110 Mendonça et al., "New Heterogeneous Catalyst for Biodiesel Production from Waste Tucumã Peels (*Astrocaryum aculeatum* Meyer): Parameters Optimization Study."
111 Barros et al., "Esterification of Lauric Acid with Butanol over Mesoporous Materials"; Mendonça et al., "Application of Calcined Waste Cupuaçu (*Theobroma grandiflorum*) Seeds as a Low-Cost Solid Catalyst in Soybean Oil Ethanolysis: Statistical Optimization."
112 de Sá Barros et al., "Pineapple (*Ananás comosus*) Leaves Ash as a Solid Base Catalyst for Biodiesel Synthesis"; Mendonça et al., "New Heterogeneous Catalyst for Biodiesel Production from Waste Tucumã Peels (*Astrocaryum aculeatum* Meyer): Parameters Optimization Study"; Piker et al., "A Green and Low-Cost Room Temperature Biodiesel Production Method from Waste Oil Using Egg Shells as Catalyst."
113 Busacca et al., "The Growing Impact of Catalysis in the Pharmaceutical Industry"; Yuryev and Liese, "Biocatalysis: The Outcast."
114 Shan et al., "Catalysts from Renewable Resources for Biodiesel Production," 2018; Abdullah et al., "A Review of Biomass-Derived Heterogeneous Catalyst for a Sustainable Biodiesel Production."
115 Roschat et al., "Biodiesel Production Based on Heterogeneous Process Catalyzed by Solid Waste Coral Fragment."

REFERENCES

Abdelhady, Hosam H., Hany A. Elazab, Emad M. Ewais, Mohamed Saber, and Mohamed S. El-Deab. "Efficient Catalytic Production of Biodiesel Using Nano-Sized Sugar Beet Agro-Industrial Waste." *Fuel* 261, (February 2020): 116481. https://doi.org/10.1016/j.fuel.2019.116481.

Abdullah, Sharifah Hanis Yasmin Sayid, Nur Hanis Mohamad Hanapi, Azman Azid, Roslan Umar, Hafizan Juahir, Helena Khatoon, and Azizah Endut. "A Review of Biomass-Derived Heterogeneous Catalyst for a Sustainable Biodiesel Production." *Renewable and Sustainable Energy Reviews* 70, no. July (2017): 1040–51. https://doi.org/10.1016/j.rser.2016.12.008.

Abol-Fotouh, Deyaa, Mohamed A. Hassan, Hassan Shokry, Anna Roig, Mohamed S. Azab, and Abd El-Hady B. Kashyout. "Bacterial Nanocellulose from Agro-Industrial Wastes: Low-Cost and Enhanced Production by *Komagataeibacter saccharivorans* MD1." *Scientific Reports* 10, no. 1 (December 2020): 3491. https://doi.org/10.1038/s41598-020-60315-9.

Abukhadra, Mostafa R., Aya S. Mohamed, Ahmed M. El-Sherbeeny, Ahmed Tawhid Ahmed Soliman, and Abd Elatty E. Abd Elgawad. "Sonication Induced Transesterification of Castor Oil into Biodiesel in the Presence of MgO/CaO Nanorods as a Novel Basic Catalyst: Characterization and Optimization." *Chemical Engineering and*

Processing - Process Intensification 154 (2020): 108024. https://doi.org/10.1016/j.cep.2020.108024.

Abukhadra, Mostafa R., and Mohamed Adel Sayed. "K+ Trapped Kaolinite (Kaol/K+) as Low Cost and Eco-Friendly Basic Heterogeneous Catalyst in the Transesterification of Commercial Waste Cooking Oil into Biodiesel." *Energy Conversion and Management* 177, no. July (2018): 468–76. https://doi.org/10.1016/j.enconman.2018.09.083.

Adepoju, T. F. "Optimization Processes of Biodiesel Production from Pig and Neem (*Azadirachta indica* a.Juss) Seeds Blend Oil Using Alternative Catalysts from Waste Biomass." *Industrial Crops and Products* 149 (2020): 112334. https://doi.org/10.1016/j.indcrop.2020.112334.

Ahmad, Shamshad, Shalini Chaudhary, Vinayak V. Pathak, Richa Kothari, and V. V. Tyagi. "Optimization of Direct Transesterification of Chlorella Pyrenoidosa Catalyzed by Waste Egg Shell Based Heterogenous Nano – CaO Catalyst." *Renewable Energy* 160 (2020): 86–97. https://doi.org/10.1016/j.renene.2020.06.010.

Aleman-Ramirez, J. L., Joel Moreira, S. Torres-Arellano, Adriana Longoria, Patrick U. Okoye, and P.J. Sebastian. "Preparation of a Heterogeneous Catalyst from Moringa Leaves as a Sustainable Precursor for Biodiesel Production." *Fuel* 284, no. August 2020 (January 2021): 118983. https://doi.org/10.1016/j.fuel.2020.118983.

Alsharifi, Mariam, Hussein Znad, Sufia Hena, and Ming Ang. "Biodiesel Production from Canola Oil Using Novel Li/TiO_2 as a Heterogeneous Catalyst Prepared via Impregnation Method." *Renewable Energy* (2017). https://doi.org/10.1016/j.renene.2017.07.117.

Anr, Reddy, A. A. Saleh, Md Saiful Islam, S. Hamdan, and Md Abdul Maleque. "Biodiesel Production from Crude Jatropha Oil Using a Highly Active Heterogeneous Nanocatalyst by Optimizing Transesterification Reaction Parameters." *Energy and Fuels* 30, no. 1 (2016): 334–43. https://doi.org/10.1021/acs.energyfuels.5b01899.

Araujo, Rayanne O., Jamal da Silva Chaar, Leandro Santos Queiroz, Geraldo Narciso da Rocha Filho, Carlos Emmerson Ferreira da Costa, Graziela C.T. da Silva, Richard Landers, Maria J.F. Costa, Alexandre A.S. Gonçalves, and Luiz K.C. de Souza. "Low Temperature Sulfonation of Acai Stone Biomass Derived Carbons as Acid Catalysts for Esterification Reactions." *Energy Conversion and Management* 196 (September 2019): 821–30. https://doi.org/10.1016/j.enconman.2019.06.059.

Araujo, Rayanne O., Vanuza O. Santos, Flaviana C.P. Ribeiro, Jamal da S. Chaar, Anderson M. Pereira, Newton P.S. Falcão, and Luiz K.C. de Souza. "Magnetic Acid Catalyst Produced from Acai Seeds and Red Mud for Biofuel Production." *Energy Conversion and Management* 228 (January 2021): 113636. https://doi.org/10.1016/j.enconman.2020.113636.

Ashok, A., T. Ratnaji, L. John Kennedy, J. Judith Vijaya, and R. Gnana Pragash. "Magnetically Recoverable Mg Substituted Zinc Ferrite Nanocatalyst for Biodiesel Production: Process Optimization, Kinetic and Thermodynamic Analysis." *Renewable Energy* 163 (2021): 480–94. https://doi.org/10.1016/j.renene.2020.08.081.

Asikin-Mijan, N., Lee, Hwei Voon, and Taufiq-Yap, Yun Hin. "Synthesis and Catalytic Activity of Hydration-Dehydration Treated Clamshell Derived CaO for Biodiesel Production." *Chemical Engineering Research and Design* (2015). https://doi.org/10.1016/j.cherd.2015.07.002.

Atadashi, Musa Idris, Aroua, Mohamed Khereddine, Raman, Abdul Aziz Abdul, Sulaiman, Nik Meriam. "The Effects of Catalysts in Biodiesel Production: A Review." *Journal of Industrial and Engineering Chemistry* 19, no. 1 (2013): 14–26. https://doi.org/10.1016/j.jiec.2012.07.009.

Athar, Moina, and Sadaf Zaidi. "A Review of the Feedstocks, Catalysts, and Intensification Techniques for Sustainable Biodiesel Production." *Journal of Environmental Chemical Engineering* (2020). https://doi.org/10.1016/j.jece.2020.104523.

Balajii, Muthusamy, and Subramaniapillai Niju. "Banana Peduncle – A Green and Renewable Heterogeneous Base Catalyst for Biodiesel Production from Ceiba Pentandra Oil." *Renewable Energy* (2019). https://doi.org/10.1016/j.renene.2019.08.062.

Barros, Suellen D.T. T, Aline V. Coelho, Elizabeth R. Lachter, Rosane A.S. S San Gil, Karim Dahmouche, Maria Isabel Pais da Silva, Andrea L.F. F Souza, et al. "Esterification of Lauric Acid with Butanol over Mesoporous Materials." *Renewable Energy* 50 (2013): 585–89. https://doi.org/10.1016/j.renene.2012.06.059.

Bet-Moushoul, Elsie, Khalil Farhadi, Yaghoub Mansourpanah, Ali Mohammad Nikbakht, Rahim Molaei, and Mehrdad Forough. "Application of CaO-Based/Au Nanoparticles as Heterogeneous Nanocatalysts in Biodiesel Production." *Fuel* (2016). https://doi.org/10.1016/j.fuel.2015.09.067.

Borah, Manash Jyoti, Ankur Das, Velentina Das, Nilutpal Bhuyan, and Dhanapati Deka. "Transesterification of Waste Cooking Oil for Biodiesel Production Catalyzed by Zn Substituted Waste Egg Shell Derived CaO Nanocatalyst." *Fuel* 242, no. May 2018 (2019): 345–54. https://doi.org/10.1016/j.fuel.2019.01.060.

Busacca, Carl A., Daniel R. Fandrick, Jinhua J. Song, and Chris H. Senanayake. "The Growing Impact of Catalysis in the Pharmaceutical Industry." *Advanced Synthesis and Catalysis* 353, no. 11–12 (2011): 1825–64. https://doi.org/10.1002/adsc.201100488.

Caliskan, Hakan. "Environmental and Enviroeconomic Researches on Diesel Engines with Diesel and Biodiesel Fuels." *Journal of Cleaner Production* (2017). https://doi.org/10.1016/j.jclepro.2017.03.168.

Changmai, Bishwajit, Putla Sudarsanam, and Lalthazuala Rokhum. "Biodiesel Production Using a Renewable Mesoporous Solid Catalyst." *Industrial Crops and Products* (2020). https://doi.org/10.1016/j.indcrop.2019.111911.

Chingakham, Ch., Asha David, and Vandana Sajith. "Fe_3O_4 Nanoparticles Impregnated Eggshell as a Novel Catalyst for Enhanced Biodiesel Production." *Chinese Journal of Chemical Engineering* (March 2019). https://doi.org/10.1016/j.cjche.2019.02.022.

Correia, Leandro Marques, Natália de Sousa Campelo, Raquel de Freitas Albuquerque, Célio Loureiro Cavalcante, Juan Antonio Cecilia, Enrique Rodríguez-Castellón, Eric Guibal, and Rodrigo Silveira Vieira. "Calcium/Chitosan Spheres as Catalyst for Biodiesel Production." *Polymer International* (2015). https://doi.org/10.1002/pi.4782.

Das, Velentina, Abhishek Mani Tripathi, Manash Jyoti Borah, Nurhan Turgut Dunford, and Dhanapati Deka. "Cobalt-Doped CaO Catalyst Synthesized and Applied for Algal Biodiesel Production." *Renewable Energy* 161 (2020): 1110–19. https://doi.org/10.1016/j.renene.2020.07.040.

Falowo, Olayomi A, Mustafa I. Oloko-Oba, and Eriola Betiku. "Biodiesel Production Intensification via Microwave Irradiation-Assisted Transesterification of Oil Blend Using Nanoparticles from Elephant-Ear Tree Pod Husk as a Base Heterogeneous Catalyst." *Chemical Engineering and Processing—Process Intensification* 140 (June 2019): 157–70. https://doi.org/10.1016/j.cep.2019.04.010.

Faraguna, Fabio, Marko Racar, Zoran Glasovac, and Ante Jukić. "Correlation Method for Conversion Determination of Biodiesel Obtained from Different Alcohols by 1H

NMR Spectroscopy." *Energy and Fuels* (2017). https://doi.org/10.1021/acs.energyfuels.6b02855.

Feyzi, Mostafa, Nahid Hosseini, Nakisa Yaghobi, and Rohollah Ezzati. "Preparation, Characterization, Kinetic and Thermodynamic Studies of MgO-La2O3 Nanocatalysts for Biodiesel Production from Sunflower Oil." *Chemical Physics Letters* 677 (2017): 19–29. https://doi.org/10.1016/j.cplett.2017.03.014.

Foroutan, Rauf, Reza Mohammadi, Hossein Esmaeili, Fatemeh Mirzaee Bektashi, and Sajad Tamjidi. "Transesterification of Waste Edible Oils to Biodiesel Using Calcium Oxide@magnesium Oxide Nanocatalyst." *Waste Management* 105 (March 2020): 373–83. https://doi.org/10.1016/j.wasman.2020.02.032.

Gebremariam, Shemelis Nigatu and Marchetti, Jorge Mario. "Economics of Biodiesel Production: Review" 168, no. May (2018): 74–84. https://doi.org/10.1016/j.enconman.2018.05.002.

Guoqing Guan, Katsuki Kusakabe, Nozomi Sakurai, Kimiko Moriyama. "Transesterification of Vegetable Oil to Biodiesel Fuel Using Acid Catalysts in the Presence of Dimethyl Ether." *Fuel* 88 (2009): 81–86. https://doi.org/10.1016/j.fuel.2008.07.021.

Gupta, Anilkumar R., and Virendra K. Rathod. "Waste Cooking Oil and Waste Chicken Eggshells Derived Solid Base Catalyst for the Biodiesel Production: Optimization and Kinetics." *Waste Management* 79 (September 2018): 169–78. https://doi.org/10.1016/j.wasman.2018.07.022.

Hadiyanto, Hady, Asha Herda Afianti, Ulul Ilma Navi'a, Nais Pinta Adetya, Widayat Widayat, and Heri Sutanto. "The Development of Heterogeneous Catalyst C/CaO/NaOH from Waste of Green Mussel Shell (Perna Varidis) for Biodiesel Synthesis." *Journal of Environmental Chemical Engineering* 5, no. 5 (October 2017): 4559–63. https://doi.org/10.1016/j.jece.2017.08.049.

Hazmi, Balkis, Umer Rashid, Yun Hin Taufiq-Yap, Mohd Lokman Ibrahim, and Imededdine Arbi Nehdi. "Supermagnetic Nano-Bifunctional Catalyst from Rice Husk: Synthesis, Characterization and Application for Conversion of Used Cooking Oil to Biodiesel." *Catalysts* 10, no. 2 (February 13, 2020): 225. https://doi.org/10.3390/catal10020225.

Jindapon, Wayu, Siyada Jaiyen, and Chawalit Ngamcharussrivichai. "Seashell-Derived Mixed Compounds of Ca, Zn and Al as Active and Stable Catalysts for the Transesterification of Palm Oil with Methanol to Biodiesel." *Energy Conversion and Management* 122 (2016): 535–43. https://doi.org/10.1016/j.enconman.2016.06.012.

Kant Bhatia, Shashi, Ravi Kant Bhatia, Jong Min Jeon, Arivalagan Pugazhendhi, Mukesh Kumar Awasthi, Dinesh Kumar, Gopalakrishnan Kumar, Jeong Jun Yoon, and Yung Hun Yang. "An Overview on Advancements in Biobased Transesterification Methods for Biodiesel Production: Oil Resources, Extraction, Biocatalysts, and Process Intensification Technologies." *Fuel* (2021). https://doi.org/10.1016/j.fuel.2020.119117.

Krishnamurthy, K.N., Somalapura Nagappa Sridhara, and Channapillekoppalu S. Ananda Kumar. "Optimization and Kinetic Study of Biodiesel Production from Hydnocarpus Wightiana Oil and Dairy Waste Scum Using Snail Shell CaO Nano Catalyst." *Renewable Energy* 146 (February 2020): 280–96. https://doi.org/10.1016/j.renene.2019.06.161.

Laca, Amanda, Adriana Laca, and Mario Díaz. "Eggshell Waste as Catalyst: A Review." *Journal of Environmental Management* 197 (2017): 351–59. https://doi.org/10.1016/j.jenvman.2017.03.088.

Lacerda, Orivaldo da Silva, Rodrigo Marinho Cavalcanti, Thaisa Moreira de Matos, Rômulo Simões Angélica, Geraldo Narciso da Rocha Filho, and Ivoneide de Carvalho Lopes Barros. "Esterification of Oleic Acid Using 12-Tungstophosphoric Supported in Flint Kaolin of the Amazonia." *Fuel* 108 (June 2013): 604–11. https://doi.org/10.1016/j.fuel.2013.01.008.

Laskar, Ikbal Bahar, Tuhin Deshmukhya, Piyali Bhanja, Bappi Paul, Rajat Gupta, and Sushovan Chatterjee. "Transesterification of Soybean Oil at Room Temperature Using Biowaste as Catalyst; an Experimental Investigation on the Effect of Co-Solvent on Biodiesel Yield." *Renewable Energy* 162 (2020): 98–111. https://doi.org/10.1016/j.renene.2020.08.011.

Leung, Dennis Yiu Cheong, Xuan Wu, and Michael Kwok Hi Leung. "A Review on Biodiesel Production Using Catalyzed Transesterification." *Applied Energy* 87, no. 4 (April 2010): 1083–95. https://doi.org/10.1016/j.apenergy.2009.10.006.

Liu, Hui, Hong shuang Guo, Xin jing Wang, Jian zhong Jiang, Hualin Lin, Sheng Han, and Su peng Pei. "Mixed and Ground KBr-Impregnated Calcined Snail Shell and Kaolin as Solid Base Catalysts for Biodiesel Production." *Renewable Energy* (2016). https://doi.org/10.1016/j.renene.2016.03.017.

Liu, Xiaoyan, Fenfen Zhu, Rongyan Zhang, Luyao Zhao, and Juanjuan Qi. "Recent Progress on Biodiesel Production from Municipal Sewage Sludge." *Renewable and Sustainable Energy Reviews* (2021). https://doi.org/10.1016/j.rser.2020.110260.

Machado, Yguatyara De Luna L., José Luís Cardozo Fonseca, Jackson Queiroz Malveira, Afonso Avelino Dantas Neto, and Tereza Neuma C. Dantas. "Study of Pressure and Temperature Influence on Rapeseed Biodiesel Oxidation Kinetics Using PetroOXY Method." *Fuel* (2020). https://doi.org/10.1016/j.fuel.2020.118771.

Madhuvilakku, Rajesh, and Shakkthivel Piraman. "Biodiesel Synthesis by TiO2-ZnO Mixed Oxide Nanocatalyst Catalyzed Palm Oil Transesterification Process." *Bioresource Technology* 150 (2013): 55–59. https://doi.org/10.1016/j.biortech.2013.09.087.

Manojkumar, Narasimhan, Chandrasekaran Muthukumaran, and Govindasamy Sharmila. "A Comprehensive Review on the Application of Response Surface Methodology for Optimization of Biodiesel Production Using Different Oil Sources." *Journal of King Saud University - Engineering Sciences* (2020). https://doi.org/10.1016/j.jksues.2020.09.012.

Mansir, Nasar, Siow Hwa Teo, Nurul-Asikin Mijan, and Taufiq-Yap Yun Hin. "Efficient Reaction for Biodiesel Manufacturing Using Bi-Functional Oxide Catalyst." *Catalysis Communications* 106201 (2020). https://doi.org/10.1016/j.catcom.2020.106201.

Martínez, Araceli, Gabriela E. Mijangos, Issis C. Romero-Ibarra, Raúl Hernández-Altamirano, and Violeta Y. Mena-Cervantes. "In-Situ Transesterification of Jatropha Curcas L. Seeds Using Homogeneous and Heterogeneous Basic Catalysts." *Fuel* 235, no. January 2018 (January 1, 2019): 277–87. https://doi.org/10.1016/j.fuel.2018.07.082.

Mendonça, Iasmin Maquiné, Flávia Lopes Machado, Cláudia Cândida Silva, Sérgio Duvoisin Junior, Mitsuo Lopes Takeno, Paulo José de Sousa Maia, Lizandro Manzato, Flávio A. de Freitas. "Application of Calcined Waste Cupuaçu (*Theobroma grandiflorum*)

Seeds as a Low-Cost Solid Catalyst in Soybean Oil Ethanolysis: Statistical Optimization." *Energy Conversion and Management* 200, no. September (November 2019): 112095. https://doi.org/10.1016/j.enconman.2019.112095.

Mendonça, Iasmin M., Orlando A.R.L. Paes, Paulo J.S. Maia, Mayane P. Souza, Richardson A. Almeida, Cláudia C. Silva, Sérgio Duvoisin, and Flávio A. de Freitas. "New Heterogeneous Catalyst for Biodiesel Production from Waste Tucumã Peels (*Astrocaryum aculeatum* Meyer): Parameters Optimization Study." *Renewable Energy* 130 (January 2019): 103–10. https://doi.org/10.1016/j.renene.2018.06.059.

Moshfegh, A. Z. "Nanoparticle Catalysts." *Journal of Physics D: Applied Physics* 42, no. 23 (2009). https://doi.org/10.1088/0022-3727/42/23/233001.

Moura, Carla Verônica Rodarte de, Adriano Gomes de Castro, Edmilson Miranda de Moura, Jose Ribeiro dos Santos, and Jose Machado Moita Neto. "Heterogeneous Catalysis of Babassu Oil Monitored by Thermogravimetric Analysis." *Energy & Fuels* 24, no. 12 (December 16, 2010): 6527–32. https://doi.org/10.1021/ef101228f.

Muhammad, Nawshad, Yasir A. Elsheikh, Muhammad Ibrahim Abdul Mutalib, Aqeel Ahmed Bazmi, Rahmat Ali Khan, Hidayatullah Khan, Sikander Rafiq, Zakaria Man, and Ihsnullah Khan. "An Overview of the Role of Ionic Liquids in Biodiesel Reactions." *Journal of Industrial and Engineering Chemistry* (2015). https://doi.org/10.1016/j.jiec.2014.01.046.

Nayebzadeh, Hamed, Mohammad Haghighi, Naser Saghatoleslami, Mohammad Tabasizadeh, and Sina Yousefi. "Fabrication of Carbonated Alumina Doped by Calcium Oxide via Microwave Combustion Method Used as Nanocatalyst in Biodiesel Production: Influence of Carbon Source Type." *Energy Conversion and Management* 171, no. February (2018): 566–75. https://doi.org/10.1016/j.enconman.2018.05.081.

Nisar, Jan, Rameez Razaq, Muhammad Farooq, Munawar Iqbal, Rafaqat Ali Khan, Murtaza Sayed, Afzal Shah, and Inayat ur Rahman. "Enhanced Biodiesel Production from Jatropha Oil Using Calcined Waste Animal Bones as Catalyst." *Renewable Energy* 101, no. August 2016 (2017): 111–19. https://doi.org/10.1016/j.renene.2016.08.048.

Nur Syazwani, Osman, Umer Rashid, and Yun Hin Taufiq Yap. "Low-Cost Solid Catalyst Derived from Waste *Cyrtopleura costata* (Angel Wing Shell) for Biodiesel Production Using Microalgae Oil." *Energy Conversion and Management* (2015). https://doi.org/10.1016/j.enconman.2015.05.075.

Oliveira, Alex De Nazaré De, Laura Rafaela Da Silva Costa, Luíza Helena De Oliveira Pires, Luís Adriano S. Do Nascimento, Rômulo S. Angélica, Carlos E.F. Da Costa, José R. Zamian, and Geraldo N. Da Rocha Filho. "Microwave-Assisted Preparation of a New Esterification Catalyst from Wasted Flint Kaolin." *Fuel* 103 (2013): 626–31. https://doi.org/10.1016/j.fuel.2012.07.017.

Pandit, Priti R., and Madhusudan Hiraman Fulekar. "Biodiesel Production from Scenedesmus Armatus Using Egg Shell Waste as Nanocatalyst." *Materials Today: Proceedings* 10 (2019): 75–86. https://doi.org/10.1016/j.matpr.2019.02.191.

Pandit, Priti R., and Madhusudan Hiraman Fulekar. "Biodiesel Production from Microalgal Biomass Using CaO Catalyst Synthesized from Natural Waste Material." *Renewable Energy* 136 (2019): 837–45. https://doi.org/10.1016/j.renene.2019.01.047.

Pessoa Junior, Wanison A.G., Mitsuo L Takeno, Francisco X Nobre, Silma De S Barros, Ingrity S.C. Sá, Edson P Silva, Lizandro Manzato, Stefan Iglauer, and Flávio A. de Freitas. "Application of Water Treatment Sludge as a Low-Cost and Eco-Friendly Catalyst in the Biodiesel Production via Fatty Acids Esterification: Process

Optimization." *Energy* 213 (December 2020): 118824. https://doi.org/10.1016/j.energy.2020.118824.

Piker, Alla, Betina Tabah, Nina Perkas, and Aharon Gedanken. "A Green and Low-Cost Room Temperature Biodiesel Production Method from Waste Oil Using Egg Shells as Catalyst." *Fuel* 182 (October 2016): 34–41. https://doi.org/10.1016/j.fuel.2016.05.078.

Qu, Tongxin, Shengli Niu, Xiangyu Zhang, Kuihua Han, and Chunmei Lu. "Preparation of Calcium Modified Zn-Ce/Al2O3 Heterogeneous Catalyst for Biodiesel Production through Transesterification of Palm Oil with Methanol Optimized by Response Surface Methodology." *Fuel* 284 (2021): 118986. https://doi.org/10.1016/j.fuel.2020.118986.

Quah, Ray Vern, Yie Hua Tan, Munjawar. Mubarak, Jibrail Kansedo, Mohammad Khalid, Ezzat Chan B. Abdullah, and Mohammad Omar Abdullah. "Magnetic Biochar Derived from Waste Palm Kernel Shell for Biodiesel Production via Sulfonation." *Waste Management* (2020). https://doi.org/10.1016/j.wasman.2020.09.016.

Ravi, Aiswarya, Baskar Gurunathan, Naveenkumar Rajendiran, Sunita Varjani, Edgard Gnansounou, Ashok Pandey, Simming You, Jegannathan Kenthorai Raman, and Praveenkumar Ramanujam. "Contemporary Approaches towards Augmentation of Distinctive Heterogeneous Catalyst for Sustainable Biodiesel Production." *Environmental Technology and Innovation* 19 (2020): 100906. https://doi.org/10.1016/j.eti.2020.100906.

Reis, Michele C., Flavio A. Freitas, Elizabeth R. Lachter, Rosane A. S. San Gil, Regina S. V. Nascimento, Rodrigo L. Poubel, and Leandro B. Borré. "Biodiesel Production From Fatty Acids of Refined Vegetable Oils by Heterogeneous Acid Catalysis and Microwave Irradiation." *Química Nova* 38, no. 10 (2015): 1307–12. https://doi.org/10.5935/0100-4042.20150163.

Rezania, Shahabaldin, Bahareh Oryani, Junboum Park, Beshare Hashemi, Krishna Kumar Yadav, Eilhann E. Kwon, Jin Hur, and Jinwoo Cho. "Review on Transesterification of Non-Edible Sources for Biodiesel Production with a Focus on Economic Aspects, Fuel Properties and by-Product Applications." *Energy Conversion and Management* 201, no. July (2019): 112155. https://doi.org/10.1016/j.enconman.2019.112155.

Roschat, Wuttichai, Mattana Kacha, Boonyawan Yoosuk, Taweesak Sudyoadsuk, and Vinich Promarak. "Biodiesel Production Based on Heterogeneous Process Catalyzed by Solid Waste Coral Fragment." *Fuel* (2012). https://doi.org/10.1016/j.fuel.2012.04.009.

Rudreshwar Balinge, Kamlesh, and J. Balakrishnan. "Mini Review on Recent Progress toward Sustainable Production of Biodiesel from Biomass." *Materials Today: Proceedings* (2020). https://doi.org/10.1016/j.matpr.2020.08.444.

de Sá Barros Silma, Wanison A.G. Pessoa Junior, Ingrity S.C. Sá, Mitsuo L. Takeno, Francisco X. Nobre, William Pinheiro, Lizandro Manzato, Stefan Iglauer, and Flávio A. de Freitas. "Pineapple (*Ananás comosus*) Leaves Ash as a Solid Base Catalyst for Biodiesel Synthesis." *Bioresource Technology* 312, no. May (September 2020): 123569. https://doi.org/10.1016/j.biortech.2020.123569.

Sahani, Shalini, and Yogesh Chandra Sharma. "Economically Viable Production of Biodiesel Using a Novel Heterogeneous Catalyst: Kinetic and Thermodynamic Investigations." *Energy Conversion and Management* 171 (2018): 969–83. https://doi.org/10.1016/j.enconman.2018.06.059.

Sakthi Vignesh, Nagamalai, Elamathi Vimali, Ramalingam Sangeetha, Muthu Arumugam, Balasubramaniem Ashokkumar, Innasimuthu Ganeshmoorthy, and Perumal Varalakshmi. "Sustainable Biofuel from Microalgae: Application of Lignocellulosic Wastes and Bio-Iron Nanoparticle for Biodiesel Production." *Fuel* 278 (2020): 118326. https://doi.org/10.1016/j.fuel.2020.118326.

Saravanan Arumugamurthy, Sakthi, Periyasamy Sivanandi, Sivakumar Pandian, Himanshu Choksi, and Deepalakshmi Subramanian. "Conversion of a Low Value Industrial Waste into Biodiesel Using a Catalyst Derived from Brewery Waste: An Activation and Deactivation Kinetic Study." *Waste Management*, 2019. https://doi.org/10.1016/j.wasman.2019.09.030.

Seffati, Kambiz, Bizhan Honarvar, Hossein Esmaeili, and Nadia Esfandiari. "Enhanced Biodiesel Production from Chicken Fat Using CaO/CuFe 2 O 4 Nanocatalyst and Its Combination with Diesel to Improve Fuel Properties." *Fuel* 235, no. September 2018 (2019): 1238–44. https://doi.org/10.1016/j.fuel.2018.08.118.

Seffati, Kambiz, Hossein Esmaeili, Bizhan Honarvar, and Nadia Esfandiari. "AC/CuFe2O4@CaO as a Novel Nanocatalyst to Produce Biodiesel from Chicken Fat." *Renewable Energy* 147 (March 2020): 25–34. https://doi.org/10.1016/j.renene.2019.08.105.

Shan, Rui, Lili Lu, Yueyue Shi, Haoran Yuan, and Jiafu Shi. "Catalysts from Renewable Resources for Biodiesel Production." *Energy Conversion and Management* 178, no. October (2018): 277–89. https://doi.org/10.1016/j.enconman.2018.10.032.

Sharma, Yogesh Chandra, and Veena Singh. "Microalgal Biodiesel: A Possible Solution for India's Energy Security." *Renewable and Sustainable Energy Reviews* (2017). https://doi.org/10.1016/j.rser.2016.08.031.

Singh, Bhupinder. "Rice Husk Ash." In *Waste and Supplementary Cementitious Materials in Concrete: Characterisation, Properties and Applications*, edited by Rafat Siddique and Paulo Cachim, 417–60. Elsevier, (2018). https://doi.org/10.1016/B978-0-08-102156-9.00013-4.

Singh, Digambar, Dilip Sharma, Shyamlal Soni, Chandrapal Singh Inda, Sumit Sharma, Pushpendra Kumar Sharma, and Amit Jhalani. "A Comprehensive Review of Physicochemical Properties, Production Process, Performance and Emissions Characteristics of 2nd Generation Biodiesel Feedstock: Jatropha Curcas." *Fuel* 285 (2021). https://doi.org/10.1016/j.fuel.2020.119110.

Soltani, Soroush, Umer Rashid, Saud Ibrahim Al-Resayes, and Imededdine Arbi Nehdi. "Recent Progress in Synthesis and Surface Functionalization of Mesoporous Acidic Heterogeneous Catalysts for Esterification of Free Fatty Acid Feedstocks: A Review." *Energy Conversion and Management* 141 (2017): 183–205. https://doi.org/10.1016/j.enconman.2016.07.042.

Sudarsanam, Putla, Elise Peeters, Ekaterina V. Makshina, Vasile I. Parvulescu, and Bert F. Sels. "Advances in Porous and Nanoscale Catalysts for Viable Biomass Conversion." *Chemical Society Reviews* 48 (2019): 2366–2421. https://doi.org/10.1039/c8cs00452h.

Suresh, T., Natesan Sivarajasekar, and K. Balasubramani. "Enhanced Ultrasonic Assisted Biodiesel Production from Meat Industry Waste (Pig Tallow) Using Green Copper Oxide Nanocatalyst: Comparison of Response Surface and Neural Network Modelling." *Renewable Energy* 164 (2021): 897–907. https://doi.org/10.1016/j.renene.2020.09.112.

Takeno, Mitsuo L., Iasmin M. Mendonça, Silma de S. Barros, Paulo J. de Sousa Maia, Wanison A. G. Pessoa Jr., Mayane P. Souza, Elzalina R. Soares, et al. "A Novel

CaO-Based Catalyst Obtained from Silver Croaker (*Plagioscion squamosissimus*) Stone for Biodiesel Synthesis: Waste Valorization and Process Optimization." *Renewable Energy* 172 (July 1, 2021): 1035–45. https://doi.org/10.1016/j.renene.2021.03.093.

Tang, Xincheng, and Shengli Niu. "Preparation of Carbon-Based Solid Acid with Large Surface Area to Catalyze Esterification for Biodiesel Production." *Journal of Industrial and Engineering Chemistry*, 2019. https://doi.org/10.1016/j.jiec.2018.09.016.

Teo, Siow Hwa, Aminul Islam, Hamid Reza Fard Masoumi, Yun Hin Taufiq-Yap, Jidon Janaun, Eng Seng Chan, and M. A. khaleque. "Effective Synthesis of Biodiesel from Jatropha Curcas Oil Using Betaine Assisted Nanoparticle Heterogeneous Catalyst from Eggshell of Gallus Domesticus." *Renewable Energy* 111 (2017): 892–905. https://doi.org/10.1016/j.renene.2017.04.039.

Thiele, Ernest W. "Relation between Catalytic Activity and Size of Particle." *Industrial and Engineering Chemistry* 31, no. 7 (1939): 916–20. https://doi.org/10.1021/ie50355a027.

Tripathi, Nimisha, Colin D. Hills, Raj S. Singh, and Christopher J. Atkinson. "Biomass Waste Utilisation in Low-Carbon Products: Harnessing a Major Potential Resource." *Npj Climate and Atmospheric Science* 2, no. 1 (December 2019): 35. https://doi.org/10.1038/s41612-019-0093-5.

Vadery, Vinu, Binitha N. Narayanan, Resmi M. Ramakrishnan, Sudha Kochiyil Cherikkallinmel, Sankaran Sugunan, Divya P. Narayanan, and Sreenikesh Sasidharan. "Room Temperature Production of Jatropha Biodiesel over Coconut Husk Ash." *Energy*, 2014. https://doi.org/10.1016/j.energy.2014.04.045.

Wang, Jiayan, Lingmei Yang, Wen Luo, Gaixiu Yang, Changlin Miao, Junying Fu, Shiyou Xing, Pei Fan, Pengmei Lv, and Zhongming Wang. "Sustainable Biodiesel Production via Transesterification by Using Recyclable Ca2MgSi2O7 Catalyst." *Fuel* 196 (2017): 306–13. https://doi.org/10.1016/j.fuel.2017.02.007.

Yuryev, Ruslan, and Andreas Liese. "Biocatalysis: The Outcast." *ChemCatChem* 2, no. 1 (2010): 103–7. https://doi.org/10.1002/cctc.200900126.

Zhou, Yan, Shengli Niu, and Jing Li. "Activity of the Carbon-Based Heterogeneous Acid Catalyst Derived from Bamboo in Esterification of Oleic Acid with Ethanol." *Energy Conversion and Management*, 2016. https://doi.org/10.1016/j.enconman.2016.02.027.

8

Two-Dimensional (2D) Layered Materials as Emerging Nanocatalysts in the Production of Biodiesel

Inbaoli A., Sujith Kumar C.S.,* and Jayaraj S.

CONTENTS

8.1 Introduction

The successful energy transition towards a carbon-neutral source and sustainability is driving the progress of novel materials for catalysis and energy [1–4]. In this regard, excitons, free carriers, plasmons, polarons and other transient properties of 2D or layered materials are areas of particular interest. Concerning the energy transition, catalysis is one of the eminent sectors for devising new solutions for sustainability and a carbon-neutral environment [5–7]. Catalysis is employed in more than one process

* *Corresponding Author* sujithcs@nitc.ac.in

DOI: 10.1201/9781003120858-8

step in the majority of industrial processes such as refineries and chemical production [8]. In addition, many pollution abatement technologies depend on catalysis for their efficacy. Limiting ourselves to biodiesel application, this chapter describes the advancements in catalytic activities in biodiesel production due to 2D layered material supported nanoparticles.

Homogeneous and heterogeneous catalysis are the conventional methods for producing biodiesel. Of these two, heterogeneous catalysts have been predominantly adopted in biodiesel production, especially in the transesterification process, due to their recovery and reusability. However, the issue with these solid catalysts is the reduction of the overall catalytic activity due to the need for a smaller surface area and greater quantity [9]. The latest advancements in nanotechnology have demonstrated that nanocatalysts are a crucial candidate in reducing the reaction time by improving the catalytic activity and also the biodiesel quality and yield. Recent research has shown that incorporating a nanocatalyst can also decrease the alcohol–oil ratio, reaction temperatures and weight of the catalyst.

Further enhancement can also be achieved by tailoring the properties such as the number of active sites, pH (acidic or basic) and other nanomaterial surface factors along with the porosity. Nevertheless, these properties are an attribute of the morphology of NPs. Therefore, researchers are analysing the possibility of manipulating the morphology of nanocatalysts through various synthesis techniques. Certain morphologies like roughened mesoporous structures on nanocrystalline, nanosponges and nanocubes facilitate enhanced active sites. These features increase the overall surface area and reactivity, favouring a high biodiesel yield [10]. Pushing the limits of catalytic activity of nanoparticles is paramount, and this is gaining immense research interest, with researchers exploring the influence of 2D materials as support for nanoparticles. Generally, it has been presumed that to enhance the catalytic activity of a nanohybrid, the chemical interactions and charge transfer between 2D layered materials (supports) and NPs have to be modulated, and one way to do this is by alteration of the NPs' surface properties, such as the active sites [11]. Furthermore, stronger interactions impact the homogeneous distribution of metal NPs as well as the stability of the catalysts [12].

8.1.1 Classification of Layered Materials

Figure 8.1 shows the most common method of classification of layered materials. The first category is the primary forms that spontaneously exist in the earthly environment, such as silicates and clays, or others developed by the chemical process. It should be noted that primary forms are also referred to as layered precursors to distinguish them from chemically synthesized materials. The primary composition class category is based on layer charge, insulator layered materials and layered redox materials. Further layer charge categories of compositional classes include neutral, positive or negative. In addition, other types of grouping are based on the layer thickness:

(i) Class I – mono-layer thickness of physical limit,
(ii) Class II – one atom thick,
(iii) Class III – few atoms thick (layers).

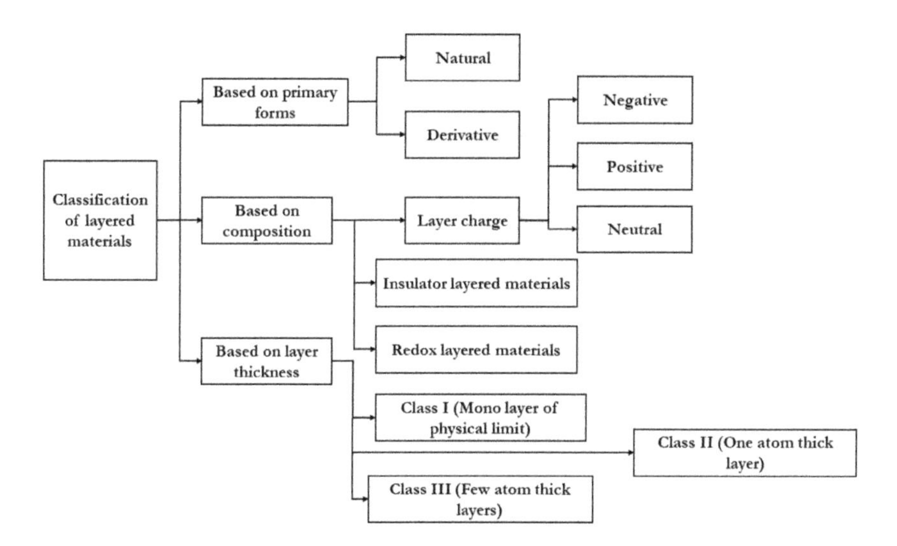

FIGURE 8.1 Classification of layered materials.

Each of the layered compositional classes has its unique practical benefits [13]. The compositional line flourished as researchers continued to exploit and discover various layered materials. Nevertheless, there have been experiments on the preparation of mixed structures containing multiple layers, such as negative oxides and positive layered double hydroxides [14].

8.1.2 Two-Dimensional Layered Materials (2DLMs)

Materials with an aspect ratio greater than 100 belong to the category called 2D materials. These are often called thin films owing to their single or few-layered nanosheets. In general, 2D layered materials exhibit diverse and unprecedented properties compared with their bulk counterpart. In order to develop novel materials with tuneable active sites and high surface areas, the delamination of layered materials with relatively mild interlayer forces is widely preferred [15–17]. These nanomaterials have unique properties compared to bulk materials and can be used to improve catalytic properties (electrocatalytic and photocatalytic). Such advanced properties pave the way for the synthesis of prolific and active catalysts, such as CO_2 conversion into value-added particulates and energy fluids [4]. Researchers have manipulated 2D nanosheets to develop various nanostructures, opening a window of exciting and new possibilities using the self-assembly prospects. This method of delaminating layered materials is known as exfoliation, and it found to increase the catalytic activity and stability. As a result, there has been an unprecedented research surge in the regime of layered materials in various sectors including electronics, gas sensors, nanophononics, optoelectronics, etc. [18–20]. Recent advances in nano-scale fabrication techniques pave the way for tailoring 2D nanostructures to deliver the desired catalytic activity for fuel synthesis and energy production. Recently, Tan et al. [21] reviewed the different varieties of 2D nanomaterials concerning energy applications. In this chapter, the catalytic activity of graphitic carbon nitride (g-C_3N_4) and hexagonal boron nitride (h-BN) is elaborated on. Some other 2D materials are:

- Metal phosphorus trichalcogenides
- Transition metal dichalcogenides (TMDs)
- Layered double hydroxides (LDHs)
- Silicates and hydroxides (clays)
- Covalent–organic frameworks (COFs)
- Metal halides
- Metal oxides and 2D zeolites
- Transition metal oxyhalides
- Perovskites and niobates
- Metal–organic frameworks (MOFs)

8.2 Synthesis of 2D Layered Materials

Currently, the laboratory methods for the synthesis of 2D layered materials include mechanical exfoliation, ultrasonication exfoliation (liquid-phase exfoliation), physical/chemical vapour deposition (PVD/CVD), chemical vapour transport (CVT) and wet chemistry. To extend the limits of 2D layered material properties, researchers have employed series technologies to adjust the morphology and energy band structure of 2D layered materials to change their physical, chemical and surface properties, such as active sites, layer thickness and number, developing hybrid heterojunction, intercalation, alloying and doping, etc.

8.2.1 Mechanical Exfoliation

In the exfoliation technique, the top-down approach is widely used to synthesize graphene from graphite. The mechanical exfoliation process includes peeling the graphene layer by layer from the solid graphite. As these adjacent graphene layers are stabilized due to an intermolecular force of attraction (van der Waals), researchers have focused on how the intermolecular attraction force can be overcome to obtain graphene. There are two different mechanical forces in mechanical exfoliation, i.e., normal and lateral forces imposed on graphite to exfoliate graphene flakes. When graphite layers are peeled apart, the normal force is imposed to overcome the intermolecular van der Waals attraction of adjacent layers. One way to do that is by using Scotch tape to create micromechanical cleavage. Later, force is employed to accelerate the relative motion adjacent to graphite layers. Lateral force takes advantage of graphite's self-lubricating ability and makes it easier to exfoliate. A schematic illustration of two routes of mechanical exfoliation is shown in Figure 8.2. It is worth mentioning that for all the exfoliation processes reported up to now, the aforementioned mechanical routes are essential for graphene production. By tailoring these two mechanical paths, it is expected that regulated graphitic exfoliation can produce

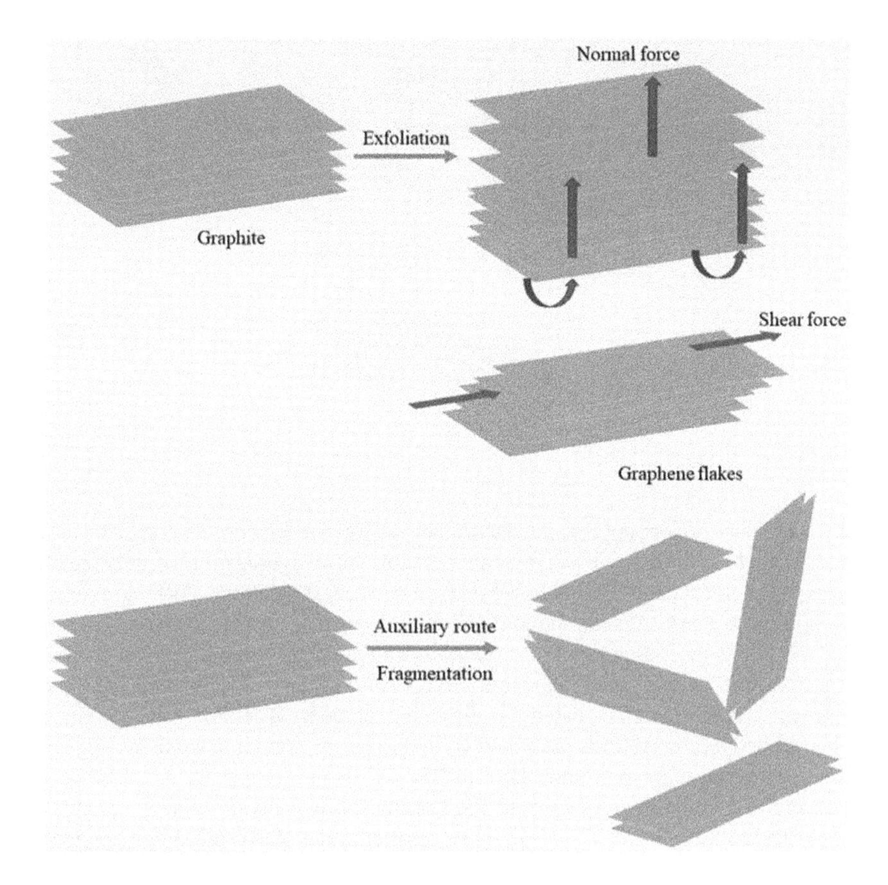

FIGURE 8.2 Mechanical exfoliation and fragmentation effect.

high-efficacy good-quality graphene. Another exfoliation method is the fragmentation effect during layer separation, as illustrated in Figure 8.2. The term fragmentation effect indicates that the forces involved in the exfoliation method break the bulk graphite particles or graphene into tiny ones. Its significance comes from the low magnitude of van der Waals force among adjacent layers in smaller graphite flakes, making them weak and quicker to exfoliate than larger ones. On the other hand, this auxiliary exfoliation has its own drawbacks, such as the reduced lateral size of graphene, which prevents the development of graphene with a large surface area [22].

8.2.2 Ultrasonic Exfoliation

One of the efficient strategies to develop single- or few-layer atomic thick sheets is ultrasonic exfoliation. This method overcomes the shortcomings of mechanical exfoliation in terms of productivity and efficacy. An illustration of ultrasonic exfoliation is shown in Figure 8.3. As illustrated in Figure 8.3, time and appropriate solvents in

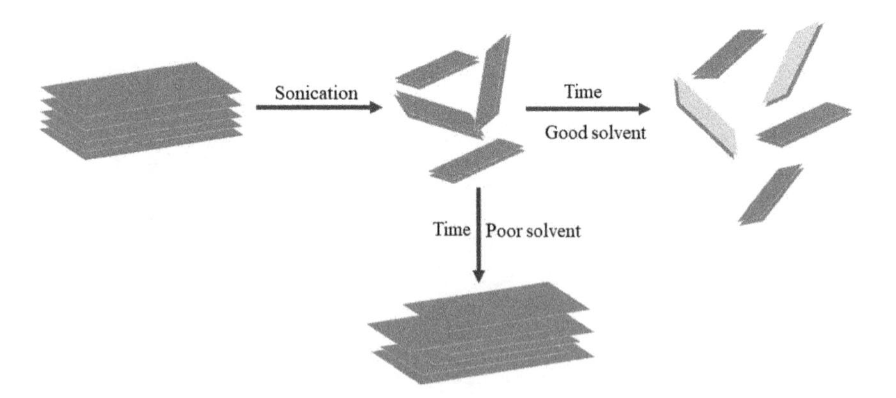

FIGURE 8.3 Ultrasonication.

sonication play paramount roles in delamination. Solvent selection is crucial because the undesirable solvent will cause sedimentation and reaggregation of delaminated layers. The solvents are selected based on their surface energies and ability to stabilize the exfoliated sheets against agglomeration and coagulation. Recently, researchers have exploited the possibility of various organic solvents as a dispersing medium for exfoliation of van der Waals solids [23–25]. Although this method has many advantages, high-purity 2D material is difficult to come by. The ultrasonic exfoliation of black phosphorous (BP) is presented with the view of providing some insights into the sonication strategy to produce layered BP.

Recently, using an organic solvent, Xie's group successfully developed pristine 2D BP via ultrasonic exfoliation [24]. Initially, distilled water of 100 mL is agitated in the presence of noble gases like argon to prevent oxidation. After that, 50 mg of bulk BP is mixed homogeneously in the water. Then, the mixture solution is sonicated for 8 hours in ice-cold water, keeping the bath temperature low. The ultrasonicated solution is subjected to centrifugation for 10 min at 1500 rpm. This process step is employed to remove the unexfoliated BP, and finally, the supernatant is collected for further use. The developed layered BP is a potential material that can replace graphene in electronic applications.

8.2.3 Chemical Vapour Transport

The chemical vapour transport (CVT) technique includes a reversible chemical reaction to grow crystals at different temperatures. By regulating the temperature and growth direction, CVT facilitates the early occurrence of intercalation. For example, Mn intercalated 2H-TaS_2 through the CVT method is summarized here. Firstly, the researchers, Li et al.[26], heated a stoichiometric mixture of Mn and TaS_2 in a vacuum quartz tube. This resulted in the formation of an intercalated polycrystalline compound Mn_xTaS_2. Next, they took 1 g of Mn_xTaS_2 compound and added 100 mg of iodine. After that, the mixture was sealed tightly and subjected to vacuum heating for 10 days. Finally, they obtained high-quality navy-blue mirror plate layered materials.

8.2.4 Wet Chemical Strategy

Due to its high productivity, comparatively low cost and the potential for mass production, the wet chemical technique is a viable strategy to synthesize all variants of 2D materials. Many 2D materials have been developed by this method, e.g., Rh, ZnO, MoS_2, MnO_2, TiO_2 and Co_3O_4 [27–30]. This technique employs sophisticated chemical routes such as hydro/solvothermal and template synthesis. In this section, a discussion on hydro/solvothermal synthesis for developing 2D nanomaterials is provided. The hydro/solvothermal method is a prevalent technique regularly used for preparing inorganic materials. Low temperature (generally in the range of 100–240 °C) and ease of changing reaction conditions are two advantages of the hydro/solvothermal approach over other forms of crystal growth. Some important factors such as temperature, reaction time and reactant ratio significantly influence the synthesis of the ultrathin nanostructure.

Similarly, using the solvothermal method, Li et al. [30] prepared ultrathin Rh nanosheets. They started the synthesis process with the preparation of solvent by mixing benzyl alcohol and formaldehyde and then dissolving $Rh(acac)_3$ and PVP in the solvent. After dissolution, the mixture is vigorously stirred for 1 h, followed by transferring it to an autoclave. The autoclave is sealed and held at 180 °C for 8 hours before being cooled to room temperature. The obtained black product is vacuum dried before being allowed to precipitate with acetone (10 mL), parted in a centrifuge and rinsed with ethanol (10 mL).

8.3 Role of 2D Layered Materials as Catalyst Support

NPs with catalytic support are crucial materials and indispensable in heterogeneous catalysis. Generally, as the nanocatalytic supporting material, mesoporous materials such as carbon, Al_2O_3, SiO_2 and zeolite, etc., have been used. Among these materials, carbon materials like CNTs, activated carbon, carbon black, and graphene, etc., exhibit immense potential as emergent support owing to their larger surface area, chemical stability at varying pH ranges and frequent attainability of the active sites. With its peculiar electronic properties and structure, graphene is considered the top contender of the carbon family. Therefore, it has attracted global attention as an outstanding support to nanocatalysts with boosted catalytic activity for a variety of transformations. As a result, a wide range of highly effective and finer metal NPs based on graphene have been successfully showcased as proficient catalysts for different electrochemical and chemical reactions [31]. After the establishment of graphene as a successful catalyst support, research around the world focused on developing other 2D materials and exploiting their effect on catalysis. The thriving characteristics of 2D materials have paved a new route for researchers to develop durable and efficient catalysts.

8.4 Catalytic Activity of Nanoparticles Supported with 2D Layered Materials

8.4.1 Graphene

Graphene has a peculiar 2D structure of sp2 bonded carbon atoms tightly arranged in a honeycomb crystal lattice that is just one atom thick. The distinctive architectural and electronic properties of graphene fulfil the fundamental demands of a perfect carrier for nanocatalysts. Increased charge transfer from catalysts to the graphene substrate improve the catalytic activity of graphene-supported materials. Zhang and colleagues successfully immobilized RhNi NPs on graphene oxide surfaces and suggested that apart from acting as a support, graphene oxide serves as a dissolving agent for NPs in the aqueous solution. The ability of graphene oxide to anchor small metal NPs is indicated by the prevalence of hydrophilic phenyl epoxide, the hydrophobic basal plane and hydroxyl groups on its surface [32]. RhNi NPs synthesized on graphene oxide have a smaller size (5.0 nm) and are better distributed on graphene surfaces than RhNi NPs, which were developed void of graphene oxide support. Obtained RhNi/GO demonstrated prolific catalytic activity for hydrocarbon decomposition. N and S dual-doped graphene loaded with Pd NPs have recently been identified as multifunctional catalyst NPs [33]. According to the authors, the outstandingly high efficacy of the Pd/NSG catalyst for methanol oxidation reaction (MOR) and formic acid oxidation reaction (FAOR) over Pd/C and Pd/G with higher anodic peak current densities and exceptional stability is due to the specific structural features and robust synergistic effect between N,S-doped graphene Pd electrocatalyst.

The attachment of PEI has greatly influenced the dispersion and size of metal NPs to the GO surface. Furthermore, the use of electron-rich functional groups on GO surfaces is a valuable tool to customize the electrical properties of GO and affect the catalytic efficacy of nanomaterials. Li et al. [34] used the co-reduction method to make NiFe NPs assisted by polyethyleneimine-decorated rGO (NiFe/PEI-rGO). The electrochemical results showed that a $Ni_{80}Fe_{20}$/PEI-rGO hybrid had higher catalytic activity against the hydrazine oxidation reaction than $Ni_{80}Fe_{20}$ NPs directly deposited on GO. Song et al. [35] also investigated the catalytic activity of ultrafine Pd NPs immobilized on diamine-alkalized reduced graphene oxide (Pd/PDA-rGO) and through catalytic dehydrogenation of formic acid facilitated hydrogen production.

8.4.2 Graphitic Carbon Nitride (g-C_3N_4)

Two-dimensional g-C_3N_4 is widely used as a catalyst support because graphitic carbon nitride is the most stable allotrope under atmospheric circumstances. . Interestingly the influence of g-C_3N_4-supported noble metals such as Au, Pd and Pt demonstrated profound catalytic activity and recovery in heterogeneous catalysis [36,37]. Recently, many research groups have prepared highly effective, reliable and durable g-C_3N_4-supported catalysts with enhanced kinetics for a variety of chemical reactions. g-C_3N_4-supported Pt NPs was developed by Shiraishi et al.[38] using reduction methods such as high-temperature hydrogen reduction (Pt/g-C_3N_4-HR) (473–873 K) and room temperature photo-reduction (Pt/g-C_3N_4-PR). Their photocatalytic studies revealed that Pt/g-C_3N_4-PR catalyst showed high activity with a ~10-fold enhancement over that of Pt/g-C_3N_4-HR catalyst. This research finding is important because both catalysts developed through two reduction methods have similar sizes but exhibited significant variation in catalytic activity.

In addition, g-C_3N_4-supported nanomaterials played a significant part as a catalyst for considered hydrogenation of a carbon–oxygen bonding pair (C=O bonds) and a carbon double bond (C=C). Homogeneously distributed g-C_3N_4-supported Pt NPs has been developed and established as a better and recyclable catalyst for furfural hydrogenation of selective furfuryl alcohol with specific selectivity[39]. Furthermore, Guo et al.[40] prepared g-C_3N_4-supported AuCo NPs (AuCo@g-C_3N_4) displaying remarkably high photocatalytic activity for AB hydrolysis and witnessed the higher catalytic activity of AuCo@g-C_3N_4.

8.4.3 Hexagonal Boron Nitride (h-BN)

Hexagonal boron nitride (h-BN) is a layered crystalline form with the same quantity of B and N atoms packed in a hexagonal lattice. Within each layer, B and N are bound by strong covalent interactions. This covalent bond among B and N atoms contributes to the high chemical, mechanical and thermal stability of hexagonal-BN (h-BN). Although h-BN has structural similarity to graphite with almost identical cell factors, it varies drastically in its electrical properties. BNNs supported by active metal NPs have been used as efficient catalysts in various chemical reactions.

In contrast, they interacted with one another by van der Waals forces. The other study, without using any surfactant or dispersing agent, used thionyl chloride ($SOCl_2$) to exfoliate commercial h-BN powder to prepare ultrathin BNNS was experimented by Sun et al. [41], who reported a high yield at ~20 wt% of h-BN powder. In addition, by using the deposition-precipitation method, immobilization of Pd NPs on the BNNS surface is achieved, which results in high catalytic activity and recyclability of Pd/BNNS catalyst towards nitro aromatics hydrogenation.

In another study, Huang and coworkers [42] developed BNNS-supported Ag, Au and Pt NPs and analysed catalytic activity for reducing 4-nitrophenol. They began with the synthesis of the layered BNNS, with direct exfoliation. This involves the ultrasonication of ethylene glycol with dispersed hexagonal-BN to facilitate uniform dispersion and, after that, immobilization of NPs. Analysis of catalytic activity showed that upon loading 12 wt% Ag in h-BN/Ag the catalytic activity is enhanced with a complete reduction of 4-nitrophenol within the span of 9 minutes. BNNS-supported Ag NPs were reported by Sen et al. [43] using a facile and eco-friendly approach deprived of further reducing agent in an aqueous solution maintained at room temperature. The obtained NPs demonstrated good catalytic activity in the reduction reaction of 4-nitrophenol. In another study, the encapsulation of Ni NPs with layered h-BN shells exhibited enhanced methanation activity with higher resistance to NP sintering and suppressed carbon deposition [44]. The data obtained from chemisorption analysis and surface science investigation established the prevalence of methanation reactions on Ni surfaces under h-BN cover. The developed NPs showed stability and enhanced catalytic activity due to the confinement effect of h-BN shells.

8.5 Prospects and Future Research Directions

Two-dimensional layered materials have revolutionized the way researchers approach the functional catalyst and catalysis process. However, the significance of 2D layered materials as multifunctional catalysts is yet to be explored. Moreover, complete knowledge in the field of 2D layered oxides will open the door to advanced application areas. Some of the potential future research works include the following:

1. The development of economically feasible highly active and functionalized heterogeneous catalysts for industrial applications.
2. Vivid analysis on the production of enzymatic biodiesel to ensure a potential option for industrial sectors.
3. Developing novel materials for catalysts supports high active sites and surface areas with tuneable pore sizes for the application of intercalation.
4. Minimizing catalyst production cost, expanding their sustainability and commercialization by exploring alternative sources such as biomass.
5. Improving 2D layered catalyst preparation routes and treatment processes for improving their efficacy to meet the industrial standards.
6. Breaching the conventional barriers such as high energy consumption, production time and overall cost by exploring the possibility of recyclability and reusability of nanocatalysts.

8.6 Conclusion

Due to its versatile structures and exciting chemical and physical properties, 2D materials are fascinating nanomaterials in the academic research area and industrial science. 2D materials display superior physical and structural properties, which are more versatile than those of the bulk material. Such disparity offers new opportunities to exploit 2D materials as catalyst supports for different chemical reactions. We briefly summarized the synthesis of 2D layered materials and their role as support for catalysts. By deliberately tuning the morphology and structural properties, the catalytic activity nanocatalyst has been enhanced. Amongst the vast number of materials, we focused on graphene, a fascinating 2D material that has captivated the research interest since its invention. This chapter discussed the influence of graphene in heterogeneous catalysis. Its distinct characteristics such as high chemical stability and surface area, excellent electrical properties and ultrahigh mechanical properties are testimony to the claim that it is a viable and suitable material for nanocatalyst support. This chapter also provides details about the catalytic behaviour of NPs supported with 2D layered materials, especially graphene. The catalytic activity of hexagonal boron nitride and nanocatalyst support has also been discussed. 2D materials have captivated the field with their elegant properties; as a result, 2D materials like graphene are an exciting topic of research in various academic areas including materials science, and chemical and physical sciences, with the interest in them continually increasing.

REFERENCES

1. Paola Lanzafame et al., "Beyond Solar Fuels: Renewable Energy-Driven Chemistry," *ChemSusChem* 10, no. 22 (2017): 4409–19, https://doi.org/10.1002/cssc.201701507.
2. Robert Schlögl, "E-Mobility and the Energy Transition," *Angewandte Chemie—International Edition* 56, no. 37 (2017): 11019–22, https://doi.org/10.1002/anie.201701633.

3. Matthias Beller, Gabriele Centi, and Licheng Sun, "Chemistry Future: Priorities and Opportunities from the Sustainability Perspective," *ChemSusChem* 10, no. 1 (2017): 6–13, https://doi.org/10.1002/cssc.201601739.
4. Damien Voiry et al., "Low-Dimensional Catalysts for Hydrogen Evolution and CO2 Reduction," *Nature Reviews Chemistry* 2, no. 1 (2018), https://doi.org/10.1038/s41570-017-0105.
5. Robert Schlögl, "Heterogeneous Catalysis," *Angewandte Chemie - International Edition* 54, no. 11 (2015): 3465–3520, https://doi.org/10.1002/anie.201410738.
6. John Meurig Thomas and Kenneth D.M. Harris, "Some of Tomorrow's Catalysts for Processing Renewable and Non-Renewable Feedstocks, Diminishing Anthropogenic Carbon Dioxide and Increasing the Production of Energy," *Energy and Environmental Science* 9, no. 3 (2016): 687–708, https://doi.org/10.1039/c5ee03461b.
7. Gabriele Centi and Siglinda Perathoner, "Catalysis: Role and Challenges for a Sustainable Energy," *Topics in Catalysis* 52, no. 8 (2009): 948–61, https://doi.org/10.1007/s11244-009-9245-x.
8. Fabrizio Cavani, Gabriele Centi, Siglinda Perathoner, Ferruccio Trifirò, *Sustainable Industrial Processes*, Wiley-VCH, Weinheim, Germany (2009).
9. Manash Jyoti Borah et al., "Transesterification of Waste Cooking Oil for Biodiesel Production Catalyzed by Zn Substituted Waste Egg Shell Derived CaO Nanocatalyst," *Fuel* 242, no. January (2019): 345–54, https://doi.org/10.1016/j.fuel.2019.01.060.
10. Elyssa G. Fawaz et al., "Study on the Catalytic Performance of Different Crystal Morphologies of HZSM-5 Zeolites for the Production of Biodiesel: A Strategy to Increase Catalyst Effectiveness," *Catalysis Science and Technology* 9, no. 19 (2019): 5456–71, https://doi.org/10.1039/c9cy01427f.
11. Gerhard Ertl, Helmut Knözinger, Jens Weitkamp, in *Handbook of Heterogeneous Catalysis* (2008). Weinheim and Wiley-VCH, New York
12. Miao Zhou et al., "Strain-Enhanced Stabilization and Catalytic Activity of Metal Nanoclusters on Graphene," *Journal of Physical Chemistry C* 114, no. 39 (2010): 16541–46, https://doi.org/10.1021/jp105368j.
13. Wieslaw J. Roth et al., "Layer like Porous Materials with Hierarchical Structure," *Chemical Society Reviews* 45, no. 12 (2016): 3400–3438, https://doi.org/10.1039/c5cs00508f.
14. Liang Li et al., "Layer-by-Layer Assembly and Spontaneous Flocculation of Oppositely Charged Oxide and Hydroxide Nanosheets into Inorganic Sandwich Layered Materials," *Journal of the American Chemical Society* 129, no. 25 (2007): 8000–8007, https://doi.org/10.1021/ja0719172.
15. Eleonora Conterosito et al., "On the Rehydration of Organic Layered Double Hydroxides to Form Low-Ordered Carbon/LDH Nanocomposites," *Inorganics* 6, no. 3 (2018): 1–16, https://doi.org/10.3390/inorganics6030079.
16. Xiao Zhang, Hongfei Cheng, and Hua Zhang, "Recent Progress in the Preparation, Assembly, Transformation, and Applications of Layer-Structured Nanodisks beyond Graphene," *Advanced Materials* 29, no. 35 (2017), https://doi.org/10.1002/adma.201701704.
17. Bálint Náfrádi, Mohammad Choucair, and László Forró, "Electron Spin Dynamics of Two-Dimensional Layered Materials," *Advanced Functional Materials* 27, no. 19 (2017), https://doi.org/10.1002/adfm.201604040.
18. Fangxu Yang et al., "2D Organic Materials for Optoelectronic Applications," *Advanced Materials* 30, no. 2 (2018), https://doi.org/10.1002/adma.201702415.

19. Shan Chen and Gaoquan Shi, "Two-Dimensional Materials for Halide Perovskite-Based Optoelectronic Devices," *Advanced Materials* 29, no. 24 (2017), https://doi.org/10.1002/adma.201605448.
20. Kan Sheng Chen et al., "Emerging Opportunities for Two-Dimensional Materials in Lithium-Ion Batteries," *ACS Energy Letters* 2, no. 9 (2017): 2026–34, https://doi.org/10.1021/acsenergylett.7b00476.
21. Chaoliang Tan et al., "Recent Advances in Ultrathin Two-Dimensional Nanomaterials," *Chemical Reviews* 117, no. 9 (2017): 6225–6331, https://doi.org/10.1021/acs.chemrev.6b00558.
22. Min Yi and Zhigang Shen, "A Review on Mechanical Exfoliation for the Scalable Production of Graphene," *Journal of Materials Chemistry A* 3, no. 22 (2015): 11700–715, https://doi.org/10.1039/c5ta00252d.
23. Jonathan N Coleman et al., "Produced by Liquid Exfoliation of Layered Materials" 331, no. February (2011): 568–72.
24. Hui Wang et al., "Ultrathin Black Phosphorus Nanosheets for Efficient Singlet Oxygen Generation," *Journal of the American Chemical Society* 137, no. 35 (2015): 11376–82, https://doi.org/10.1021/jacs.5b06025.
25. Joohoon Kang et al., "Solvent Exfoliation of Electronic-Grade, Two-Dimensional Black Phosphorus," *ACS Nano* 9, no. 4 (2015): 3596–3604, https://doi.org/10.1021/acsnano.5b01143.
26. Lun Jun Li et al., "Influence of the Low Mn Intercalation on Magnetic and Electronic Properties of 2H-TaS2 Single Crystals," *Journal of Magnetism and Magnetic Materials* 323, no. 21 (2011): 2536–41, https://doi.org/10.1016/j.jmmm.2011.04.002.
27. Yu Chen et al., "Ultrasmall Fe3O4 Nanoparticle/MoS2 Nanosheet Composites with Superior Performances for Lithium Ion Batteries," *Small* 10, no. 8 (2014): 1536–43, https://doi.org/10.1002/smll.201302879.
28. Chao Hu et al., "Large-Scale, Ultrathin and (001) Facet Exposed TiO2 Nanosheet Superstructures and Their Applications in Photocatalysis," *Journal of Materials Chemistry A* 2, no. 7 (2014): 2040–43, https://doi.org/10.1039/c3ta14343k.
29. Ziqi Sun et al., "Generalized Self-Assembly of Scalable Two-Dimensional Transition Metal Oxide Nanosheets," *Nature Communications* 5, no. May (2014): 1–9, https://doi.org/10.1038/ncomms4813.
30. Haohong Duan et al., "Ultrathin Rhodium Nanosheets," *Nature Communications* 5 (2014): 1–8, https://doi.org/10.1038/ncomms4093.
31. Xiaobin Fan, Guoliang Zhang, and Fengbao Zhang, "Multiple Roles of Graphene in Heterogeneous Catalysis," *Chemical Society Reviews* 44, no. 10 (2015): 3023–35, https://doi.org/10.1039/c5cs00094g.
32. Jun Wang et al., "Rhodium-Nickel Nanoparticles Grown on Graphene as Highly Efficient Catalyst for Complete Decomposition of Hydrous Hydrazine at Room Temperature for Chemical Hydrogen Storage," *Energy and Environmental Science* 5, no. 5 (2012): 6885–88, https://doi.org/10.1039/c2ee03344e.
33. Xin Zhang et al., "Palladium Nanoparticles Supported on Nitrogen and Sulfur Dual-Doped Graphene as Highly Active Electrocatalysts for Formic Acid and Methanol Oxidation," *ACS Applied Materials and Interfaces* 8, no. 17 (2016): 10858–65, https://doi.org/10.1021/acsami.6b01580.
34. Jing Li et al., "Polyethyleneimine Decorated Graphene Oxide-Supported Ni 1-XFex Bimetallic Nanoparticles as Efficient and Robust Electrocatalysts for Hydrazine

Fuel Cells," *Catalysis Science and Technology* 3, no. 12 (2013): 3155–62, https://doi.org/10.1039/c3cy00487b.

35. Fu Zhan Song et al., "Diamine-Alkalized Reduced Graphene Oxide: Immobilization of Sub-2 Nm Palladium Nanoparticles and Optimization of Catalytic Activity for Dehydrogenation of Formic Acid," *ACS Catalysis* 5, no. 9 (2015): 5141–44, https://doi.org/10.1021/acscatal.5b01411.
36. Yong Wang et al., "Highly Selective Hydrogenation of Phenol and Derivatives over a Pd@carbon Nitride Catalyst in Aqueous Media," *Journal of the American Chemical Society* 133, no. 8 (2011): 2362–65, https://doi.org/10.1021/ja109856y.
37. Dongshun Deng et al., "Palladium Nanoparticles Supported on Mpg-C3N4 as Active Catalyst for Semihydrogenation of Phenylacetylene under Mild Conditions," *Green Chemistry* 15, no. 9 (2013): 2525–31, https://doi.org/10.1039/c3gc40779a.
38. Yasuhiro Shiraishi et al., "Platinum Nanoparticles Strongly Associated with Graphitic Carbon Nitride as Efficient Co-Catalysts for Photocatalytic Hydrogen Evolution under Visible Light," *Chemical Communications* 50, no. 96 (2014): 15255–58, https://doi.org/10.1039/c4cc06960a.
39. Xiufang Chen et al., "Highly Selective Hydrogenation of Furfural to Furfuryl Alcohol over Pt Nanoparticles Supported on G-C3N4 Nanosheets Catalysts in Water," *Scientific Reports* 6, no. April (2016): 1–13, https://doi.org/10.1038/srep28558.
40. Lin-tong Guo et al., "Lin-Tong Guo, Yi-Yu Cai, Jie-Min Ge, Ya-Nan Zhang, Ling-Hong Gong, Xin-Hao Li, * Kai-Xue Wang, Qi-Zhi Ren, * Juan Su, and Jie-Sheng Chen," 2015.
41. Wenliang Sun et al., "High-Yield Production of Boron Nitride Nanosheets and Its Uses as a Catalyst Support for Hydrogenation of Nitroaromatics," *ACS Applied Materials and Interfaces* 8, no. 15 (2016): 9881–88, https://doi.org/10.1021/acsami.6b01008.
42. Caijin Huang et al., "Stable Colloidal Boron Nitride Nanosheet Dispersion and Its Potential Application in Catalysis," *Journal of Materials Chemistry A* 1, no. 39 (2013): 12192–97, https://doi.org/10.1039/c3ta12231j.
43. Heng Shen et al., "Facile in Situ Synthesis of Silver Nanoparticles on Boron Nitride Nanosheets with Enhanced Catalytic Performance," *Journal of Materials Chemistry A* 3, no. 32 (2015): 16663–69, https://doi.org/10.1039/c5ta04188k.
44. Lijun Gao et al., "Enhanced Nickel-Catalyzed Methanation Confined under Hexagonal Boron Nitride Shells," *ACS Catalysis* 6, no. 10 (2016): 6814–22, https://doi.org/10.1021/acscatal.6b02188.

9

Size-Dependent Catalytic Properties of Nanomaterials, Their Suitability in Terms of Efficiency, Cost-Effectiveness and Sustainability

Sunita Singh and Jitamanyu Chakrabarty*

CONTENTS

9.1 Introduction: Nanomaterials and Their Alluring Features

The advent of nano-based materials has hugely contributed in various fields of science. One of the remarkable applications of nanomaterials is catalysis science [1,2]. Nanoparticles (NPs) are particles with a size in the range of 1–100 nanometres, i.e. 10^{-9} metres [3]. The large surface area of NPs directly influences the reaction rate and catalytic activity [4,5]. Also, the fine-tuning of nanocatalysts with regard to their composition (type of core–shell, bimetallic or use of supporting material), size and shape has provided greater selectivity [6,7]. Previous studies have established that tailored nanostructures have the potential to meet the stringent requirements of catalysis, disabling the shortcomings of homogeneous and heterogeneous catalysts [8–10]. Figure 9.1 presents the basic differences in bulk catalysis and catalysis shown by nanoscale materials.

* *Corresponding Author* jitamanyu.chakrabarty@ch.nitdgp.ac.in

DOI: 10.1201/9781003120858-9

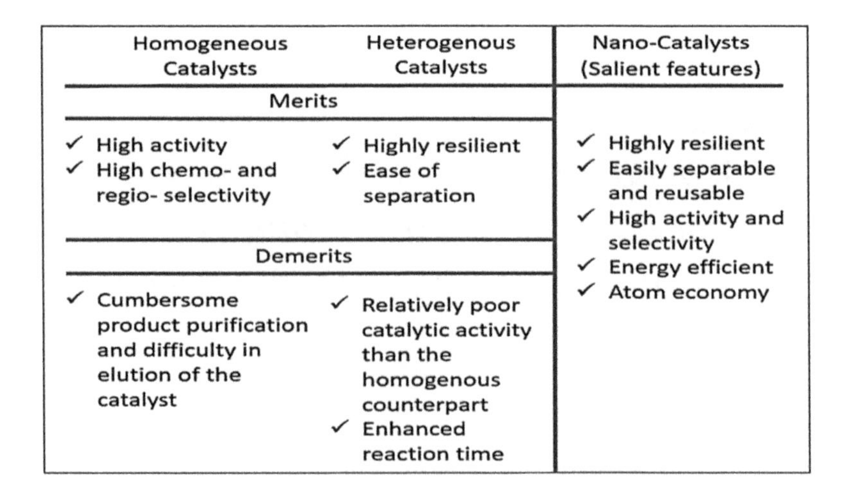

	Homogeneous Catalysts	Heterogenous Catalysts	Nano-Catalysts (Salient features)
Merits	✓ High activity ✓ High chemo- and regio- selectivity	✓ Highly resilient ✓ Ease of separation	✓ Highly resilient ✓ Easily separable and reusable ✓ High activity and selectivity ✓ Energy efficient ✓ Atom economy
Demerits	✓ Cumbersome product purification and difficulty in elution of the catalyst	✓ Relatively poor catalytic activity than the homogenous counterpart ✓ Enhanced reaction time	

FIGURE 9.1 Basic difference in bulk catalysis and catalysis shown by nanoscale materials.

9.2 Synthesis and Characterization of Nanomaterials

The synthesis of stabilized NPs may be done by various process that are categorized under two major headings as: (i) top-down technologies and (ii) bottom-up technologies. Figure 9.2 presents the scheme of nanomaterial synthesis, and also shows the synthetic routes for nanomaterials.

Nanomaterials are characterized using various sophisticated instruments to characterize and analyse the actual shape, size, surface structure, chemical composition, valency, electron band gap, light emission and absorption, light scattering, bonding environment and diffraction properties [11]. A list of the characterization techniques of nanomaterials with their utilities has been summarized in Table 9.1.

9.3 Nanomaterials: Surface Chemistry and Catalytic Activity

NPs are recognized as a prominent industrial catalyst having multiple applications ranging from manufacturing of chemicals to the conversion and storage of energy. The heterogeneity of NPs and their individual variance in shape and size result in their flexible nature and particle-specific activity as a catalyst. Several materials and elements such as silica, clays, iron, aluminium and TiO_2 have been used as nanocatalysts for several years by many researchers [12–18]. It has already been established that the properties of NPs that vary with their shape and structure affect the catalytic performance of the material [19–23]. The surface chemistry of the nanocatalyst is also hugely affected by the chemical and structural parameters of the nanomaterial [22,24,25]. Therefore, further investigation needs to be carried out in order to explore in what ways the catalytic performance of an NP is affected by the physical properties and also for the influence of the fabrication parameters affecting those physical properties

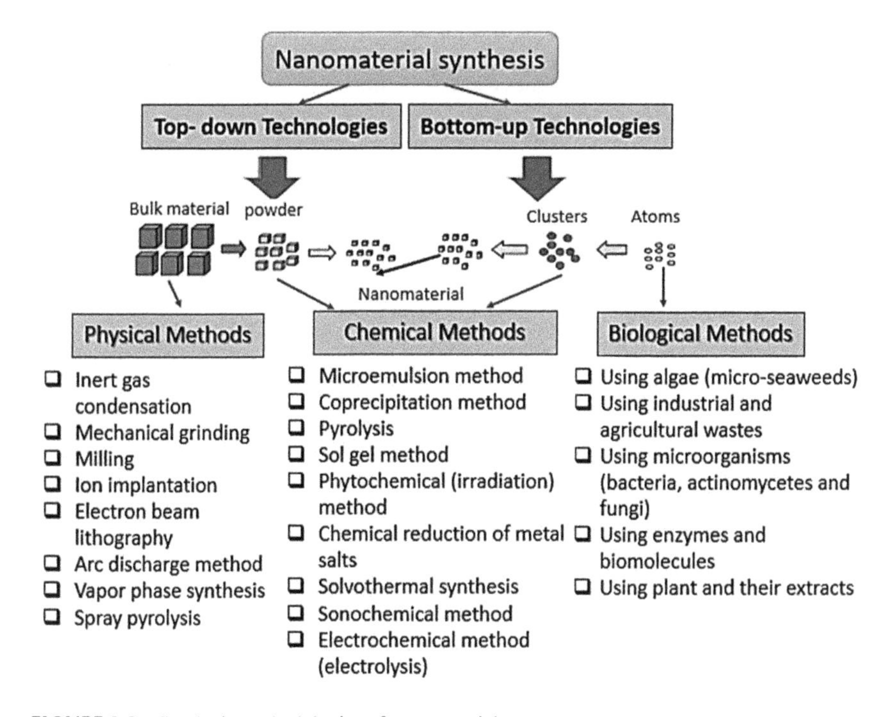

FIGURE 9.2 Synthetic methodologies of nanomaterials.

TABLE 9.1

List of the Characterization Techniques and Their Utility

S. No.	Techniques	Utility
I	**Optical (imaging) probe characterization techniques**	
1	Confocal laser-scanning microscopy (CLSM)	Imaging/ultrafine morphology
2	Scanning near-field optical microscopy (SNOM)	Rastered images
3	Two-photon fluorescence microscopy (2PFM)	Fluorophores/biological systems
4	Dynamic light scattering (DLS)	Particle sizing
5	Brewster angle microscopy (BAM)	Gas-liquid interface imaging
II	**Electron probe characterization techniques**	
1	Scanning electron microscopy (SEM)	Imaging/topology/morphology
2	Electron probe microanalysis (EPM)	Particle size/local chemical analysis
3	Transmission electron microscopy (TEM)	Imaging/particle size and shape
4	High-resolution transmission electron microscopy (HRTEM)	Imaging structure and chemical analysis
5	Low-energy electron diffraction (LEED)	Surface/adsorbate bonding
6	Electron energy-loss spectroscopy (EELS)	Inelastic electron interaction

(continued)

TABLE 9.1 (Continued)

List of the Characterization Techniques and Their Utility

S. No.	Techniques	Utility
7	Auger electron spectroscopy (AES)	Chemical surface analysis
III	**Scanning probe characterization techniques**	
1	Atomic force microscopy (AFM)	Imaging/topology/surface structure
2	Chemical force microscopy (CFM)	Chemical/surface analysis
3	Magnetic force microscopy (MFM)	Magnetic material analysis
4	Scanning tunnelling microscopy (STM)	Imaging/topology/surface
5	Atomic probe microscopy (APM)	Three-dimensional imaging
6	Field ion microscopy (FIM)	Chemical profiles/atomic spacing
7	Atomic probe tomography (APT)	Position-sensitive lateral location of atoms
IV	**Photon (spectroscopic) probe characterization techniques**	
1	Ultraviolet photoemission spectroscopy (UPS)	Surface analysis
2	UV-visible spectroscopy (UVVS)	Chemical analysis
3	Atomic absorption spectroscopy (AAS)	Chemical analysis
4	Inductively coupled plasma spectroscopy (ICPS)	Elemental analysis
5	Fluorescence spectroscopy (FS)	Elemental analysis
6	Localized surface plasmon resonance (LSPR)	Nano-sized particle analysis
V	**Ion-particle probe characterization techniques**	
1	Rutherford back scattering (RBS)	Quantitative–qualitative elemental analysis
2	Small angle neutron scattering (SANS)	Surface characterization
3	Nuclear reaction analysis (NRA)	Depth profiling of solid thin film
4	Raman spectroscopy (RS)	Vibration analysis
5	X-ray diffraction (XRD)	Crystal structure
6	Energy dispersive X-ray spectroscopy (EDX)	Elemental analysis
7	Small angle X-ray scattering (SAXS)	Surface analysis/particle sizing (1–100 nm)
8	Cathodoluminescence (CLS)	Characteristic emission
9	Nuclear magnetic resonance spectroscopy (NMR)	Analysis of odd number of nuclear species
VI	**Thermodynamic characterization techniques**	
1	Thermal gravimetric analysis (TGA)	Mass loss vs. temperature
2	Differential thermal analysis (DTA)	Reaction heat capacity
3	Differential scanning calorimetry (DSC)	Reaction heat phase changes
4	Nano calorimetry (NC)	Latent heats of fusion
5	Brunauer-Emmette-Teller method (BET)	Surface area analysis
6	Sears method (Sears)	Colloid size, specific surface area

to be assessed. This comprehension will further aid in the rational design of better performing catalysts which would be highly active and selective, and have high resilience. This will enable the chemical reactions occurring in industries to become more resource-efficient, consuming less energy and resulting in less waste; thus, helping to counter the environmental impact.

9.3.1 Size-Dependent Catalytic Properties of Nanomaterials

Various types of nanocatalysts have already been synthesized and employed for catalytic applications such as magnetic, nano mixed metal oxides, core–shell, nano-supported catalysts, graphene-based nanocatalysts, etc. [26–32]. The nanocatalysts have shown superior performance and a wider range of applicability than most of the prevailing catalysts in the concerned fields and have continued to play a significant role in various petrochemical processes including energy conversion, oil refining, chemical transformations, as well as environmental remediation [33–35]. It has been observed that the fundamental insights into the interrelationship between the structure and the chemical factors have been a crucial parameter in designing a catalyst with better activity, selectivity and durability [36–39].

Previously, few studies have been carried out to determine the impact of the size or shape of the catalyst on the catalytic performance [38–42]. Figure 9.3 shows the correlation of the size and shape of the particle on the catalytic surface and the catalytic performance.

Researchers have studied the size-dependent catalytic activity and selectivity of a series of metal NPs with the same composition and shape. Guo et al. (2018) analysed the hydrolytic dehydrogenation reaction of ammonia borane using monodispersed nickel NPs as catalyst and studied the size-dependent activity of the catalyst [43]. Zhou et al. (2017) analysed the size-dependent activity and dynamics of NPs of gold at the single molecular level [44], Suchomel et al. (2018) synthesized size-controlled Au NPs and studied the variance in catalytic activity depending on size [45]. Latha et al.

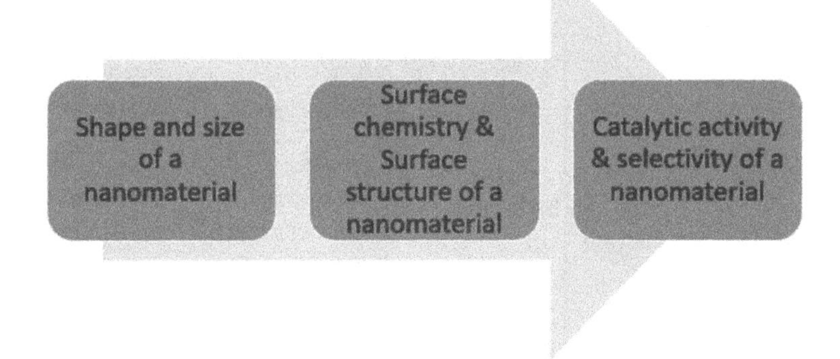

FIGURE 9.3 Correlation of the size and shape of the particle on the catalytic surface and the catalytic performance.

(2017) assessed the catalytic properties of gold NPs influenced by variations in their size mediated by the leaf extract of *Justicia adhatoda* [46]. Li et al. (2020) studied the catalytic performance of Pd NPs with carbon support in the dehydrogenation reaction of formic acid and scrutinized the difference in catalytic activity based on the variable size of the NPs [47] and Guisbiers et al. (2011) studied the size-dependent melting and catalytic properties of Pt-Pd NPs [48].

9.3.2 Size-Dependent Electronic and Structural Parameters of the Surface of a Metal Catalyst

The reaction in a heterogeneous catalysis occurs at the surface or the interface between the solid nanocatalyst and the liquid/gaseous reactant [49]. Nanocatalysts have shown impressively high activity in various chemical and biological reactions, which is mainly attributed to the much higher surface area-to-volume ratio of the NPs [50]. The fraction of atoms present at the topmost surface layer and at the corners and edges of the nanoparticle depends on the atomic size of the nanoparticle and it also contributes to the specific surface area of the catalyst [22]. With a decrease in the size of the atoms in the coordinated sites, the electronic state, crystallographic shapes, packing and density of a metal nanoparticle may vary or even transform from a metallic state to a molecular state [22]. Figure 9.4 presents the effect of the intrinsic properties of nanomaterials on their catalytic activity.

For example, the oxidation of CO at room temperature (RT) occurs on the under co-ordinated Pt atoms of the restructured Pt (557) [22,51]. The restructured terrace of Pt (557) caused by relatively higher CO pressure gives Pt NPs of 2.2 nm size [22,52]. These Pt atoms with the coordination number (CN) 7 are highly under coordinated at the edge of the triangular nanoclusters; but actively participate in CO oxidation even at RT [22, 52]. Meanwhile, these Pt atoms on the terrace of Pt (111) with CN=9 are inactive for CO oxidation at RT [22]. Therefore, it becomes evident that the catalytic performance is clearly determined and dependent on the co-ordination number of the chemical environment [22, 51].

Similarly, researchers have suggested that the band structure of a metal nanoparticle of 1–2 nm could also show a molecular state instead of exhibiting a metallic state [52].

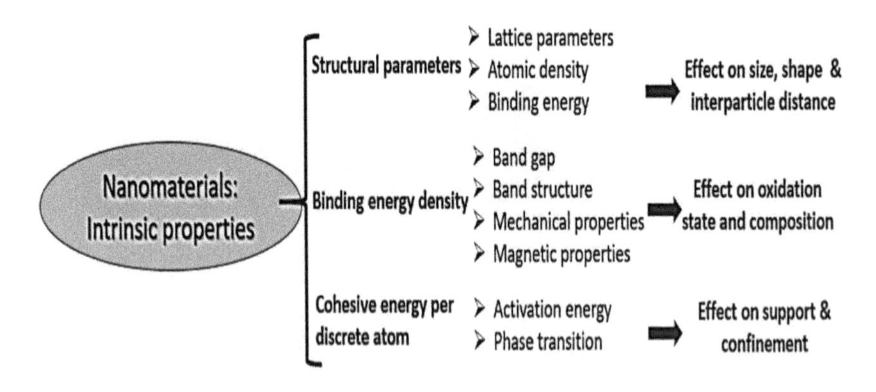

FIGURE 9.4 Effects of intrinsic properties of nanomaterials on their catalytic activity.

For instance, according to the view point of an electronic state, an Au nanoparticle which is smaller than 1 nm behaves like a molecule rather than displaying a metallic state. The molecular electronic state of a smaller metal nanoparticle (1–2 nm) intrinsically exhibits different catalytic activity than the larger sized nanoparticle [22, 52]. This case of the electronic factor was experimentally demonstrated for the first time by Goodman and co-workers when they carried out the oxidation of CO on nanoclusters of Au with three times the atomic layer thickness which are supported on titanium dioxide (TiO_2) [53]. By analysing Au L_{III} XANES, the experiment concluded that the white lines of the supported NPs of Au that measure 3 nm have an average co-ordination number of 9.5, 1 nm has an average co-ordination number of 6 and 0.5–1 nm has an average co-ordination number of 3.6, which demonstrates that the Au nanoparticles with a smaller size possess higher electron density and a narrower band located closer to the fermi level. This clearly illustrates that the electronic environment is size-dependent [53].

9.3.3 Size-Dependent Adsorption and Activation Energy

The adsorptions of the reactant molecule, dissociated species, the intermediate and the product molecule are the crucial steps of the catalytic event at the molecular level [22,53]. Studies have suggested that the adsorption energy of the adsorbates on the metal nanoparticle surface is size-dependent [54, 55]. Further computational studies also have concluded that the activation energy needed for chemisorption of the adsorbates on the Pt nanocatalyst decreases with size [56]. The size dependency of the activation energy results from the cohesive energy, which also depends on size and this in turn is correlated to the fraction of under co-ordinated atoms present on the surface of the nanocatalyst [22,56].

The size of the atoms significantly influences the performance of the catalyst and has a complex impact over the catalytic performance [22]. It has been found that a lower co-ordination numbered metal atom exhibits high activity if an intramolecular chemical bond of the reactant species dissociates and is the slowest step, i.e. the rate-determining step of the catalytic reaction [22,57,58]. This is because lower co-ordination numbered metal atoms display stronger adsorption energies for the reactant by decreasing the barrier of activation energy, thus easing the dissociation of reactant and hence accelerating the catalytic event. Therefore, for a reaction where the rate-determining step of the reaction is the dissociation step, the catalytic activity would be controlled by the size of the atoms.

Contrary to the trend of increased catalytic efficiency with a decrease in size, there are some reactions where the catalytic activity decreases with a decreasing size of the nanoparticle. This happens when lower co-ordination numbered metal atoms strongly bind with the dissociated species after readily dissociating a chemical bond (like the O–O bond of an O_2 molecule) [22]. This strong binding of the dissociated species, like an oxygen atom, slows down the coupling of the molecule with another species in the catalytic cycle or sometimes poisons the catalytic surface to some extent. Thus, by modulating various features of the individual nanocomponents like their shape, size, surface composition, spatial distribution along with the chemical and thermal stability, the nanomaterial can be modified to possess newer properties and hence, enhanced catalytic performance.

The adsorption of small adsorbates containing hydrogen and oxygen on the Pt nanocatalyst with varying co-ordination numbers is illustrated here. Extensive study has been done on the adsorption of these species on Pt_n (where, n = 201, 769, 28) having sizes in the range of 0.7–1.7 nm. The adsorption energy of the adsorbates, including *O, $*O_2$, *OOH, *OH, $*H_2O_2$ and $*H_2O$, decreases linearly with the increase in the co-ordination number from 3 to 9. Similarly, some transition metal catalysts such as Co, Ni, Cu, Rh, Pd, Ag, Ir and Au exhibit a linear correlation of the adsorption energy with that of the co-ordination number [10]. Another example is the oxidation of CO catalysed via Au NPs. In this case, on increasing the CN of Au atoms (Au–Au), stronger binding of the Co molecule and O atom over Au atoms was observed. Au atoms of an (111) surface with CN=9 display enhanced binding as compared to the Au nanoparticles with CN=4. According to the calculations, the adsorption energies of CO on an Au atom for CN values of 9, 7, 6, 5 and 4 are 0.12 eV, –0.23 eV, –0.48 eV, –0.56 eV and –0.83 eV, respectively. Adsorption energies of O on an Au atom are –0.23 eV, –0.24 eV, –0.29 eV, –0.27 eV and –0.59 eV with CN values of 9, 7, 6, 5 and 4, respectively [59].

9.4 Nanomaterials: Suitability in Terms of Efficiency, Cost-Effectiveness and Sustainability

The introduction and application of nanoscience in various fields has led to intricate insights for a better understanding and results of the processes. The synthesis and use of nanocatalysts, specifically in chemical and biological sciences, have provided more opportunities and improved methods of achieving the desired results. The salient features and wide range of applications of nanocatalysts, like better performance and selectivity, easy recovery and reusability, have been recognized as superior key features over the other catalysts employed in the concerned fields.

Nanostructured catalysts hold numerous potential advantages such as upgraded catalytic efficiency along with the reusability feature, because of which they have attracted the attention of substantial industrial and academic research in recent times. The high catalytic efficiency of nanomaterials is imparted due to their nanoscale size range, shape and surface area-to-volume ratio. This exceptionally large ratio of surface area to volume, which is caused by electronic and structural changes on the surface, makes them different to the bulk materials [60]. Metal NPs are also used as supports in various reactions and they have become the centre of economic and scientific interest because of their vital applications in several oil-refining processes, chemical manufacturing techniques and environmental catalysis such as selective oxidation and hydrogenation reactions under mild conditions [61]. It has been discovered that the conversion of bulk materials into their nano-size components improves the functionalities of the materials [62]. Therefore, on the nano-scale conversion of conventional catalysts, notable strengthening of their catalytic performance occurs.

The following are some examples of catalysis performed using nano-range metals.

Bi and Lu (2008) studied the production of H2 from formaldehyde at RT using nanoscale copper as catalyst, assessed the catalytic activity and also reported the plausible pathway of the reaction for the conversion of reactant to the desired product (shown in Scheme 9.1) [63].

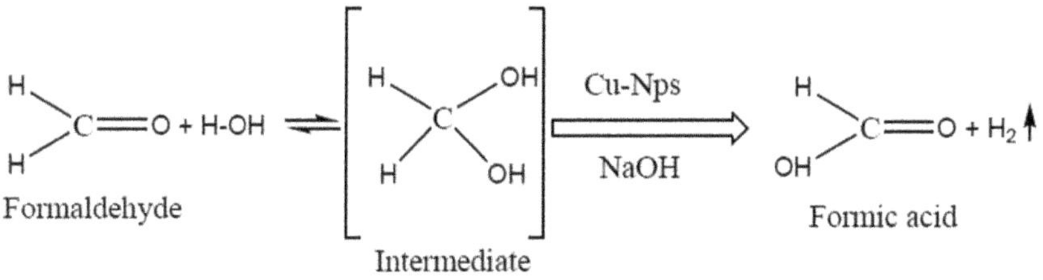

$$\underset{\text{Formaldehyde}}{H_2C{=}O} + H\text{-}OH \rightleftharpoons \underset{\text{Intermediate}}{\left[H_2C(OH)_2\right]} \xrightarrow[\text{NaOH}]{\text{Cu-Nps}} \underset{\text{Formic acid}}{H(OH)C{=}O} + H_2\uparrow$$

SCHEME 9.1 Production of hydrogen by using nanoscale copper as catalyst.

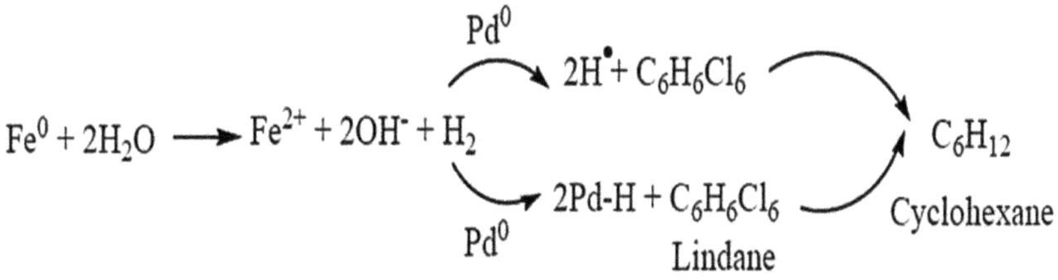

$$Fe^0 + 2H_2O \rightarrow Fe^{2+} + 2OH^- + H_2$$
$$H_2 \xrightarrow{Pd^0} 2H^\bullet + C_6H_6Cl_6 \rightarrow C_6H_{12}$$
$$H_2 \xrightarrow{Pd^0} 2Pd\text{-}H + \underset{\text{Lindane}}{C_6H_6Cl_6} \rightarrow \underset{\text{Cyclohexane}}{C_6H_{12}}$$

SCHEME 9.2 De-chlorination of lindane using nanoscale Fe(0)-Pd(0) bimetallic nanocatalyst.

Nagpal et al. (2010) reported the reductive dechlorination of lindane (γ-hexachlorocyclohexane) in aqueous medium catalysed by nanoscale Fe-Pd bimetallic particles. The presence of Pd on the surface of nanoscale iron particles increased the catalytic activity of the Fe-Pd bimetallic system to enhance the dechlorination of lindane. Dechlorination occurs via adsorption of chlorinated compounds (lindane) on the particle surface. Palladium on the surface acts as a collector of hydrogen gas that is produced by reduction of the water molecules in the presence of nanoscale iron (iron corrosion reaction). Nanoscale palladium (Pd^0) reacts with hydrogen gas to form either metal hydride or hydrogen radicals. Both are highly reactive towards C–Cl bonds and finally replace all chlorine atoms from lindane to form cyclohexane (shown in Scheme 9.2) [64].

Gawande et al. (2013) reported the use of surface-functionalized nano-magnetite supported NPs. These NPs act as a bridge between heterogeneous and homogeneous catalysis. Magnetite-supported metal nanocatalysts have been successfully used in organic synthesis for a variety of important reactions (shown in Scheme 9.3)[65]. Successful use of various nanocatalysts have also been done in several organic transformations, as shown in Table 9.2.

For a catalyst to be sustainable, it must embody the key sustainability features, like environmentally friendly nature of raw constituents, high production yield, selectivity, reusability and easy disposal [78]. The catalyst must show prominent improved properties compared with existing technologies, while also maintaining the economic demands. Nanocatalysts have the capability to exhibit all the described properties and fulfil the requirements of a suitable heterogeneous catalyst and catalytic support in many reactions. The application of nanocatalysts has drastically changed the face of heterogeneous catalysis because of their highly sustainable features, such

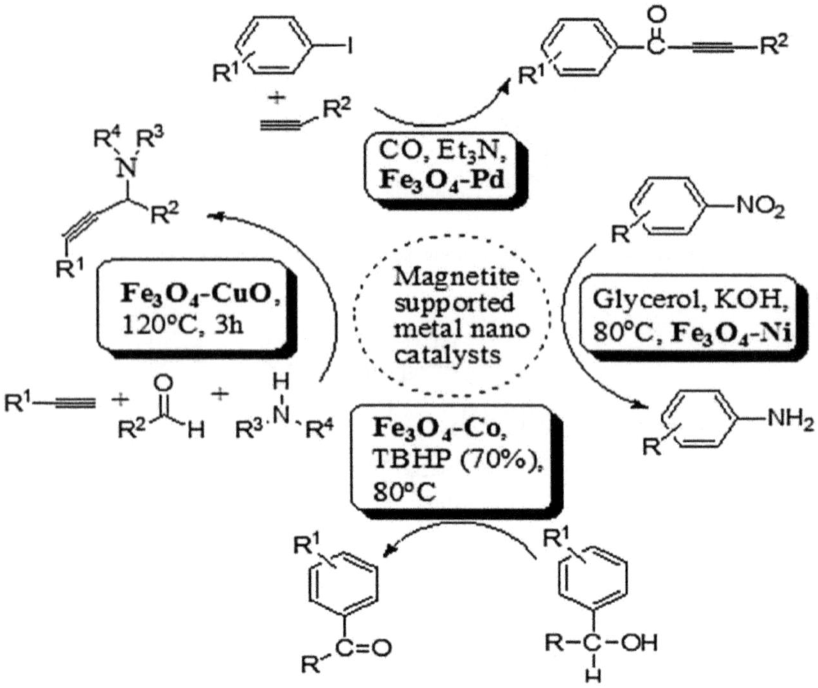

SCHEME 9.3 Some organic reactions catalysed using magnetite-supported metal nanocatalyst.

TABLE 9.2

Some Selective Nano-Catalysed Reactions Which Highlight the Application of Nano-catalysts in Organic Synthesis

Nanocatalyst	Reaction	References
Gold	Synthesis of 1,2,3-triazole	[66]
Gold	Deoxygenation of epoxides	[67]
Calcium oxide	Synthesis of highly substituted pyridines	[68]
Calcium oxide	Photodegradation of 2,4,6-trinitrophenol	[69]
Palladium	Hiyama coupling reactions of benzyl halides	[70]
Palladium	C-C coupling reactions	[71]
Copper	Synthesis of phenols, anilines, thiophenols from aryl halides	[72]
Copper	Synthesis of some aromatic aldehydes and acids	[73]

TABLE 9.2 (Continued)

Some Selective Nano-Catalysed Reactions Which Highlight the Application of Nanocatalysts in Organic Synthesis

Nanocatalyst	Reaction	References
Zinc oxide	One-pot synthesis of caumarins through Knoevenagel condensation	[74]
Zinc oxide	Synthesis of pyranopyrazoles	[75]
Silver	Diels-Alder cycloadditions of 2-hydroxychalcones	[76]
Silver	One-pot synthesis of benzofurans	[77]

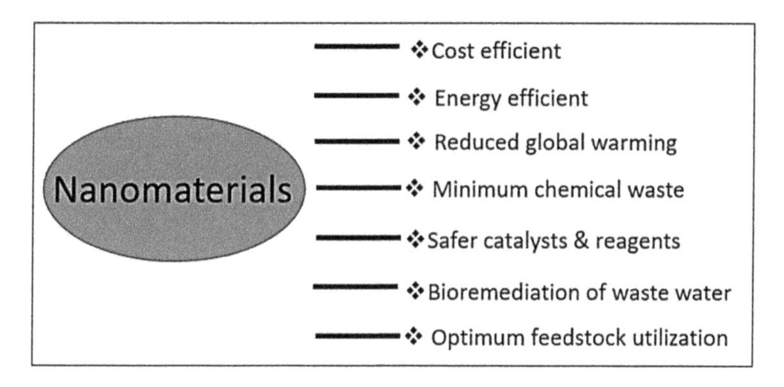

FIGURE 9.5 Various benefits of nanomaterials.

as recyclability and recovery, environmentally benign synthesis, energy and cost efficiency as well as their low toxicity towards living organisms [78]. Figure 9.5 presents various benefits of nanomaterials.

Nanotechnology exclusively offers technologies to establish new industries that will be mainly based on cost-efficient economies and will contribute to sustainable development. Nanotechnology offers stringent methodologies having the potential to significantly diminish the impact of the production, use and storage of energy [79]. Strict efforts and policies have been issued by various countries to focus on the judicious use of energy [79]. For example, one of the prominent themes of the 7th Framework Program of the European Union (FP7) has been energy, where the main focus will be to achieve a sustainable energy-based economy by accelerating the research and developments in the cost-effective technologies [79]. Similarly, from the 'Roadmap Report Concerning the Use of Nanomaterials in the Energy Sector' according to the 6th Framework Program of the European Union (FP6), the application of the domain of energy conversion will be primarily focussed on the use of solar energy (mostly focusing on photovoltaic technology for local supply), conversion of hydrogen and thermoelectric devices [79].

Recently, a tin-based anode nanobattery with the application of nano-alloy instead of the graphite electrode has been commercialized by Sony Corporation with the trade name Nexelion [79]. Similarly, a breakthrough has been announced in lithium-ion batteries by Toshiba Corporation that reduces the recharge time of the battery. This new nanobattery has the capacity to recharge up to 80% of the battery's energy in just 1 min, which is approximately 60 times faster than the commonly used typical lithium-ion batteries and also combines this speed-up recharge time with performance enhancements in the energy density [79].

9.5 Concluding Remarks

With a rapidly growing world population and the resulting energy demands, the production, transformation and application of sustainable energy with minimized environmental impact has been a major concern in order to maintain easy access to energy. The emergence of novel multifunctional nanomaterials has overcome a major part of the confinements of the technology and provided numerous alternatives to further explore and implicate the field of non-renewable energies. For the transition from a carbon-based energy economy towards more sustainable generation, several technological advancements are required in the areas of transportation, transformation, storage, and then the final use of energy. However, due to the varied functionalities of nanomaterials, their application provides efficient solutions at feasible cost. Also, with the unprecedented control over the structure, size and organization of the matter in nanotechnology, unique properties can be obtained which contribute to overcoming many of the challenges faced.

REFERENCES

[1] M.M. Norhasri, M. Hamidah, A.M. Fadzil, Applications of using nano material in concrete: A review, *Construction and Building Materials* 133(2017) 91–97.

[2] Z.-Y. Zhou, N. Tian, J.-T. Li, I. Broadwell, S.-G. Sun, Nanomaterials of high surface energy with exceptional properties in catalysis and energy storage, *Chemical Society Reviews* 40(7) (2011) 4167–4185.

[3] S. Pawar, Science of nanomaterials, *Progress and Prospects in Nanoscience Today* 1 (2020) 1–414.

[4] C. Wang, J. Tuninetti, Z. Wang, C. Zhang, R. Ciganda, L. Salmon, S. Moya, J. Ruiz, D. Astruc, Hydrolysis of ammonia-borane over Ni/ZIF-8 nanocatalyst: high efficiency, mechanism, and controlled hydrogen release, *Journal of the American Chemical Society* 139(33) (2017) 11610–11615.

[5] T.N. Pingel, M. Jørgensen, A.B. Yankovich, H. Grönbeck, E. Olsson, Influence of atomic site-specific strain on catalytic activity of supported nanoparticles, *Nature Communications* 9(1) (2018) 1–9.

[6] S.B. Kalidindi, B.R. Jagirdar, Nanocatalysis and prospects of green chemistry, *ChemSusChem* 5(1) (2012) 65–75.

[7] S. Prabhudev, M. Bugnet, C. Bock, G.A. Botton, Strained lattice with persistent atomic order in Pt3Fe2 intermetallic core–shell nanocatalysts, *ACS Nano* 7(7) (2013) 6103–6110.
[8] E. Farnetti, R. Di Monte, J. Kašpar, Homogeneous and heterogeneous catalysis, *Inorganic and Bio-Inorganic Chemistry* 2(6) (2009) 50–86.
[9] F. Zaera, Nanostructured materials for applications in heterogeneous catalysis, *Chemical Society Reviews* 42(7) (2013) 2746–2762.
[10] A. Verma, M. Shukla, I. Sinha, Introductory chapter: salient features of nanocatalysis, *Nanocatalysts*, IntechOpen (2019). DOI: http: //dx.doi.org/10.5772/intechopen.78514
[11] M. Larramendy, S. Soloneski, *Green Nanotechnology: Overview and Further Prospects*, BoD–Books on Demand (2016).
[12] E.G. Garrido-Ramírez, B.K. Theng, M.L. Mora, Clays and oxide minerals as catalysts and nanocatalysts in Fenton-like reactions—A review, *Applied Clay Science* 47(3–4) (2010) 182–192.
[13] V. Vimonses, Development of multifunctional nanomaterials and adsorption-photocatalysis hybrid system for wastewater reclamation, (2011).
[14] Z. Guowu, Synthesis of integrated nanocatalysts with mesoporous silica/silicate and microporous MOFs, (2016).
[15] E. Ruiz-Hitzky, P. Aranda, M. Akkari, N. Khaorapapong, M. Ogawa, Photoactive nanoarchitectures based on clays incorporating TiO2 and ZnO nanoparticles, *Beilstein Journal of Nanotechnology* 10(1) (2019) 1140–1156.
[16] A. Balasubramanyam, N. Sailaja, M. Mahboob, M. Rahman, S.M. Hussain, P. Grover, In vivo genotoxicity assessment of aluminium oxide nanomaterials in rat peripheral blood cells using the comet assay and micronucleus test, *Mutagenesis* 24(3) (2009) 245–251.
[17] A. Mukherjee, I. Mohammed Sadiq, T. Prathna, N. Chandrasekaran, Antimicrobial activity of aluminium oxide nanoparticles for potential clinical applications, *Science Against Microbial Pathogens: Communicating Current Research and Technological Advances* 1 (2011) 245–251.
[18] G. Cheraghian, M.P. Wistuba, Ultraviolet aging study on bitumen modified by a composite of clay and fumed silica nanoparticles, *Scientific Reports* 10(1) (2020) 1–17.
[19] S.W. Kang, Y.W. Lee, Y. Park, B.-S. Choi, J.W. Hong, K.-H. Park, S.W. Han, One-pot synthesis of trimetallic Au@ PdPt core–shell nanoparticles with high catalytic performance, *ACS Nano* 7(9) (2013) 7945–7955.
[20] M. Liu, Z. Zhao, X. Duan, Y. Huang, Nanoscale Structure Design for High-Performance Pt-Based ORR Catalysts, *Advanced Materials* 31(6) (2019) 1802234.
[21] W. Zhan, J. Wang, H. Wang, J. Zhang, X. Liu, P. Zhang, M. Chi, Y. Guo, Y. Guo, G. Lu, Crystal structural effect of AuCu alloy nanoparticles on catalytic CO oxidation, *Journal of the American Chemical Society* 139(26) (2017) 8846–8854.
[22] S. Cao, F.F. Tao, Y. Tang, Y. Li, J. Yu, Size-and shape-dependent catalytic performances of oxidation and reduction reactions on nanocatalysts, *Chemical Society Reviews* 45(17) (2016) 4747–4765.
[23] B.R. Cuenya, F. Behafarid, Nanocatalysis: size-and shape-dependent chemisorption and catalytic reactivity, *Surface Science Reports* 70(2) (2015) 135–187.

[24] N. Sharma, H. Ojha, A. Bharadwaj, D.P. Pathak, R.K. Sharma, Preparation and catalytic applications of nanomaterials: a review, *RSC Advances* 5(66) (2015) 53381–53403.

[25] K. Hemalatha, G. Madhumitha, A. Kajbafvala, N. Anupama, R. Sompalle, S. Mohana Roopan, Function of nanocatalyst in chemistry of organic compounds revolution: an overview, *Journal of Nanomaterials* 2013 (2013). DOI: https://doi.org/10.1155/2013/341015.

[26] S. Olveira, S.P. Forster, S. Seeger, Nanocatalysis: academic discipline and industrial realities, *Journal of Nanotechnology* 2014 (2014). DOI: https://doi.org/10.1155/2014/324089.

[27] R.S. Varma, Nano-catalysts with magnetic core: sustainable options for greener synthesis, *Sustainable Chemical Processes* 2(1) (2014) 11.

[28] H. Hildebrand, K. Mackenzie, F.-D. Kopinke, Novel nano-catalysts for wastewater treatment, *Global NEST Journal* 10(1) (2008) 47–53.

[29] P.N. Kapoor, A.K. Bhagi, R.S. Mulukutla, K.J. Klabunde, Mixed metal oxide nanoparticles, *Dekker Encyclopedia Nanoscience Nanotechnology* (2004) 6(2) 2007–2015.

[30] M.B. Gawande, A. Goswami, T. Asefa, H. Guo, A.V. Biradar, D.-L. Peng, R. Zboril, R.S. Varma, Core–shell nanoparticles: synthesis and applications in catalysis and electrocatalysis, *Chemical Society Reviews* 44(21) (2015) 7540–7590.

[31] J. Fan, Y. Gao, Nanoparticle-supported catalysts and catalytic reactions–a mini-review, *Journal of Experimental Nanoscience* 1(4) (2006) 457–475.

[32] M. Hu, Z. Yao, X. Wang, Graphene-based nanomaterials for catalysis, *Industrial & Engineering Chemistry Research* 56(13) (2017) 3477–3502.

[33] H. Lyu, B. Gao, F. He, C. Ding, J. Tang, J.C. Crittenden, Ball-milled carbon nanomaterials for energy and environmental applications, *ACS Sustainable Chemistry & Engineering* 5(11) (2017) 9568–9585.

[34] M. Khalil, B.M. Jan, C.W. Tong, M.A. Berawi, Advanced nanomaterials in oil and gas industry: design, application and challenges, *Applied Energy* 191 (2017) 287–310.

[35] P. Rao, Nanocatalysis: applications in the chemical industry, (2010).

[36] D. Astruc, *Nanoparticles and catalysis*, John Wiley & Sons (2008).

[37] M.A. Newton, Dynamic adsorbate/reaction induced structural change of supported metal nanoparticles: heterogeneous catalysis and beyond, *Chemical Society Reviews* 37(12) (2008) 2644–2657.

[38] N. Wang, W. Qian, W. Chu, F. Wei, Crystal-plane effect of nanoscale CeO 2 on the catalytic performance of Ni/CeO 2 catalysts for methane dry reforming, *Catalysis Science & Technology* 6(10) (2016) 3594–3605.

[39] S. Chaemchuen, Z. Luo, K. Zhou, B. Mousavi, S. Phatanasri, M. Jaroniec, F. Verpoort, Defect formation in metal–organic frameworks initiated by the crystal growth-rate and effect on catalytic performance, *Journal of Catalysis* 354 (2017) 84–91.

[40] D. Li, C. Wang, D. Tripkovic, S. Sun, N.M. Markovic, V.R. Stamenkovic, Surfactant removal for colloidal nanoparticles from solution synthesis: the effect on catalytic performance, *ACS Catalysis* 2(7) (2012) 1358–1362.

[41] G.-H. Wang, W.-C. Li, K.-M. Jia, B. Spliethoff, F. Schüth, A.-H. Lu, Shape and size controlled α-Fe2O3 nanoparticles as supports for gold-catalysts: synthesis and influence of support shape and size on catalytic performance, *Applied Catalysis A: General* 364(1–2) (2009) 42–47.

[42] A.M. Henning, J. Watt, P.J. Miedziak, S. Cheong, M. Santonastaso, M. Song, Y. Takeda, A.I. Kirkland, S.H. Taylor, R.D. Tilley, Gold–palladium core–shell

nanocrystals with size and shape control optimized for catalytic performance, *Angewandte Chemie* 125(5) (2013) 1517–1520.

[43] K. Guo, H. Li, Z. Yu, Size-dependent catalytic activity of monodispersed nickel nanoparticles for the hydrolytic dehydrogenation of ammonia borane, *ACS Applied Materials & Interfaces* 10(1) (2018) 517–525.

[44] F. Fu, Y. Li, Z. Yang, G. Zhou, Y. Huang, Z. Wan, X. Chen, N. Hu, W. Li, L. Huang, Molecular-level insights into size-dependent stabilization mechanism of gold nanoparticles in 1-Butyl-3-methylimidazolium tetrafluoroborate ionic liquid, *Journal of Physical Chemistry C* 121(1) (2017) 523–532.

[45] P. Suchomel, L. Kvitek, R. Prucek, A. Panacek, A. Halder, S. Vajda, R. Zboril, Simple size-controlled synthesis of Au nanoparticles and their size-dependent catalytic activity, *Scientific Reports* 8(1) (2018) 1–11.

[46] D. Latha, C. Arulvasu, P. Prabu, V. Narayanan, Photocatalytic activity of biosynthesized silver nanoparticle from leaf extract of *Justicia adhatoda* (2017).

[47] L. Di, J. Zhang, M. Craven, Y. Wang, H. Wang, X. Zhang, X. Tu, Dehydrogenation of formic acid over Pd/C catalysts: insight into the cold plasma treatment, *Catalysis Science & Technology* 10(18) (2020) 6129–6138.

[48] G. Guisbiers, G. Abudukelimu, D. Hourlier, Size-dependent catalytic and melting properties of platinum-palladium nanoparticles, *Nanoscale Research Letters* 6(1) (2011) 396.

[49] T. Bligaard, J.K. Nørskov, Heterogeneous catalysis. In Anders Nilsson, Lars G.M. Pettersson and Jens K. Nørskov, Eds., *Chemical Bonding at Surfaces and Interfaces*, Elsevier (2008), pp. 255–321.

[50] C. Huang, X. Chen, Z. Xue, T. Wang, Effect of structure: a new insight into nanoparticle assemblies from inanimate to animate, *Science Advances* 6(20) (2020) eaba1321.

[51] M. Fernández-García, J.A. Rodriguez, *Nanomaterials: Inorganic and Bioinorganic Perspectives*. (2007) Chemistry Department Brookhaven National Laboratory, New York.

[52] J.A. Adekoya, K.O. Ogunniran, T.O. Siyanbola, E.O. Dare, N. Revaprasadu, Band structure, morphology, functionality, and size-dependent properties of metal nanoparticles, noble and precious metals—properties. In Seehra, M.S., and Bristow, A.D., Eds., *Nanoscale Effects and Applications* (2018) pp. 15–42. London, IntechOpen DOI:10.5772/intechopen.69142.

[53] M. Chen, D. Goodman, The structure of catalytically active gold on titania, *Science* 306(5694) (2004) 252–255.

[54] I.V. Yudanov, A. Genest, S. Schauermann, H.-J. Freund, N. Rösch, Size dependence of the adsorption energy of CO on metal nanoparticles: a DFT search for the minimum value, *Nano Letters* 12(4) (2012) 2134–2139.

[55] S. Lee, C. Fan, T. Wu, S.L. Anderson, Cluster size effects on CO oxidation activity, adsorbate affinity, and temporal behavior of model Au n/TiO 2 catalysts, *Journal of Chemical Physics* 123(12) (2005) 124710.

[56] A.D. Allian, K. Takanabe, K.L. Fujdala, X. Hao, T.J. Truex, J. Cai, C. Buda, M. Neurock, E. Iglesia, Chemisorption of CO and mechanism of CO oxidation on supported platinum nanoclusters, *Journal of the American Chemical Society* 133(12) (2011) 4498–4517.

[57] Y. Wang, W. Qiu, E. Song, F. Gu, Z. Zheng, X. Zhao, Y. Zhao, J. Liu, W. Zhang, Adsorption-energy-based activity descriptors for electrocatalysts in energy storage applications, *National Science Review* 5(3) (2018) 327–341.

[58] R. Jinnouchi, R. Asahi, Predicting catalytic activity of nanoparticles by a DFT-aided machine-learning algorithm, *Journal of Physical Chemistry Letters* 8(17) (2017) 4279–4283.

[59] K. Sawabe, T. Koketsu, J. Ohyama, A. Satsuma, A Theoretical insight into enhanced catalytic activity of au by multiple twin nanoparticles, *Catalysts* 7(6) (2017) 191.

[60] Y. Li, G.A. Somorjai, Nanoscale advances in catalysis and energy applications, *Nano Letters* 10(7) (2010) 2289–2295.

[61] M.J. Ndolomingo, N. Bingwa, R. Meijboom, Review of supported metal nanoparticles: synthesis methodologies, advantages and application as catalysts, *Journal of Materials Science* (2020) 1–47.

[62] J. Jeevanandam, A. Barhoum, Y.S. Chan, A. Dufresne, M.K. Danquah, Review on nanoparticles and nanostructured materials: history, sources, toxicity and regulations, *Beilstein Journal of Nanotechnology* 9(1) (2018) 1050–1074.

[63] Y. Bi, G. Lu, Nano-Cu catalyze hydrogen production from formaldehyde solution at room temperature, *International Journal of Hydrogen Energy* 33(9) (2008) 2225–2232.

[64] V. Nagpal, A.D. Bokare, R.C. Chikate, C.V. Rode, K.M. Paknikar, Reductive dechlorination of γ-hexachlorocyclohexane using Fe–Pd bimetallic nanoparticles, *Journal of Hazardous Materials* 175(1–3) (2010) 680–687.

[65] M.B. Gawande, P.S. Branco, R.S. Varma, Nano-magnetite (Fe 3 O 4) as a support for recyclable catalysts in the development of sustainable methodologies, *Chemical Society Reviews* 42(8) (2013) 3371–3393.

[66] M. Boominathan, N. Pugazhenthiran, M. Nagaraj, S. Muthusubramanian, S. Murugesan, N. Bhuvanesh, Nanoporous titania-supported gold nanoparticle-catalyzed green synthesis of 1, 2, 3-triazoles in aqueous medium, *ACS Sustainable Chemistry & Engineering* 1(11) (2013) 1405–1411.

[67] A. Noujima, T. Mitsudome, T. Mizugaki, K. Jitsukawa, K. Kaneda, Gold nanoparticle-catalyzed environmentally benign deoxygenation of epoxides to alkenes, *Molecules* 16(10) (2011) 8209–8227.

[68] J. Safaei-Ghomi, M. Ghasemzadeh, M. Mehrabi, Calcium oxide nanoparticles catalyzed one-step multicomponent synthesis of highly substituted pyridines in aqueous ethanol media, *Scientia Iranica* 20(3) (2013) 549–554.

[69] A. Imtiaz, M.A. Farrukh, M. Khaleeq-ur-Rahman, R. Adnan, Micelle-assisted synthesis of Al2O3· CaO nanocatalyst: optical properties and their applications in photodegradation of 2, 4, 6-trinitrophenol, *Scientific World Journal* 2013 (2013).

[70] D. Srimani, A. Bej, A. Sarkar, Palladium nanoparticle catalyzed Hiyama coupling reaction of benzyl halides, *Journal of Organic Chemistry* 75(12) (2010) 4296–4299.

[71] A. Balanta, C. Godard, C. Claver, Pd nanoparticles for C–C coupling reactions, *Chemical Society Reviews* 40(10) (2011) 4973–4985.

[72] H.-J. Xu, Y.-F. Liang, Z.-Y. Cai, H.-X. Qi, C.-Y. Yang, Y.-S. Feng, CuI-nanoparticles-catalyzed selective synthesis of phenols, anilines, and thiophenols from aryl halides in aqueous solution, *Journal of Organic Chemistry* 76(7) (2011) 2296–2300.

[73] P.K. Tandon, S.B. Singh, M. Srivastava, Synthesis of some aromatic aldehydes and acids by sodium ferrate in presence of copper nano-particles adsorbed on K 10 montmorillonite using microwave irradiation, *Applied Organometallic Chemistry* 21(4) (2007) 264–267.

[74] B.V. Kumar, H.S.B. Naik, D. Girija, B.V. Kumar, ZnO nanoparticle as catalyst for efficient green one-pot synthesis of coumarins through Knoevenagel condensation, *Journal of Chemical Sciences* 123(5) (2011) 615–621.

[75] S.U. Tekale, S.S. Kauthale, K.M. Jadhav, R.P. Pawar, Nano-ZnO catalyzed green and efficient one-pot four-component synthesis of pyranopyrazoles, *Journal of Chemistry* 2013 (2013).

[76] H. Cong, C.F. Becker, S.J. Elliott, M.W. Grinstaff, J.A. Porco Jr, Silver nanoparticle-catalyzed Diels–Alder cycloadditions of 2′-hydroxychalcones, *Journal of the American Chemical Society* 132(21) (2010) 7514–7518.

[77] J. Safaei-Ghomi, M.A. Ghasemzadeh, Silver iodide nanoparticle as an efficient and reusable catalyst for the one-pot synthesis of benzofurans under aqueous conditions, *Journal of Chemical Sciences* 125(5) (2013) 1003–1008.

[78] J.F. Jenck, F. Agterberg, M.J. Droescher, Products and processes for a sustainable chemical industry: a review of achievements and prospects, *Green Chemistry* 6(11) (2004) 544–556.

[79] E. Serrano, G. Rus, J. Garcia-Martinez, Nanotechnology for sustainable energy, *Renewable and Sustainable Energy Reviews* 13(9) (2009) 2373–2384.

10

Utilization of Biodiesel By-Products in Various Industrial Applications

Rituraj Dubey* **and Laxman Singh**

CONTENTS

10.1 Introduction

In the current era, the power of a nation is measured on the basis of the oil reserves it has. The world is looking for more than a 50% increase in the energy required by the end of the next decade. The reason behind this is the rapid growth in economic development, industrialization, human population, transportation of goods and services and technology demands. Additionally, the shipping sector depends completely on fossil fuels (Sani et al., 2014). In this way, it can be claimed that almost the last stocks of conventional fuel reserves are being used nowadays due to enormous consumption of oil and, finally, this has stimulated researchers to work on unconventional sources of energy. The most popular unconventional sources of energy are geothermal, ocean, wind, hydropower and biofuels. These sources, other than biofuels, can be generated and harnessed only in specific locations, and hence the need of the hour is

* *Corresponding Author* riturajdubey0@gmail.com

DOI: 10.1201/9781003120858-10

to replace the petroleum-based products with biofuels. Liquid biofuel like biodiesel have got established itself as an emerging option of renewable fuel which can be easily produced and transported around the world (Noshadi et al., 2012). Still, biodiesel is not in high demand due to its production cost. However, the production cost of biodiesel can be reduced just by enhancing the utilization of the main by-product *viz.* crude glycerol (Zheng et al., 2008). Biodiesel is considered to be environment friendly due to lower smoke and particulate's production, lower carbon monoxide and hydrocarbon's emissions, no sulphur dioxide's emissions and higher cetane number in comparison to the conventional petroleum-based fuels because the carbon in the fats which is the source of biodiesel production is originated from carbon dioxide present in the air (Lotero et al., 2005; Ma & Hanna, 1999).

According to the American Society for Testing and Materials, currently known as ASTM International, 'biodiesel' is the term given to a mixture of long-chain mono-alkylic fatty esters derived from vegetable oils or animal fats, designated B100. It can be used alone or blended with diesel oil, indicated as 'Bx' where 'x' denotes the percentage of biodiesel present in the blended mixture, and hence B100 means pure biodiesel (Abbaszaadeh et al., 2012; Fukuda et al., 2001; Knothe et al., 1997). Biodiesel is a composition of methyl and/or ethyl esters of lauric, palmitic, stearic, oleic and other long-chain fatty acids and is synthesized by transesterification of unsaturated fats with methanol or ethanol in the presence of a catalyst, such as sodium hydroxide. Plants like palm, sunflower, soybean, etc., and animal products including chicken, pork, fish, hamburgers, etc., are used as the source of unsaturated fats (Antolín et al., 2002; Encinar et al., 2007; Huber et al., 2006).

A significant drawback of the mass production of biodiesel is that it creates a large amount of by-products which are not usable in the biodiesel generation process. The transesterification reaction involved in biodiesel production gives the main by-product, namely crude glycerol (Kolesárová et al., 2011). Another by-product, oil-cake, is also produced during the process of extraction of oil from oil seed plants. When animal fat is extracted from cooking-waste during biodiesel production, it yields a used-oil sediment by-product containing fragments of meat, bone and bread. In addition, the cleaning process in biodiesel mass production yields a significant amount of wastewater, while refining of crude glycerol yields a large amount of methanol. When a waterless technique has been used for the washing process the use of ion exchange resin-styrene plastic beads come into effect and yields a used ion-exchange resin by-product. Similarly, another waterless procedure uses magnesium silicate which also comes out as a by-product. Subsequently several micro by-products *viz.* ion-exchange resin, magnesium silicate and used oil sediment and macro by-products *viz.* wastewater, methanol and oil-cake, are also produced (Kolesárová et al., 2011; Mandolesi de Araújo et al., 2013). The complete process of biodiesel mass-production is shown in Figure 10.1. In this chapter, we discuss the industrial applications of all the by-products produced during the biodiesel mass production.

10.2 Main By-Product: Crude Glycerol

The main by-product [about 10% (w/w)] in biodiesel's production is glycerol. As per some reports, the anticipated mass production of biodiesel will reach 110–120 billion

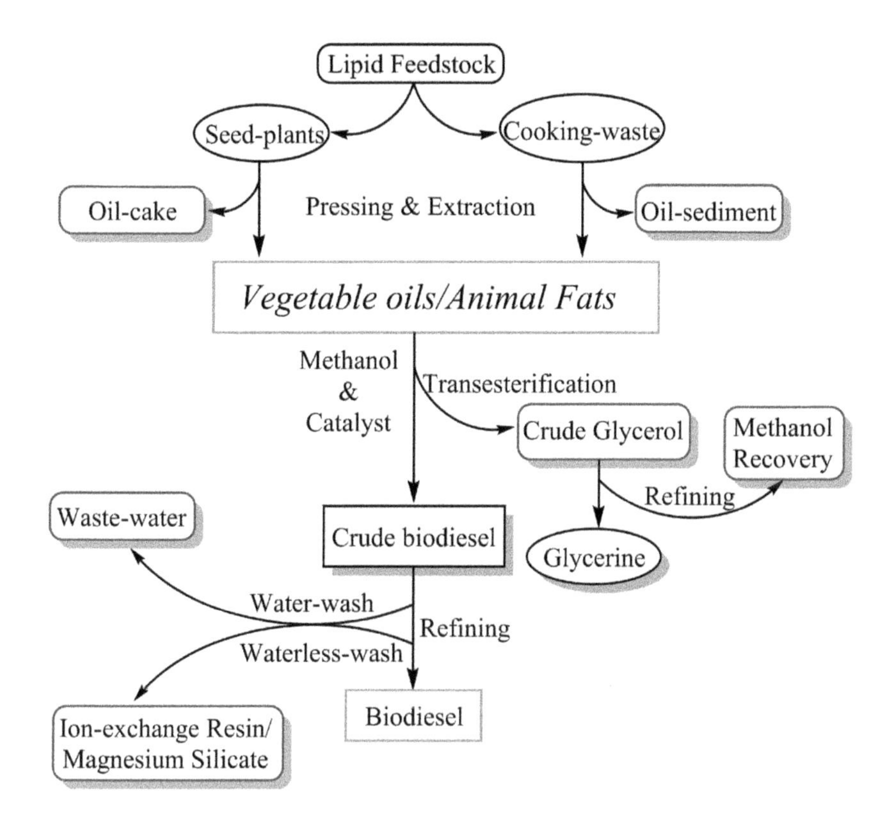

FIGURE 10.1 Complete process of biodiesel production along with its all by-products.

gallons between 2020 and 2025, which will indirectly yield 11–12 billion gallons of crude glycerol (Kaur et al., 2020; Ogunkunle & Ahmed, 2019). Hence the exploration for industrial applications of glycerol is essential, not only for increment in its value but also for sustainable mass production of biodiesel. Chemically, glycerol is 1,2,3-propanetriol and it is known to be the oldest human-isolated organic molecule since its usage has been reported in soap production in 2800 BC. Glycerol can be utilized in pharmaceutical, cosmetic, chemical, food, polymer and other allied industries, either directly as a major raw material or as a precursor to another major raw material (Anitha et al., 2016; Ciriminna et al., 2014; Kaur et al., 2020; Luo et al., 2016; Zhu et al., 2013) as described below.

10.2.1 Direct Applications

The direct applications of glycerol in industries are shown in Figure 10.2. The main use of glycerine was traditionally in the soap industry (for softening and moisturization purposes). Currently, it is widely used in the pharmaceutical industry to dissolve drugs, to provide the required amount of humidity to medicine pills, to increase the viscosity of cough syrups, to carry antibiotics and antiseptics in ear infection medicines and to

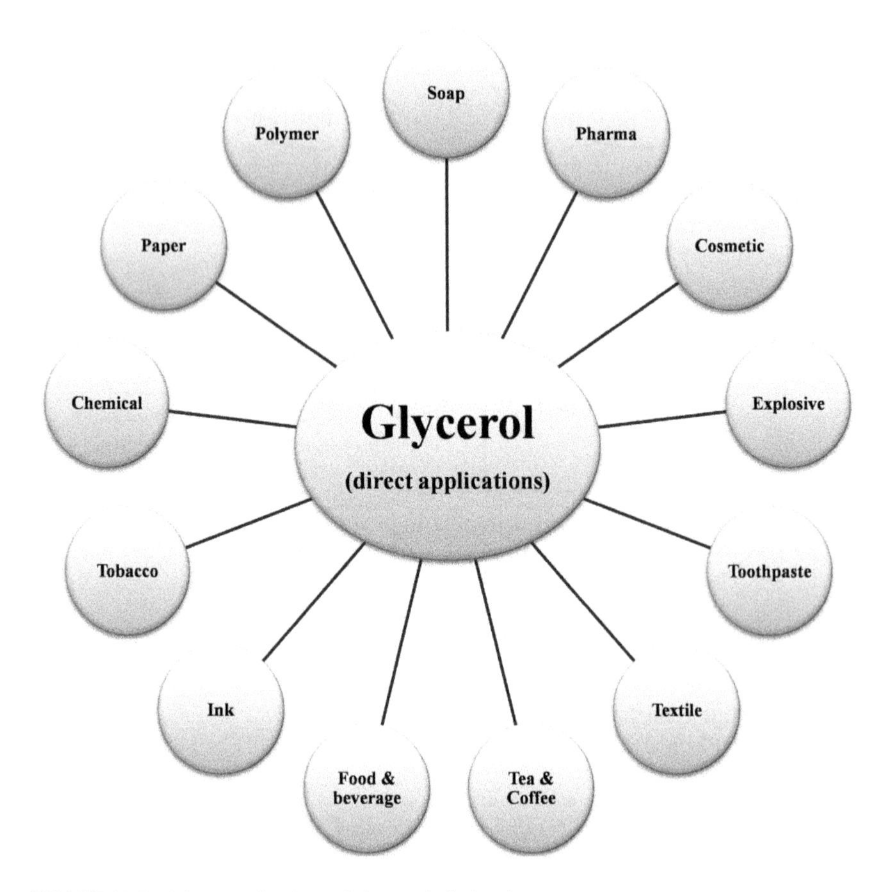

FIGURE 10.2 Direct applications of glycerol in industries.

plasticize medicine capsules as well as to manufacture ointments, creams and lotions (Singhabhandhu & Tezuka, 2010). It is used in the chemical industry to dissolve several molecules that are in daily use, such as mercury chloride, tannins, halogens like iodine and bromine, phenol and alkaloids (Tan et al., 2013). It is also utilized in the cosmetic industry to provide lubrication, smoothness and moisturization to dermatology products (Singhabhandhu & Tezuka, 2010) due to its moisturizing and emollient properties. It is used widely in the toothpaste industry to keep the toothpaste soft and wet (Tan et al., 2013). It is also used in the food, beverage, tea and coffee industries as a solvent, softener, sweetener and preservative agent (Antolín et al., 2002). Further, the use of glycerol in the tobacco industry is to provide the optimized moisture content and to preserve freshness in order to eliminate the unpleasant irritating taste in tobacco. It is used in the paper manufacturing industry as a plasticizer and lubricant, while in the textile industry it is used as lubricant, softener and protective filler (Tan et al., 2013). Glycerol is used as a raw material in the polymer industries for the production of dendrimers and hyperbranched polyesters (Haag et al., 2000). It is also used in the explosive (for manufacturing nitroglycerin), chemical (for the fabrication of urethane foams, alkydic resins and cellophane) and ink industries.

10.2.2 Indirect Applications (as a Precursor Molecule)

The conversion of glycerol to other chemicals like poly(hydroxyalkanoates), ethanol, butanol, citric acid, lactic acid, 1,3-dihydroxyacetone, glyceric acid, succinic acid, oxalic acid, 2,3-butanediol, 1,3-propanediol, dichloro-2-propanol, mannitol, erythritol, arabitol, acrolein, 1,2-propanediol, glycerol carbonate, monoglyceride, diacylglycerol and triglyceride (Figure 10.3) may create several industrial applications of glycerol as a precursor molecule (Bagnato et al., 2017; Fan et al., 2010) as detailed below.

Poly(hydroxyalkanoates) are produced from crude glycerol through biological conversion (Ashby et al., 2004). Poly(hydroxyalkanoates) are used as bacterial polyesters and have been utilized further as a substitute for non-biodegradable petrochemically produced polymers (Yang et al., 2012). *Polyhydroxybutyrate* is the most useful member of the poly(hydroxyalkanoate) group and hence it is produced

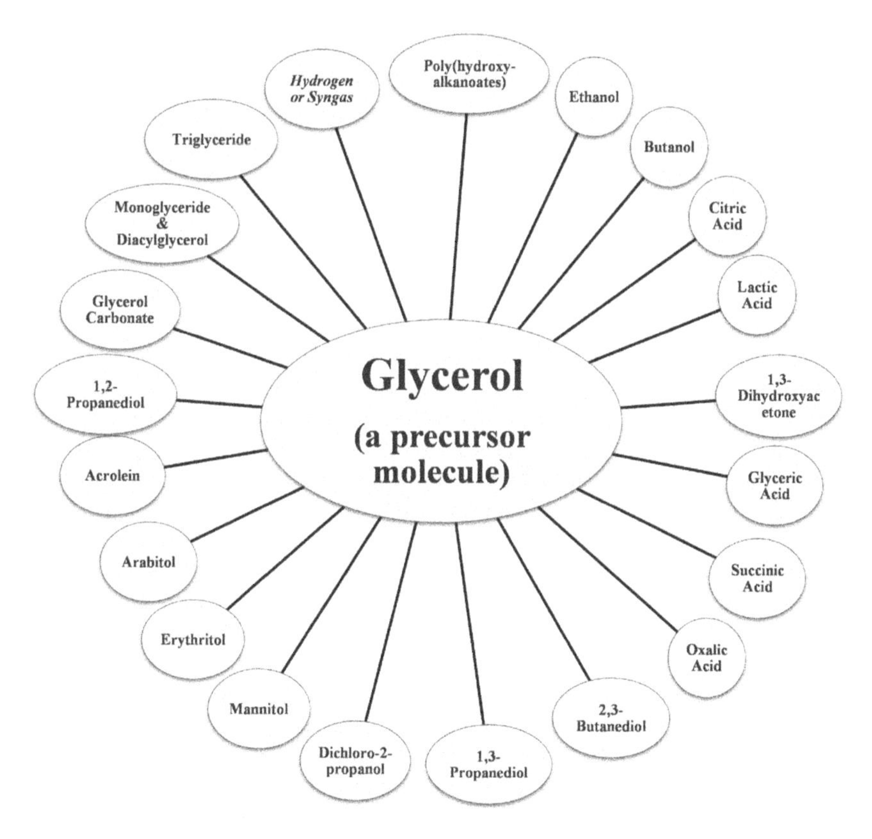

FIGURE 10.3 Indirect applications of glycerol as a precursor molecule to give several molecules of industrial importance.

from crude glycerol with the help of *P. denitrificans* and *C. necator* JMP 134. However, the presence of NaCl in crude glycerol would reduce its production and hence crude glycerol from different manufacturers should be mixed before cell growth (Mothes et al., 2007). In addition, for large-scale production of polyhydroxybutyrate crude glycerol should be subjected to *C. necator* DSM 545 fermentation, however, the impact of hindrance of NaCl could not be eliminated completely (Cavalheiro et al., 2009). Hindrance of NaCl has been overcome by fermenting the crude glycerol based on *Z. denitrificans* MW1 to produce polyhydroxybutyrate (Ibrahim & Steinbüchel, 2009). With the help of microbial consortia, methanol present in the crude glycerol can be transformed to polyhydroxybutyrate with a very high yield of 20.9 tons/year/10 million gallons biodiesel (Dobroth et al., 2011). Similarly, *P. oleovorans* NRRL B-14682 can also be used for its production from crude glycerol (Ashby et al., 2011). Polyhydroxybutyrate is used in the polymer industry (Prabisha et al., 2015).

Ethanol is synthesized from crude glycerol through bio-conversion using various microbes such as *E. coli*, *Z. mobilis*, *K. oxytoca*, *K. cryocrescens* and *S. cerevisiae* (Hong et al., 2010). It is used in the food, fuel, pharmaceutical (in many cough and cold liquids as a solvent and/or preservative) and chemical industries (Nitayavardhana & Khanal, 2010). In the food industry, it is used as a main part of beer, wine or brandy-like alcoholic beverages. In plum pudding and fruit cake production, it is used for even distribution of food colour as well as for enhancement of the flavour of food extracts. It is used in varnish in the fuel industry. It has a higher octane number than gasoline and hence is used for 10% blending with gasoline in the United States as a fuel. In the chemical industry, it is used as a common ingredient in many cosmetics and beauty products due to its astringent quality and as a preservative and binding agent for lotion ingredients and also in hairspray as it also has adherence quality. It is also a common ingredient in many hand sanitizers due to its effectiveness against bacteria, fungi and viruses.

Butanol is produced from crude glycerol using *C. pasteurianum* and/or *C. acetobutylicum* for the bioconversion process (Taconi et al., 2009). It is used in the paint (in the formulation of paints and lacquers), automotive fuel, plastics, polymers, lubricants, brake fluids, synthetic rubber and resin manufacturing industries (Harvey & Meylemans, 2011). It is also used as a powerful industrial solvent for cleaning and polishing products. It is also used as an artificial food flavouring in theUnited States In addition, it is used in products such as shampoo, shaving products and soaps as it is almost insoluble in water, but soluble in almost all other organic solvents. This is also why it is used as a universal solvent for dyes, natural and synthetic resins, paints, lacquers and varnishes, gums, and vegetable oils and alkaloids.

Citric acid ($C_6H_8O_7$; 2-hydroxy-1,2,3-propane tricarboxylic acid) is considered to be one of the most versatile, naturally produced and highly consumed chemicals in the world (Soccol et al., 2006). It can be produced from crude glycerol using *Y. lipolytica* ACA-DC 50109 (Papanikolaou & Aggelis, 2003) and other strains of *Y. lipolytica* (Imandi et al., 2007; Kamzolova et al., 2011; Rymowicz et al., 2006). With the help of *Y. lipolytica* Wratislavia AWG7 strain, crude glycerol yields a similar concentration of citric acid (131.5 g/L) to that yielded from pure glycerol (139 g/L). Meanwhile, the fermentation of crude glycerol with *Y. lipolytica* Wratislavia K1 gives comparatively low concentrations of citric acid (~87–89 g/L) (Rywiska et al., 2013).

However, *Y. lipolytica* Wratislavia K1 was found to be the best for the conversion of crude glycerol into erythritol (Rymowicz et al., 2009). Additionally, *Y. lipolytica* LGAM S (7)1 and *Y. lipolytica* N15 have shown good results for the conversion of crude glycerol to citric acid, even up to 98 g/L (Kamzolova et al., 2011; Papanikolaou et al., 2002). It is used mainly in the food and soft drink industries (70%). It can also be utilized as an additive in cosmetics, toiletries and pharmaceuticals (Papanikolaou et al., 2002).

The conversion of glycerol into lactic acid/lactic acid salts and esters is not an easy task as only one strain, AC-521, out of eight strains of *E. coli* has given suitable results with very high productivity up to 85.8 g/L (Hong et al., 2009). The *salts and esters of lactic acid* have been found to be useful in the industrial, food, cosmetic and pharmaceutical, and polymer (in polyester resins and polyurethane) industries (Castro-Aguirre et al., 2016; Datta & Henry, 2006).

One-pot electrocatalytic oxidation of glycerol can yield *1,3-dihydroxyacetone* (Ciriminna et al., 2006) but conversion of crude glycerol into 1,3-dihydroxyacetone was reported *via* fermentation of the alga *S. limacinum* on crude glycerol up to a very high level (Pyle et al., 2008). 1,3-Dihydroxyacetone is used in the cosmetics industry as an artificial browning agent and in the agriculture, food and pharmaceutical industries (in the production of fungicides) (Petersen et al., 2003).

Glyceric acid is a naturally occurring organic acid with a molecular formula of $C_3H_6O_4$ (2R, 3-hydroxy propanoic acid) (Handa et al., 1986). Its chemical synthesis is very costly and hence restricts it usage value. Chemically, it is synthesized either by a microbial fermentation process or by metallic oxidation of glycerol (Bianchi et al., 2005). It is used in the chemical, cosmetic and pharmaceutical industries (Habe et al., 2009; Rosseto et al., 2008; Sile et al., 2013).

Succinic acid or *butanedioic acid* has a molecular formula of C_4H_6O and is chemically a dicarboxylic acid (Millard et al., 1996). It is also known as amber acid due to its first ever production using amber as a source. In this case, too, the chemical synthesis is quite expensive which is why the bioconversion route is followed. Chemically, it is synthesized from butane with the help of maleic anhydride while biochemically, it is produced from crude glycerol by a fermentation process with the help of *M. succiniciproducens*, *C. glutamicum*, recombinant *E. coli*, *A. succiniciproducens*, *Lactobacillus*, *A. succinogenes* and *S. cerevisiae* and other carbon sources and it is widely available in food market (Beauprez et al., 2010; Jiang et al., 2013; Song & Lee, 2006). It is in used in the food, pharmaceutical (in the formulation of antibiotics, amino acids, vitamins and green solvents) and polymer (for production of biodegradable plastics) industries (Harry-O'kuru et al., 2015). The various dicarboxylic acid derivatives of succinic acid are also used widely in various chemical conversions (Zeikus et al., 1999).

Oxalic acid is synthesized from crude glycerol *via* biotransformation using *A. niger* NRRL and *A. niger* LFMB (André et al., 2010). It is used in paper (in the pretreatment process of total mechanical pulping), manufacture (as a bleaching agent in the wood and textile industries) and detergent industries as it is a good reducing agent and hence has been found to be useful in removing tarnish, rust, ink stains or calcium deposits, etc. (Musiał et al., 2011). It is also used as a hole scavenger in wastewater treatment (Lucchetti et al., 2017) and as a miticide against the *Varroa* mite parasite by beekeepers.

2,3-Butanediol is synthesized from crude glycerol using *B. amyloliquefaciens* in the fermentation process (Yang et al., 2013). It is used in the polymer (to produce plastics), automotive (to produce anti-freeze solutions), rubber (to produce 1,3-butadiene), methyl ethyl ketone production, and pharmaceutical and cosmetics industries (to produce diacetyl and the precursors of polyurethane) (Ji et al., 2011).

1,3-Propanediol is synthesized biochemically from crude glycerol through anaerobic fermentation in fed-batch cultures of *K. pneumoniae* (Mu et al., 2006). However, the yield in this case is very low in comparison to the yield achieved from methanol esterification of soybean oil by alkali- (51.3 g/L) and lipase-catalysis (53 g/L). To overcome this, surface methodology has been used and achieved an optimized yield of 1,3-propanediol (13.8 g/L) (Oh et al., 2008). Additionally, a high yield of 56 g/L of 1,3-propanediol has been achieved *via* bioconversion of *Jatropha* biodiesel into crude glycerol and then into 1,3-propanediol by *K. pneumoniae* ATCC 15380 (Hiremath et al., 2011). In addition, with the help of *C. butyricum* (strains F2b, VPI 3266 and VPI 1718) 1,3-propanediol can be produced from crude glycerol through a synthetic medium and biochemical conversion route and it has been observed that the yield differs from commercial glycerol to crude glycerol as the initial substrate (Chatzifragkou et al., 2011; González-Pajuelo et al., 2004; Papanikolaou & Aggelis, 2003; Papanikolaou et al., 2008). It has a variety of applications in the cosmetic, food, lubricant, paint and pharmaceutical industries. It is also used in the polymer and textile industries to produce polytrimethylene terephthalate, polyurethanes, polyesters, polybutylene terephthalate, etc. (Johnson & Rehmann, 2016).

Dichloro-2-propanol is synthesized from glycerol on the basis of a second-order nucleophilic substitution mechanism using an acetic acid catalyst *via* an intermediate 3-chloro-1,2-propanediol (Luo et al., 2009). Another report showed the same synthesis with the use of heteropolyacid as a replacement for acetic acid and yielded better results (Lee et al., 2008). It is a precursor to epichlorohydrin, which is used as an intermediate in epoxide resin production and in the production of glycerine and several other chemicals used by pharmaceutical industries (Ma et al., 2007).

Mannitol is produced from crude glycerol using *C. magnoliae* with a yield of 0.51 g/g (Khan et al., 2009). It is used in the food and pharmaceutical industries (Li et al., 2016). Additionally, with the help of *Y. lipolytica* (Wratislavia K1) simultaneous production of *erythritol* and citric acid has been observed (Rymowicz et al., 2009). *Erythritol* is used in food industries (Saran et al., 2015). *Arabitol* can be produced using *D. hansenii* SBP-1 from crude glycerol up to a yield of 0.5 g/g (Koganti et al., 2011). *Arabitol* is also used in food industries (Koganti et al., 2011).

Acrolein, the simplest unsaturated aldehyde, is synthesized from glycerol in the presence of acid catalyst in a supercritical aqueous condition (673 K, 34.5 MPa) up to a very high yield of 74 mol% (Watanabe et al., 2007). Another study has been reported at relatively very low pressure (0.85 bar) and at low temperature (533 K) in the presence of activated carbon catalyst or H_3PO_4 to give quite a good yield of 66.8% (Yan & Suppes, 2009). It is mostly used in the paper industry as a slimicide. Further, it is also used in detergent, acrylic acid ester synthesis and super-absorber polymer synthesis industries (Yang et al., 2012).

1,2-Propanediol is synthesized from glycerol *via* its hydrogenolysis using Cu/MgO catalysts (Yuan et al., 2010). It is used in cosmetics (as an additive), paint (as a protective agent for paints), tobacco (as a humectant), polymer (in the production of polyester resins as feedstock), lubricant (as an antifreezing agent), colourings and flavours (as solvent), nutrition products (as a less toxic alternative), hydraulic fluids (as a constituent) and pharmaceutical industries. It is also used in fibre manufacturing, general manufacturing, solvents, emulsifiers and plasticizers (as a precursor molecule) (Rajkhowa et al., 2017; Villa et al., 2015).

With the introduction of carbon dioxide (CO_2) in the presence of a catalyst, *glycerol carbonate* can be synthesized from carboxylation of glycerine. It is used in the pharmaceutical industry (as a chemical precursor), polymer industry (to produce varnish coatings), paint industry (to produce coatings, paints, electrolytes, solvents, lubricants, adhesives and surfactants), and in cosmetics industry and in the components of gas separation membranes, respectively (Zheng et al., 2008). It is a stable and colourless liquid and has been found to be a very useful solvent for several polymers including nylon, cellulose acetate, polyacrylonitrile and nitrocellulose (Zheng et al., 2008).

With the assistance of biological conversion, crude glycerol can be easily converted into *lipids* which can be further utilized as a sustainable biodiesel feedstock for various direct industrial applications. It is known that culturing and algal growth of *S. limacinum SR21* and *C. curvatus* to yield lipids is inversely proportional to the concentrations of glycerol as its higher concentration suppresses cell growth. In this way, the highest cellular lipid content (73.3%) can be achieved with optimal glycerol concentrations of 25 and 35 g/L for untreated and treated crude glycerol, respectively, in the case of culturing of yellow grease-derived crude glycerol (Liang et al., 2010b). The presence of methanol may harm negatively the growth of *S. limacinum* SR21. On the other hand, in the case of *C. curvatus* yeast, only 52% of monounsaturated fatty acid concentrated lipid production was achieved, even after a two-stage fed-batch process with an almost negligible impact of the presence of methanol on cell growth (Liang et al., 2010a). Another report showed that *R. glutinis* TISTR 5159 cultured crude glycerol also produced the highest lipid content of 10.05 g/L with a lipid yield of 60.7% (Saenge et al., 2011). Similarly, *C. protothecoides* converted crude glycerol to lipid with a yield of 0.31 g lipids/g substrate (O'Grady & Morgan, 2011). *Monoglyceride* and *diacylglycerol* are used in the polymer (in polyesters and plasticizers), explosive, food (in preservatives and emulsifiers) and ink industries (Nascimento et al., 2019). *Triglyceride* is used in the cosmetics (as a moisturizer), energy (as fuel additives, and as an anti-fuel oil detonation agent), food (as additives) and tobacco (as cigarette filters) industries (Nascimento et al., 2019).

The conversion of glycerol to *hydrogen* or *syngas via* steam reforming and/or gasification with air or oxygen can be used in energy sector industries to generate fuels (Fan et al., 2010; Slinn et al., 2008; Yoon et al., 2010).

10.3 Macro By-Products

10.3.1 WasteWater

Biodiesel wastewater contains nitrates as predominant contaminants and hence it cannot be used without remediation because of the hazardous nature of nitrates to human health. For remediation purposes, biodiesel wastewater is subjected to photocatalytic reduction in which, with the assistance of solar energy nitrates, it can be

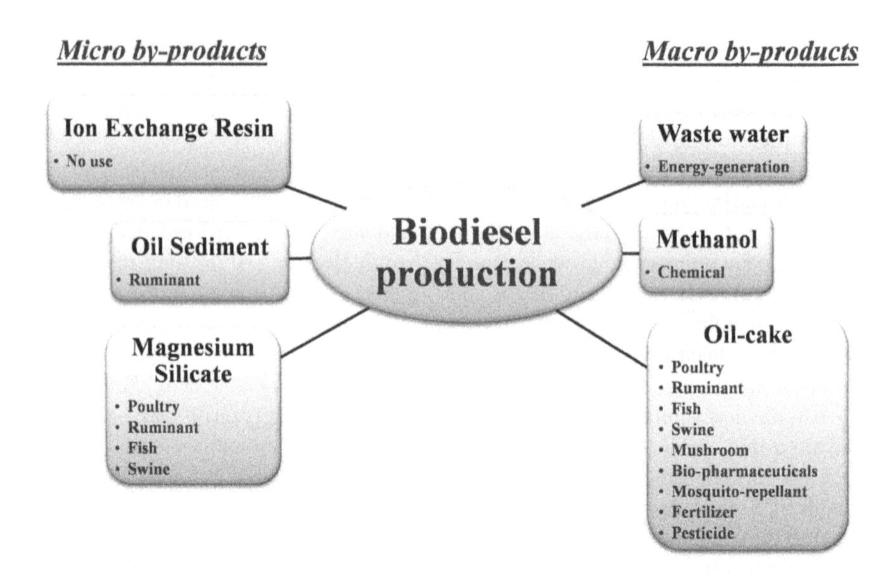

FIGURE 10.4 Industrial application of micro and macro by-products of biodiesel production.

easily converted into nitrogen gas, a non-toxic final by-product (Lucchetti et al., 2017). Various organic molecules like formic acid, ethanol and oxalic acid are used as hole scavengers in this photolytic reaction. Interestingly, other biodiesel by-products, *viz.* methanol and glycerol, have also been utilized as organic hole scavengers. The utilization of glycerol as nitrates, nitrite and phosphorus remover has also been reported in several other research works (Akunna et al., 1993; Bernat et al., 2015; Costa et al., 2018; Yang et al., 2018). Very polar organic hole scavengers lead to the highest photocatalytic reduction rates. Several other results have confirmed that glycerol can yield complete denitrification of biodiesel wastewater (Bodík et al., 2009; Cyplik et al., 2013; Guerrero et al., 2012). After this, biodiesel wastewater can be utilized in energy-generation industries as it can be used to provide a carbon source to produce biogas, biohydrogen and biomethane fuels (Jaruwat et al., 2010; Phukingngam et al., 2011) (Figure 10.4).

10.3.2 Oil-Cake

Since oil-cake is produced from oil seeds which are mainly composed of proteins, fibres and minerals through an extraction process, it has finally been found useful as a protein-rich food for poultry, ruminants, fish and swine industries (Ramachandran et al., 2007; Zhou et al., 2016). Due to the presence of nitrogen, phosphorous and potassium, oil-cake can be used in fertilizer industries as a suitable organic nitrogenous fertilizer (Mazzoncini et al., 2015). Additionally, the presence of a series of biologically active compounds, mainly isothiocyanates, especially in those oil-cakes extracted from Brassicaceae oilseeds, means that it can also be used in pesticide industries (Mazzoncini et al., 2015). Oil-cake may also be used in the mushroom industry and biopharmaceutical industry to produce antibiotics and other biochemicals due to its nutritional value (Ramachandran et al., 2007; Vidyarthi et al., 2002). However, oil-cakes originating from non-edible oil plants, including *Jatropha* and *Karanja*, can be used only in the mosquito-repellent manufacturing industry due to the presence of phorbol esters and karinjin, respectively (Pant et al., 2016) (Figure 10.4).

10.3.3 Methanol

During the transesterification reaction methanol is used in excess. This excess methanol is further separated from the reaction mixture through distillation of the polar phase (crude glycerol) followed by reuse in the next biodiesel production process cycles which in general not considered to be a by-product of the biodiesel mass production (Cao et al., 2008). Nevertheless, methanol is used mostly in the chemical industry either directly or as a precursor molecule (Ali et al., 2015; Dalena et al., 2018) (Figure 10.4).

10.4 Micro By-Products

10.4.1 Ion-Exchange Resin Sediment

Ion-exchange resin is used to remove impurities like inorganic salts and free ions during biodiesel production. It is also used to purify glycerol in order to achieve high-grade glycerol and in this way it yields sediment (waste). Hence, ion-exchange resin sediment (waste) is treated as a micro by-product as only one pound of beads is produced per 900 gallons of biodiesel produced (Mangayil et al., 2015). Since it is used as a catalyst it must be reused, principally, but its reuse is of a similar expense to that of using new beads and hence it tends not to be reused. Conclusively, it can be stated that it has no further industrial applications (Figure 10.4).

10.4.2 Magnesium Silicate Sediment

Magnesium silicate is also widely used as an adsorbent material in the dry washing process (a waterless procedure) in which contaminants are removed from crude biodiesel. Its most popular brand name is Magnesol. Magnesium silicate sediment is a by-product that is also categorized as a micro by-product of biodiesel mass production due to the production of only about 1% of the weight of the biodiesel. However, it is non-reusable, like ion-exchange resins, and hence generally it is disposed of in a landfill due to its non-poisonous nature (Mangayil et al., 2015). Due to the available high nutritional values of biodiesel absorbed contaminants it can be used as a compost and animal feed additive, specifically, in the poultry, ruminant, fish and swine industries (Figure 10.4).

10.4.3 Oil Sediment

When biodiesel is produced from waste cooking oil, it yields an oil sediment which contains meat, bone and/or bread fragments. In some cases, sludge containing meat, bone and/or bread fragments is used for biodiesel production and eventually yields oil sediments from the rendering process. Similar to magnesium silicate contaminated residues, these oil sediment residues can also be used in the poultry, ruminants, fish and swine industries as animal feed (Figure 10.4).

10.5 Conclusion

The increasing demand for oil resources, which are limited, and the high cost of petroleum products and increasing demand for reduction of environmental pollution have inspired researchers to look into further enhancement of biodiesel mass production. In this scenario, utilization of by-products of biodiesel production plays a vital role in reducing the costs associated with biodiesel. Glycerol has been found to be the main by-product, while wastewater, oil-cake and methanol are considered as macro by-products and ion exchange resins, magnesium silicate and oil sediments are categorized as micro by-products that cannot be neglected. In this chapter, we have explained in depth the utilization of all these biodiesel production by-products. In this way, it can be stated that effective utilization of the main by-product, *viz.* crude glycerol, is one of the most important tasks in the commercialization and cost-reduction of biodiesel. Not only the direct use of glycerol but also its indirect use as a precursor molecule (as a feedstock), as explained in this chapter, has proved its utility as a carbon source for the production of fuels and as an energy source for the production of ethanol, butanol, citric acid, glyceric acid, etc. and high-value chemicals. Thus, it may also be stated that the industrial utilization of glycerol has reduced our dependence on non-renewable energy (oil) resources. Utilization of macro by-products like wastewater in energy generation, and oil-cake in the agriculture, bio-pharmaceutical, fertilizer, mosquito-repellent and pesticide industries has also benefitted the cost-cutting and commercialization of biodiesel. In addition, the utilization of micro by-products, like oil sediments and magnesium silicate, in the ruminant, swine, poultry and fishery industries has also add to the fuel cost reductions. Despite obtaining fruitful results it can be said that there remain several aspects that require further improvisation of the technologies involved in the optimization of reaction parameters, production yields, fermentation conditions, and so on.

REFERENCES

Abbaszaadeh, A., Ghobadian, B., Omidkhah, M.R., Najafi, G. 2012. Current biodiesel production technologies: a comparative review. *Energy Conversion and Management,* **63**, 138–148.

Akunna, J.C., Bizeau, C., Moletta, R. 1993. Nitrate and nitrite reductions with anaerobic sludge using various carbon sources: glucose, glycerol, acetic acid, lactic acid and methanol. *Water Research*, **27**(8), 1303–1312.

Ali, K.A., Abdullah, A.Z., Mohamed, A.R. 2015. Recent development in catalytic technologies for methanol synthesis from renewable sources: a critical review. *Renewable and Sustainable Energy Reviews,* **44**, 508–518.

André, A., Diamantopoulou, P., Philippoussis, A., Sarris, D., Komaitis, M., Papanikolaou, S. 2010. Biotechnological conversions of bio-diesel derived waste glycerol into added-value compounds by higher fungi: production of biomass, single cell oil and oxalic acid. *Industrial Crops and Products*, **31**(2), 407–416.

Anitha, M., Kamarudin, S.K., Kofli, N.T. 2016. The potential of glycerol as a value-added commodity. *Chemical Engineering Journal,* **295**, 119–130.

Antolín, G., Tinaut, F.V., Briceño, Y., Castaño, V., Pérez, C., Ramírez, A.I. 2002. Optimisation of biodiesel production by sunflower oil transesterification. *Bioresource Technology*, **83**(2), 111–114.

Ashby, R.D., Solaiman, D.K.Y., Foglia, T.A. 2004. Bacterial poly(hydroxyalkanoate) polymer production from the biodiesel co-product stream. *Journal of Polymers and the Environment*, **12**(3), 105–112.

Ashby, R.D., Solaiman, D.K.Y., Strahan, G.D. 2011. Efficient utilization of crude glycerol as fermentation substrate in the synthesis of poly(3-hydroxybutyrate) Biopolymers. *Journal of the American Oil Chemists' Society*, **88**(7), 949–959.

Bagnato, G., Iulianelli, A., Sanna, A., Basile, A. 2017. Glycerol production and transformation: a critical review with particular emphasis on glycerol reforming reaction for producing hydrogen in conventional and membrane reactors. *Membranes (Basel)*, **7**(2) 1–31.

Beauprez, J., De Mey, M., Soetaert, W. 2010. Microbial succinic acid production: natural versus metabolic engineered producers. *Process Biochemistry,* **45**, 1103–1114.

Bernat, K., Kulikowska, D., Żuchniewski, K. 2015. Glycerine as a carbon source in nitrite removal and sludge production. *Chemical Engineering Journal,* **267**, 324–331.

Bianchi, C., Canton, P., Dimitratos, N., Porta, F., Prati, L. 2005. Selective oxidation of glycerol with oxygen using mono and bimetallic catalysts based on Au, Pd and Pt metals. *Catalysis Today,* **102**, 203–212.

Bodík, I., Blšťáková, A., Sedláček, S., Hutňan, M. 2009. Biodiesel waste as source of organic carbon for municipal WWTP denitrification. *Bioresource Technology*, **100**(8), 2452–2456.

Cao, P., Dubé, M.A., Tremblay, A.Y. 2008. Methanol recycling in the production of biodiesel in a membrane reactor. *Fuel*, **87**(6), 825–833.

Castro-Aguirre, E., Iñiguez-Franco, F., Samsudin, H., Fang, X., Auras, R. 2016. Poly(lactic acid)—Mass production, processing, industrial applications, and end of life. *Advanced Drug Delivery Reviews*, **107**, 333–366.

Cavalheiro, J.M.B.T., de Almeida, M.C.M.D., Grandfils, C., da Fonseca, M.M.R. 2009. Poly(3-hydroxybutyrate) production by *Cupriavidus necator* using waste glycerol. *Process Biochemistry*, **44**(5), 509–515.

Chatzifragkou, A., Papanikolaou, S., Dietz, D., Doulgeraki, A., Nychas, G.-J., Zeng, A.-P. 2011. Production of 1,3-propanediol by Clostridium butyricum growing on biodiesel-derived crude glycerol through a non-sterilized fermentation process. *Applied Microbiology and Biotechnology,* **91**, 101–12.

Ciriminna, R., Palmisano, G., Della Pina, C., Rossi, M., Pagliaro, M. 2006. One-pot electrocatalytic oxidation of glycerol to DHA. *Tetrahedron Letters,* **47**, 6993–6995.

Ciriminna, R., Pina, C.D., Rossi, M., Pagliaro, M. 2014. Understanding the glycerol market. *European Journal of Lipid Science and Technology*, **116**(10), 1432–1439.

Costa, D., Gomes, A., Fernandes, M., Bortoluzzi, R., Magalhães, M., Skoronski, E. 2018. Using natural biomass microorganisms for drinking water denitrification. *Journal of Environmental Management,* **217**, 520–530.

Cyplik, P., Juzwa, W., Marecik, R., Powierska-Czarny, J., Piotrowska-Cyplik, A., Czarny, J., Drożdżyńska, A., Chrzanowski, Ł. 2013. Denitrification of industrial wastewater: influence of glycerol addition on metabolic activity and community shifts in a microbial consortium. *Chemosphere*, **93**(11), 2823–2831.

Dalena, F., Senatore, A., Marino, A., Gordano, A., Basile, M., Basile, A. 2018. Methanol production and applications: an overview. In: *Methanol* (Eds.) A. Basile, F. Dalena, Elsevier, pp. 3–28.

Datta, R., Henry, M. 2006. Lactic acid: recent advances in products, processes and technologies — a review. *Journal of Chemical Technology & Biotechnology*, **81**(7), 1119–1129.

Dobroth, Z.T., Hu, S., Coats, E.R., McDonald, A.G. 2011. Polyhydroxybutyrate synthesis on biodiesel wastewater using mixed microbial consortia. *Bioresource Technology*, **102**(3), 3352–9.

Encinar, J.M., González, J.F., Rodríguez-Reinares, A. 2007. Ethanolysis of used frying oil. Biodiesel preparation and characterization. *Fuel Processing Technology*, **88**(5), 513–522.

Fan, X., Burton, R., Zhou, Y. 2010. Glycerol (byproduct of biodiesel production) as a source for fuels and chemicals–mini review. *Open Fuels & Energy Science Journal,* **3**, 17–22.

Fukuda, H., Kondo, A., Noda, H. 2001. Biodiesel fuel production by transesterification of oils. *Journal of Bioscience and Bioengineering*, **92**(5), 405–416.

González-Pajuelo, M., Andrade, J.C., Vasconcelos, I. 2004. Production of 1,3-propanediol by *Clostridium butyricum* VPI 3266 using a synthetic medium and raw glycerol. *Journal of Industrial Microbiology and Biotechnology*, **31**(9), 442–446.

Guerrero, J., Tayà, C., Guisasola, A., Baeza, J.A. 2012. Glycerol as a sole carbon source for enhanced biological phosphorus removal. *Water Research*, **46**(9), 2983–2991.

Haag, R., Sunder, A., Stumbé, J.-F. 2000. An approach to glycerol dendrimers and pseudo-dendritic polyglycerols. *Journal of the American Chemical Society*, **122**(12), 2954–2955.

Habe, H., Fukuoka, T., Kitamoto, D., Sakaki, K. 2009. Biotechnological production of D-glyceric acid and its application. *Applied Microbiology and Biotechnology*, **84**(3), 445–52.

Handa, S., Sharma, A., Chakraborti, K.K. 1986. Natural products and plants as liver protecting drugs. *Fitoterapia,* **57**, 307–351.

Harry-O'kuru, R.E., Gordon, S.H., Klokkenga, M. 2015. Bio-generation of succinic acid by fermentation of *Physaria fendleri* seed polysaccharides. *Industrial Crops and Products,* **77**, 116–122.

Harvey, B.G., Meylemans, H.A. 2011. The role of butanol in the development of sustainable fuel technologies. *Journal of Chemical Technology & Biotechnology*, **86**(1), 2–9.

Hiremath, A., Kannabiran, M., Rangaswamy, V. 2011. 1,3-Propanediol production from crude glycerol from Jatropha biodiesel process. *New Biotechnology,* **28**, 19–23.

Hong, A.-A., Cheng, K.-K., Peng, F., Zhou, S., Sun, Y., Liu, C.-M., Liu, D.-H. 2009. Strain isolation and optimization of process parameters for bioconversion of glycerol to lactic acid. *Journal of Chemical Technology and Biotechnology,* **84**, 1576–1581.

Hong, W.-K., Kim, C.-H., Heo, S.-Y., Luo, L.H., Oh, B.-R., Seo, J.-W. 2010. Enhanced production of ethanol from glycerol by engineered *Hansenula polymorpha* expressing pyruvate decarboxylase and aldehyde dehydrogenase genes from *Zymomonas mobilis*. *Biotechnology Letters*, **32**(8), 1077–1082.

Huber, G.W., Iborra, S., Corma, A. 2006. Synthesis of transportation fuels from biomass: chemistry, catalysts, and engineering. *Chemical Reviews*, **106**(9), 4044–4098.

Ibrahim, M.H.A., Steinbüchel, A. 2009. Poly(3-Hydroxybutyrate) production from glycerol by *Zobellella denitrificans* MW1 via high-cell-density fed-batch fermentation and simplified solvent extraction. *Applied and Environmental Microbiology,* **75**(19), 6222.

Imandi, S., Bandaru, V., Somalanka, S., Rao, G. 2007. Optimization of medium constituents for the production of citric acid from byproduct glycerol using Doehlert experimental design. *Enzyme and Microbial Technology,* **40**, 1367–1372.

Jaruwat, P., Kongjao, S., Hunsom, M. 2010. Management of biodiesel wastewater by the combined processes of chemical recovery and electrochemical treatment. *Energy Conversion and Management*, **51**(3), 531–537.

Ji, X.J., Huang, H., Ouyang, P.K. 2011. Microbial 2,3-butanediol production: a state-of-the-art review. *Biotechnology Advances*, **29**(3), 351–64.

Jiang, M., Dai, W., Xi, Y., Mingke, W., Kong, X., Jiangfeng, m., Zhang, M., Chen, K., Wei, P. 2013. Succinic acid production from sucrose by *Actinobacillus succinogenes* NJ113. *Bioresource Technology*, **153C**, 327–332.

Johnson, E.E., Rehmann, L. 2016. The role of 1,3-propanediol production in fermentation of glycerol by *Clostridium pasteurianum. Bioresources Technology,* **209**, 1–7.

Kamzolova, S., Dedyukhina, E., Anastassiadis, S., Golovchenko, N., Morgunov, I. 2011. Citric acid production by yeast grown on glycerol-containing waste from biodiesel industry. *Food Technology and Biotechnology,* **49**, 65–74.

Kaur, J., Sarma, A.K., Jha, M.K., Gera, P. 2020. Valorisation of crude glycerol to value-added products: perspectives of process technology, economics and environmental issues. *Biotechnology Reports,* **27**, e00487.

Khan, A., Bhide, A., Gadre, R. 2009. Mannitol production from glycerol by resting cells of *Candida magnoliae. Bioresources Technology*, **100**(20), 4911–3.

Knothe, G., Dunn, R.O., Bagby, M.O. 1997. Biodiesel: the use of vegetable oils and their derivatives as alternative diesel fuels. In: *Fuels and Chemicals from Biomass*, Saha, B.C., Ed. Vol. 666, American Chemical Society, pp. 172–208.

Koganti, S., Kuo, T.M., Kurtzman, C.P., Smith, N., Ju, L.K. 2011. Production of arabitol from glycerol: strain screening and study of factors affecting production yield. *Appl Microbiol Biotechnol*, **90**(1), 257–67.

Kolesárová, N., Hutňan, M., Bodík, I., Špalková, V. 2011. Utilization of biodiesel by-products for biogas production. *Journal of Biomedicine and Biotechnology,* **2011**, 126798.

Lee, S., Park, D., Kim, H., Lee, J., Jung, J., Woo, S., Song, W., Kwon, M., Song, I. 2008. Direct preparation of dichloropropanol (DCP) from glycerol using heteropolyacid (HPA) catalysts: a catalyst screen study. *Catalysis Communications,* **9**, 1920–1923.

Li, H., Ding, X., Zhao, Y.-C., Han, B.-H. 2016. Preparation of mannitol-based ketal-linked porous organic polymers and their application for selective capture of carbon dioxide. *Polymer,* **89**, 112–118.

Liang, Y., Cui, Y., Trushenski, J., Blackburn, J.W. 2010a. Converting crude glycerol derived from yellow grease to lipids through yeast fermentation. *Bioresource Technology*, **101**(19), 7581–7586.

Liang, Y., Sarkany, N., Cui, Y., Blackburn, J.W. 2010b. Batch stage study of lipid production from crude glycerol derived from yellow grease or animal fats through microalgal fermentation. *Bioresour Technol*, **101**(17), 6745–50.

Lotero, E., Liu, Y., Lopez, D.E., Suwannakarn, K., Bruce, D.A., Goodwin, J.G. 2005. Synthesis of biodiesel via acid Catalysis. *Industrial & Engineering Chemistry Research*, **44**(14), 5353–5363.

Lucchetti, R., Onotri, L., Clarizia, L., Natale, F.D., Somma, I.D., Andreozzi, R., Marotta, R. 2017. Removal of nitrate and simultaneous hydrogen generation through photocatalytic reforming of glycerol over "in situ" prepared zero-valent nano copper/P25. *Applied Catalysis B: Environmental*, **202**, 539–549.

Luo, X., Ge, X., Cui, S., Li, Y. 2016. Value-added processing of crude glycerol into chemicals and polymers. *Bioresource Technology,* **215**, 144–154.

Luo, Z.-H., 罗正鸿, You, X.-Z., Li, H.-R. 2009. Direct preparation kinetics of 1,3-dichloro-2-propanol from glycerol using acetic acid catalyst. *Industrial & Engineering Chemistry Research*, **48** (1) 446–452.

Ma, F., Hanna, M.A. 1999. Biodiesel production: a review. *Journal Series* #12109. *Bioresource Technology*, **70**(1), 1–15.

Ma, L., Zhu, J.W., Yuan, X.Q., Yue, Q. 2007. Synthesis of epichlorohydrin from dichloropropanols: kinetic aof the process. *Chemical Engineering Research and Design*, **85**(12), 1580–1585.

Mandolesi de Araújo, C.D., de Andrade, C.C., de Souza e Silva, E., Dupas, F.A. 2013. Biodiesel production from used cooking oil: a review. *Renewable and Sustainable Energy Reviews,* **27**, 445–452.

Mangayil, R., Aho, T., Karp, M., Santala, V. 2015. Improved bioconversion of crude glycerol to hydrogen by statistical optimization of media components. *Renewable Energy,* **75**, 583–589.

Mazzoncini, M., Antichi, D., Tavarini, S., Silvestri, N., Lazzeri, L., D'Avino, L. 2015. Effect of defatted oilseed meals applied as organic fertilizers on vegetable crop production and environmental impact. *Industrial Crops and Products,* **75**, 54–64.

Millard, C., Chao, Y.-P., Liao, J., Donnelly, M. 1996. Enhanced Production of succinic acid by overexpression of phosphoenolpyruvate carboxylase in *Escherichia coli. Applied and Environmental Microbiology,* **62**, 1808–10.

Mothes, G., Schnorpfeil, C., Ackermann, J.U. 2007. Production of PHB from crude glycerol. *Engineering in Life Sciences*, **7**(5), 475–479.

Mu, Y., Teng, H., Zhang, D.J., Wang, W., Xiu, Z.L. 2006. Microbial production of 1,3-propanediol by *Klebsiella pneumoniae* using crude glycerol from biodiesel preparations. *Biotechnology Letters*, **28**(21), 1755–9.

Musiał, I., Cibis, E., Rymowicz, W. 2011. Designing a process of kaolin bleaching in an oxalic acid enriched medium by *Aspergillus niger* cultivated on biodiesel-derived waste composed of glycerol and fatty acids. *Applied Clay Science*, **52**(3), 277–284.

Nascimento, J.A.C., Pinto, B.P., Calado, V.M.A., Mota, C.J.A. 2019. Synthesis of solketal fuel additive from acetone and glycerol using CO2 as switchable catalyst. *Frontiers in Energy Research,* **7**, 58.

Nitayavardhana, S., Khanal, S.K. 2010. Innovative biorefinery concept for sugar-based ethanol industries: production of protein-rich fungal biomass on vinasse as an aquaculture feed ingredient. *Bioresour Technol*, **101**(23), 9078–85.

Noshadi, I., Amin, N.A.S., Parnas, R.S. 2012. Continuous production of biodiesel from waste cooking oil in a reactive distillation column catalyzed by solid heteropolyacid: optimization using response surface methodology (RSM). *Fuel,* **94**, 156–164.

O'Grady, J., Morgan, J.A. 2011. Heterotrophic growth and lipid production of *Chlorella protothecoides* on glycerol. *Bioprocess and Biosystems Engineering*, **34**(1), 121–125.

Ogunkunle, O., Ahmed, N.A. 2019. A review of global current scenario of biodiesel adoption and combustion in vehicular diesel engines. *Energy Reports,* **5**, 1560–1579.

Oh, B.-R., Seo, J.-W., Choi, M.H., Kim, C.H. 2008. Optimization of culture conditions for 1,3-propanediol production from crude glycerol by *Klebsiella pneumoniae* using response surface methodology. *Biotechnology and Bioprocess Engineering*, **13**(6), 666–670.

Pant, M., Sharma, S., Dubey, S., Naik, S.N., Patanjali, P.K. 2016. Utilization of biodiesel by-products for mosquito control. *Journal of Bioscience and Bioengineering*, **121**(3), 299–302.

Papanikolaou, S., Aggelis, G. 2003. Modelling aspects of the biotechnological valorization of raw glycerol: production of citric acid by *Yarrowia lipolytica* and 1,3-propanediol by *Clostridium butyricum. Journal of Chemical Technology and Biotechnology,* **78**, 542–547.

Papanikolaou, S., Fakas, S., Fick, M., Chevalot, I., Galiotou-Panayotou, M., Komaitis, M., Marc, I., Aggelis, G. 2008. Biotechnological valorisation of raw glycerol discharged after bio-diesel (fatty acid methyl esters) manufacturing process: production of 1,3-propanediol, citric acid and single cell oil. *Biomass and Bioenergy*, **32**(1), 60–71.

Papanikolaou, S., Muniglia, L., Chevalot, I., Aggelis, G., Marc, I. 2002. *Yarrowia lipolytica* as a potential producer of citric acid from raw glycerol. *Journal of Applied Microbiology*, **92**(4), 737–44.

Petersen, A.B., Na, R., Wulf, H.C. 2003. Sunless skin tanning with dihydroxyacetone delays broad-spectrum ultraviolet photocarcinogenesis in hairless mice. *Mutation Research/Fundamental and Molecular Mechanisms of Mutagenesis*, **542**(1–2), 129–38.

Phukingngam, D., Chavalparit, O., Somchai, D., Ongwandee, M. 2011. Anaerobic baffled reactor treatment of biodiesel-processing wastewater with high strength of methanol and glycerol: reactor performance and biogas production. *Chemical Papers*, **65**(5), 644–651.

Prabisha, T.P., Sindhu, R., Binod, P., Sankar, V., Raghu, K.G., Pandey, A. 2015. Production and characterization of PHB from a novel isolate *Comamonas* sp. from a dairy effluent sample and its application in cell culture. *Biochemical Engineering Journal,* **101**, 150–159.

Pyle, D.J., Garcia, R.A., Wen, Z. 2008. Producing docosahexaenoic acid (DHA)-rich algae from biodiesel-derived crude glycerol: effects of impurities on DHA production and algal biomass composition. *Journal of Agricultural and Food Chemistry*, **56**(11), 3933–3939.

Rajkhowa, T., Marin, G.B., Thybaut, J.W. 2017. A comprehensive kinetic model for Cu catalyzed liquid phase glycerol hydrogenolysis. *Applied Catalysis B: Environmental,* **205**, 469–480.

Ramachandran, S., Singh, S.K., Larroche, C., Soccol, C.R., Pandey, A. 2007. Oil cakes and their biotechnological applications – A review. *Bioresource Technology*, **98**(10), 2000–2009.

Rosseto, R., Tcacenco, C., Ranganathan, R., Hajdu, J. 2008. Synthesis of phosphatidylcholine analogues derived from glyceric acid: a new class of biologically active phospholipid compounds. *Tetrahedron letters,* **49**, 3500–3503.

Rymowicz, W., Rywińska, A., Marcinkiewicz, M. 2009. High-yield production of erythritol from raw glycerol in fed-batch cultures of *Yarrowia lipolytica. Biotechnol Lett,* **31**(3), 377–80.

Rymowicz, W., Rywińska, A., Żarowska, B., Juszczyk, P. 2006. Citric acid production from raw glycerol by acetate mutants of *Yarrowia lipolytica*. *Chemical Papers*, **60**(5), 391–394.

Rywiska, A., Rymowicz, W., Larowska, B., Wojtatowicz, M. 2013. Biosynthesis of citric acid from glycerol by acetate mutants of *Yarrowia lipolytica* in fed-batch fermentation. *Food Technology and Biotechnology* 47 (1) 1–6.

Saenge, C., Cheirsilp, B., Suksaroge, T.T., Bourtoom, T. 2011. Potential use of oleaginous red yeast *Rhodotorula glutinis* for the bioconversion of crude glycerol from biodiesel plant to lipids and carotenoids. *Process Biochemistry*, **46**(1), 210–218.

Saran, S., Mukherjee, S., Dalal, J., Saxena, R.K. 2015. High production of erythritol from *Candida sorbosivorans* SSE-24 and its inhibitory effect on biofilm formation of *Streptococcus mutans*. *Bioresource technology,* **198**, 31–38.

Sile, E., Chornaja, S., Dubencovs, K., Zhizhkun, S., Serga, V., Kulikova, L., Palcevskis, E. 2013. Selective liquid phase oxidation of glycerol to glyceric acid over novel supported Pt catalysts. *Journal of the Serbian Chemical Society,* **78**, 1359–1372.

Singhabhandhu, A., Tezuka, T. 2010. A perspective on incorporation of glycerin purification process in biodiesel plants using waste cooking oil as feedstock. *Energy*, **35**(6), 2493–2504.

Slinn, M., Kendall, K., Mallon, C., Andrews, J. 2008. Steam reforming of biodiesel by-product to make renewable hydrogen. *Bioresource Technology*, **99**(13), 5851–5858.

Soccol, C., Vandenberghe, L., Rodrigues, C., Pandey, A. 2006. New perspectives for citric acid production and application. *Food Technology and Biotechnology*, **44** (2) 141–149.

Song, H., Lee, S.Y. 2006. Production of succinic acid by bacterial fermentation. *Enzyme and Microbial Technology,* **39**, 352–361.

Taconi, K.A., Venkataramanan, K.P., Johnson, D.T. 2009. Growth and solvent production by *Clostridium pasteurianum* ATCC® 6013™ utilizing biodiesel-derived crude glycerol as the sole carbon source. *Environmental Progress & Sustainable Energy*, **28**(1), 100–110.

Tan, H.W., Abdul Aziz, A.R., Aroua, M.K. 2013. Glycerol production and its applications as a raw material: a review. *Renewable and Sustainable Energy Reviews,* **27**, 118–127.

Vidyarthi, A.S., Tyagi, R.D., Valero, J.R., Surampalli, R.Y. 2002. Studies on the production of B. thuringiensis based biopesticides using wastewater sludge as a raw material. *Water Research*, **36**(19), 4850–4860.

Villa, A., Dimitratos, N., Chan-Thaw, C.E., Hammond, C., Prati, L., Hutchings, G.J. 2015. Glycerol oxidation using gold-containing catalysts. *Accounts of Chemical Research*, **48**(5), 1403–1412.

Watanabe, M., Iida, T., Aizawa, Y., Aida, T.M., Inomata, H. 2007. Acrolein synthesis from glycerol in hot-compressed water. *Bioresour Technol*, **98**(6), 1285–90.

Yan, W., Suppes, G. 2009. Low-pressure packed-bed gas-phase dehydration of glycerol to acrolein. *Industrial & Engineering Chemistry Research*, **48**(7), 3279–3283.

Yang, F., Hanna, M.A., Sun, R. 2012. Value-added uses for crude glycerol—a byproduct of biodiesel production. *Biotechnology for Biofuels*, **5**(1), 13.

Yang, G., Wang, D., Yang, Q., Zhao, J., Liu, Y., Wang, Q., Zeng, G., Li, X., Li, H. 2018. Effect of acetate to glycerol ratio on enhanced biological phosphorus removal. *Chemosphere,* **196**, 78–86.

Yang, T.-W., Rao, Z.-M., Zhang, X., Xu, M.-J., Xu, Z.-H., Yang, S.-T. 2013. Fermentation of biodiesel-derived glycerol by *Bacillus amyloliquefaciens*: effects of co-substrates on 2,3-butanediol production. *Applied Microbiology and Biotechnology*, **97**(17), 7651–7658.

Yoon, S.J., Choi, Y.-C., Son, Y.-I., Lee, S.-H., Lee, J.-G. 2010. Gasification of biodiesel by-product with air or oxygen to make syngas. *Bioresource Technology*, **101**(4), 1227–1232.

Yuan, Z., Wang, J., Wang, L., Xie, W., Chen, P., Hou, Z., Zheng, X. 2010. Biodiesel derived glycerol hydrogenolysis to 1,2-propanediol on Cu/MgO catalysts. *Bioresource Technology*, **101**(18), 7088–7092.

Zeikus, J., Jain, M., Elankovan, P. 1999. Biotechnology of succinic acid production and markets for derived industrial products. *Applied Microbiology and Biotechnology,* **51**, 545–552.

Zheng, Y., Chen, X., Shen, Y. 2008. Commodity chemicals derived from glycerol, an important biorefinery feedstock. *Chemical Reviews*, **108**(12).

Zhou, X., Beltranena, E., Zijlstra, R.T. 2016. Effects of feeding canola press-cake on diet nutrient digestibility and growth performance of weaned pigs. *Animal Feed Science and Technology,* **211**, 208–215.

Zhu, S., Gao, X., Zhu, Y., Zhu, Y., Zheng, H., Li, Y. 2013. Promoting effect of boron oxide on Cu/SiO2 catalyst for glycerol hydrogenolysis to 1,2-propanediol. *Journal of Catalysis*, **303**(Complete), 70–79.

11

A Life Cycle Assessment of Biodiesel Production

Mariany Costa Deprá, Patrícia Arrojo da Silva, Paola Lasta, Leila Queiroz Zepka, and Eduardo Jacob-Lopes*

CONTENTS

11.1 Introduction

The growing global demand for energy and the need for clean and renewable fuel sources represent the primary factors driving the exponential growth of the biofuels market. In terms of knowledge, the global biodiesel market was valued at approximately USD 32.4 billion in 2018. Currently, future estimates suggest that this segment will reach USD 44.2 billion by 2024, recording a compound annual growth rate of around 5% during 2019–2024 (Markets and Markets, 2020).

It is known, however, that this market increase is associated with the incorrect management of surplus energy demands since these contributed considerably to the enormous environmental burden. Consequently, increased consumer environmental

* *Corresponding Author* ejacoblopes@gmail.com

DOI: 10.1201/9781003120858-11

awareness of the benefits of biodiesel has led to its widespread adoption worldwide (Prasad et al., 2020).

Indeed, the overexploitation of energy and the depletion of energy reserves, including fossil fuels, required exploration of alternative resources. However, it is important to note that fossil fuel reserves still represent an abundant source of energy. And therefore, this scenario is expected to change in the near future, considering the critical aspects involving endorsed sustainability of the processes associated with extensive energy exploration. Thus, the need for interest in pointing out alternative approaches to renewable resources is increased in the current scenario (Rebello et al., 2020).

Undeniably, the latent potential of energy production from biomass, in general, has received substantial attention in recent decades. However, the path of consolidation of this energy bioproduct still presents several parameters to be overcome. Among them, the choice of feedstock for exploration seems to be, by far, the main problem that haunts bioenergetic processes (Chopra et al., 2020). This is because conventional biodiesel production from agricultural cultures was not considered a fully sustainable approach in the long term, due to its possible adverse effects of deforestation, depletion of biodiversity and potential competition with food resources or land destined for agriculture. In contrast, the capacity of biodiesel production from oils extracted from different sources, such as jatropha, castor, hemp, and karanja, as well as unconventional sources, such as food waste, poultry fat, used cooking oil, and microalgae have been investigated, and considered to be promising (Khanum et al., 2020).

However, despite biodiesel's advantageous attributes, there remain some gaps to be overcome before it is accepted as an appropriate fuel (Hosseinzadeh-Bandbafha et al., 2020). This is because the unit operations involved in the process, that is, production, harvesting, extraction, conversion, use and access to energy, are at the base of many of the significant challenges associated with sustainability, safety and environmental quality. Thus, a series of factors must be considered when selecting the best viable alternative (Prasad et al., 2020).

In order to assist in bioprocess performance analysis, as well as industrial decision-making, the life cycle assessment (LCA) tool has emerged as a robust systematic method, used to measure the potential impacts of the life cycle phases of a given product, from its production to the end of its life (Deprá et al., 2020). In addition, LCA aims to provide a more detailed and lucid picture of the actual environmental compensations in the process. Its application has been widely used in the evaluation of biofuels. However, a range of theoretical premises is considered, making it complex to collate results and different realities (Branco-Vieira et al., 2020).

Because of the above, this chapter aims to compile details of the main raw materials used for the production of first-, second- and third-generation biodiesels, as well as identification of the different subprocesses and their respective adverse environmental effects. Also, the importance of this compilation lies in providing an essential view on the relationship of energy generation processes as well as their role in environmental burdens, which we believe to be useful for the scientific community. Finally, the perspective beyond sustainability was also addressed, as this theme also helps to understand the influence of market relations and public policies involved in the consolidation of clean energies.

11.2 First- and Second-Generation Biodiesel

Biofuels can be classified according to the raw material used. For primary classification, first-generation biodiesel is composed of edible oils and has advantages for biodiesel production (such as soybean, canola, rapeseed, palm, sunflower, etc.). From another perspective, second-generation biofuels use non-food raw materials, such as lignocellulosic biomass, including jatropha oil (Singh et al., 2019).

However, first-generation biofuels require a significant amount of cultivation area, water and fertilizers, resulting in concerns about environmental impacts and carbon balances, and providing limits for its production (Kumar et al., 2020). In addition, the dispute between food versus fuel becomes more serious due to the increase in food prices, which is a result of the increase in the biofuel production (Naik et al., 2010).

Subsequently, second-generation biofuels, produced from woody biomass, are more energy-efficient and more flexible compared to other raw materials. Also, the ability to use cellulosic biomass has a lower cost and better environmental performance. In this way, studies on second-generation biofuels are still under development and they are not commercially available (Havlík et al., 2011).

In this sense, next, we address the main characteristics that involve the production parameters of first- and second-generation biodiesel as raw material, lipid and fatty acid content, yield, cultivation time, soil preparation, process for obtaining biodiesel and environmental impacts.

11.2.1 Soybean

Soybean (*Glycine max*) is a perennial crop in the Leguminosae family, with a height of between 0.5 to 1.2 m, it is a perennial crop with a 3-month harvest, and it is widely cultivated in several countries (Singh et al., 2020). Among them, the production of countries such as the United States, Brazil and Argentina (Bergmann et al., 2013) stands out, making this feedstock the core biodiesel source in these countries. However, Brazil has gained extra attention, given that the data indicate that about 70% of biodiesel production is attributed to soybean, corresponding to 5.4 billion USD in 2018 (da Silva César et al., 2019; Woyann et al., 2019).

In terms of biodiesel, it is essential to approach issues such as content and lipid profile. In this way, this culture presents about 18–21% oil content (Bergmann, et al., 2013). In addition, this fraction is characterized by having mainly compounds such as palmitic acid, stearic acid, oleic acid, linoleic acid and linolenic acid (Liu et al., 2008) (Table 11.1).

Additionally, the general process of converting oil into biodiesel fundamentally requires the reaction of triglycerides with methanol in the presence of sodium hydroxide which acts as a catalyst, producing a mixture of methyl esters and glycerine. This reaction requires heat and a strong catalyst, such as potassium hydroxide. The refined oil is mixed with the preparation of the methanol-potassium hydroxide transesterification reaction, resulting in methyl ester (biofuel) (Requena et al., 2011). However, it is important to highlight that soybean biodiesel production includes some unitary steps such as dehulling, extraction and degumming, which are inherent characteristics of this process (Figure 11.1).

TABLE 11.1
Yield of Raw Materials Applied to First- and Second-Generation Biodiesel

Raw material	Average agricultural yield (ton/ha)	Average biofuel yield (L/ha)
Soybean	2.24	550
Sunflower	1.37	950
Rapeseed	1.98	1200
Jatropha	4.00	2800
Palm	14.10	4500

Source: Adapted from Rocha et al. 2019.

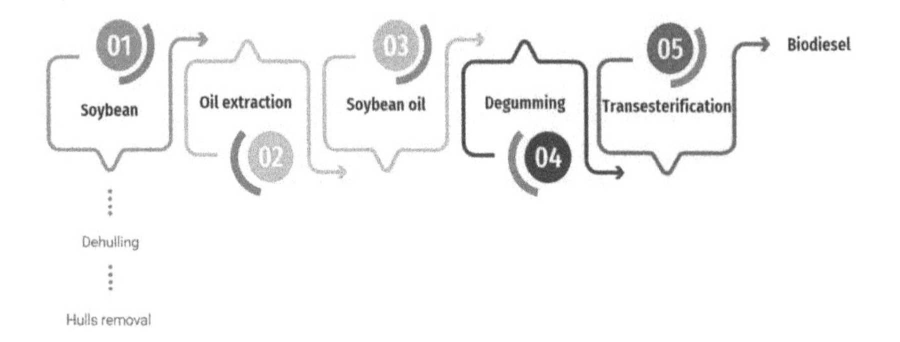

FIGURE 11.1 Simplified flowchart of the process for obtaining biodiesel soybean. (Adapted from Severo et al. 2019.)

However, under environmental aspects, some environmental burdens are found for soybean cultivation on a large scale. Among them, traditional planting and soil preparation can cause erosion and a reduction of soil organic matter. However, conventional preparation results in less erosion and oxidation of organic matter, but requires greater use of herbicides due to the difficult control of weeds (Milazzo et al., 2013). Therefore, depending on management practices, the slope and local climate, soil loss can reach 19–30 tons per hectare. Consequently, soil preparation can influence the category of land use and change. Thus, no-tillage helps to reduce soil erosion, and is more efficient in terms of energy, labour and machinery (Tomei and Upham, 2009).

Concurrently, herbicides and transoceanic transport influence the eutrophication and acidification categories. Terrestrial ecotoxicity (pesticide use, and includes the stages of production and transport) is mainly due to heavy metal emissions to the soil because of the production of chemical fertilizers, which contain heavy metals as contaminants (Xue et al., 2012).

The cultivation of soybeans in relation to global warming potential is dominated by N_2O emissions in the agricultural phase, during the degradation of crop residues (for example, straw), biological nitrogen fixation and in the burning of petroleum fuels (Da Silva et al., 2010). In parallel, the critical points for water use are for plantations

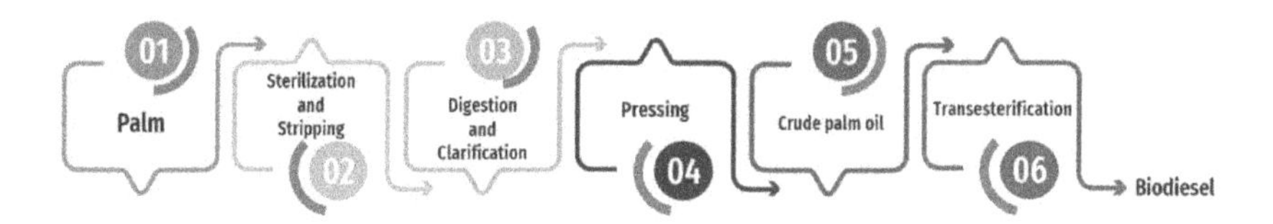

FIGURE 11.2 Simplified flowchart of the process for obtaining biodiesel palm. (Adapted from Severo, et al., 2019.)

that require irrigation, this being the case in regions with water scarcity where this is necessary for crops (Schmidt, 2015).

11.2.2 Palm

The palm (*Elaeis guineensis*) is a perennial plant of the Arecaceae family (Bergmann, et al., 2013) that is suitable for the generation of biodiesel due to the large oil yield per hectare (see Table 11.1) (Silalertruksa and Gheewala, 2012). However, the quantity and quality of the oil produced depend on the variety of palm, soil type, age of the trees and fruit management (Baryeh, 2001).

Additionally, the palm is a non-branched monoecious tree, which grows between 20 and 30 m in height and can live up to 200 years (Baryeh, 2001). Over the years, the trees become very tall, making harvesting difficult, requiring replanting. The palm begins to bear fruit after 2.5–3 years (Silalertruksa and Gheewala, 2012).

Palm oil can be extracted from two parts of the fruit – the pulp that contains 20–22% oil and the seed that contains 38–55% oil (Bergmann et al., 2013). In addition, it has a high resistance to oxidation due to its significant saturated fatty acids content (de Almeida et al., 2015). This fraction is characterized by having mainly compounds such as myristic acid, palmitic acid, stearic acid, oleic acid, linoleic acid and linolenic acid (Montoya et al., 2013).

In terms of cultivation, the palm needs soil levelling, ploughing and holes for planting seedlings, as well as the application of fertilizers. Usually, no insecticides are used, but herbicides such as glyphosate are used in crops for weed control (Saswattecha et al., 2016).

The palm biodiesel production process follows the process flow chart shown in Figure 11.2. The main steps that characterize this process consist of (i) sterilization and pickling, (ii) digestion, clarification and pressing, and (iii) transesterification. The bunches of fresh fruit are removed to separate the fruit from the stem. Then, the crude palm oil is drawn from the fruit, and then a fibre cake is produced, from which the oil is extracted from the core of these fractions. Water use is high when extracting palm oil, resulting in an additional environmental impact. Finally, biodiesel is obtained from the oil transesterification (Saswattecha et al., 2016; Severo et al., 2019).

Additionally, palm oil is acidic, generating a serious problem for the generation of biodiesel, requiring processing within 24–48 hours after harvesting to avoid a decrease in oil quality. The loss of oil quality is associated with enzymatic deterioration, that in the presence of sodium hydroxide in the transesterification reaction, can form soap instead of biodiesel (Bergmann, et al., 2013).

Notably, the main critical points in the palm biodiesel life cycle are global warming, eutrophication, acidification, toxicity and loss of biodiversity (Manik et al., 2013).

However, the impact processes related to planting are related to energy use and fertilizer production. Thus, the use of pesticides contributed to a lower impact due to the integrated biological pest management. In general, fertilizer use is the most polluting process in the system, followed by transport and emissions from the boiler (Sumiani and Sune, 2007).

In addition, in many regions, the manufacturing of palm oil is a serious problem in terms of changes in land use and deforestation. In contrast, there are pollution problems generated by palm oil generation that exacerbates a number of environmental problems, for example, methane emissions in wastewater and nitrous oxide emissions in fertilizer applications, such as nitrogen in the cultivation phase. Also, these fertilizer applications (nitrogen and phosphorus) can cause eutrophication in surface waters (Saswattecha et al., 2015).

Finally, various air pollutants are generated during fuel combustion, such as sulphur dioxide, non-methane volatile organic compounds, nitrogen oxides, particulate matter and carbon monoxide, which are harmful to human health and can cause problems with acidification, eutrophication and photochemical ozone formation (Saswattecha et al., 2015).

11.2.3 Rapeseed

Rapeseed (*Brassica napus*) is part of the Brassicaceae family, having the characteristics of 150 cm height and presenting fruits with 12–18 seeds of 1.1–1.8 mm diameter. This is a suitable crop for the production of edible vegetable oils, animal feed and biodiesel (Jaime et al., 2018), and it is considered to be the second most-produced oilseed, after soybean oil. In addition, the main producers are the European Union, Canada, China and India and, therefore, this culture has been suggested for the production of biodiesel (Aldhaidhawi et al., 2017).

The lipid content of its cultivated seeds varies from 38% to 45% of oil (Jaime et al., 2018). The amount of seed oil and its yield is influenced by the genotype, environment and their interaction. Therefore, the two most accurate measures for the adoption of the rapeseed cultivar are seed production and oil content (Moghaddam and Pourdad, 2011). In addition, rapeseed is characterized by compounds such as erucic acid, oleic acid, linoleic acid and α-linolenic acid (El-Beltagi and Mohamed, 2010). In terms of yield, rapeseed produces more oil compared to soybean and sunflower (see Table 11.1).

Rapeseed cultivation includes several stages, such as soil preparation, fertilization, sowing, weed control and harvesting. The straw is generally ploughed back into the field, an agricultural management activity with various benefits, such as nutrient return and cycling, construction of soil organic matter and the prevention of soil erosion (Malça et al., 2014). Nevertheless, cultivation is employed in two practices: conventional tillage (with soil tillage) and no-till (without soil tillage). Conventional cultivation involves inversion of the soil and an intensive sequence of soil preparation, such as ploughing and harrowing. In no-till, the crop is sown in untouched soil. Stubble is retained on the surface and weed control is achieved by greater use of herbicides (Iriarte et al., 2011).

Figure 11.3 describes the main stages in the process of obtaining rapeseed biodiesel, which consists of mechanical classification (for straw removal), extraction and

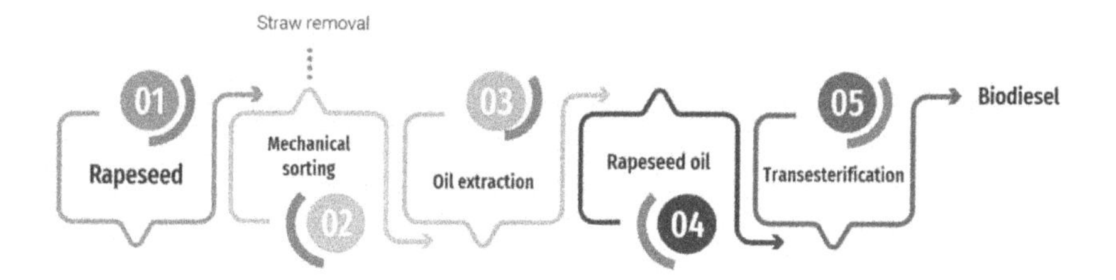

FIGURE 11.3 Flowchart of the process for obtaining biodiesel rapeseed. (Adapted from Severo, et al. 2019.)

transesterification. Large-scale oil extraction is generally preceded by grinding and cooking the seeds to assist the oil extraction process. However, during the mechanical pressing of the seeds, a protein-rich rape cake is also generated. Chemical extraction uses a solvent derived from petroleum (hexane). However, in solvent extraction, the oil undergoes a distillation process to recover hexane, which is recycled back to the oil extraction process. In addition, in the transesterification reaction, the triglyceride molecules in the oil react with methanol in an alkaline-containing catalyst to improve the biodiesel yield (Malça et al., 2014).

From an environmental burden perspective, rapeseed cultivation entails the application of fertilizers, pesticides, agrochemicals and also the occupation of arable land. Thus, ammonia emissions to air and nitrate to water can contribute to eutrophication. In addition, the use of agricultural machinery contributes to acidification, ozone depletion and global warming potential (González-García, et al., 2013).

Notably, carbon dioxide is a contributor to greenhouse gas emissions in relation to raw materials and energy input. However, pesticides are the main source of greenhouse gas emissions for primary energy. In the case of energy inputs, direct and indirect emissions are included (Malca and Freire, 2009). Direct emissions arise from land use, changes in the carbon stock due to the conversion of land use and the burning of fossil fuels. On other hand, indirect emissions are due to the manufacture of inputs, extraction, transportation and the transformation of crude fossil fuels and electricity generation (Malca and Freire, 2010).

11.2.4 Sunflower

The sunflower (*Helianthus annuus*) is an annual plant belonging to the Asteraceae family with a great flower that grows in the most diverse of soil types. The flower stem can reach up to 3 m in height, with the flower head reaching 30 cm in diameter containing the seeds (Alaboudi et al., 2018). Among the countries that stand out for the production of this crop, Turkey is considered the world's largest producer (Saydut et al., 2016).

It is worth mentioning that the lipid content of sunflower seeds contains about 40–50% oil, and its high levels of oleic/linoleic acid make the oil more resistant to oxidation and, therefore, more suitable to obtain high oxidative stability at high temperatures, which is useful for the production of biodiesel (Marvey, 2008). In addition, the sunflower is also characterized by other compounds such as palmitic and

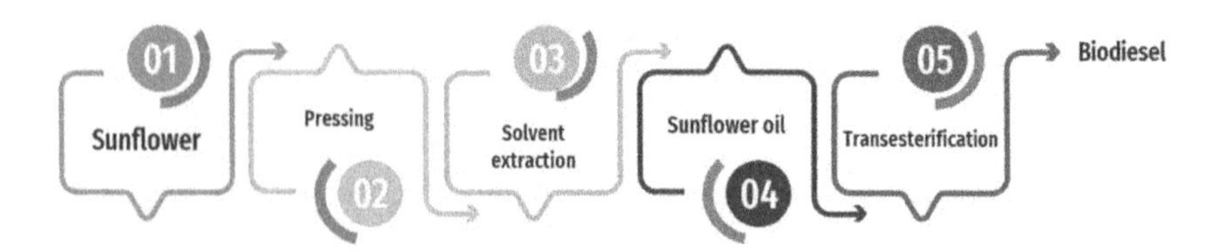

FIGURE 11.4 Flowchart of the process for obtaining biodiesel sunflower. (Adapted from Severo et al. 2019.)

stearic acids (Georgogianni et al., 2008). In terms of productivity, sunflower produces more biodiesel than soybean (Table 11.1).

Sunflower originates from subtropical and temperate zones, is widely adaptable, and more tolerant to drought than most other grain crops. It grows best on soils with a high water-holding capacity but is easily adaptable to a variety of soil conditions (Marvey, 2008). Sunflower production is often carried out under rainfed conditions worldwide, that is, without irrigation other than rain. Seed production varies depending on field operations, used fertilizers and soil composition (Bastianoni et al., 2008).

To maximize the use of this natural resource, the proper sowing period is very important, as it ensures good seed germination, as well as the timely appearance of seedlings and ideal development of the root system (Balalić et al., 2012). Some precautions must be taken with the preparation of the soil, doses and distribution of fertilizers, sowing operations (quantity of seeds per ha) and pesticide use (Spugnoli et al., 2012).

As shown in Figure 11.4, the main stages in the process of obtaining biodiesel from sunflower consist of pressing, extraction and transesterification. Raw material pressing is done in a screw press or expeller, which results in a higher oil content (Raß et al., 2008). Oil extraction from sunflower seeds is carried out by pressing, followed by solvent extraction. This combination produces a higher oil yield, therefore, the advantage of this process is to allow pressing and subsequent extraction by solvent, as they are quite difficult to process using the direct extraction method (Kartika et al., 2010). The transesterification of sunflower oil to obtain biodiesel consists of replacing triglyceride glycerol with a short-chain alcohol in the presence of a catalyst.

For the expansion of the sunflower growing area, important environmental impacts may occur due to changes in the carbon stock from the change in land use. In addition, it is of utmost importance to evaluate the parameters of the environmental impact of sunflower productivity resulting from higher levels of fertilization and the adoption of irrigation rather than rainfed agriculture (Figueiredo et al., 2017).

Biodiesel production from sunflower requires elevated fossil fuel demand, resulting in environmental burdens on the energy resources sector. Also, global warming, acidification and eutrophication potential are a comprehensive set of environmental issues related to sunflower cultivation, which include carbon, nitrogen and phosphorus emissions due to fertilization and land expansion (Figueiredo et al., 2017). The production of sunflower oil from sunflower seeds produced in conventional agriculture has a greater impact on resource consumption and human health due to the emissions used in the production of fertilizers needed in the agricultural phase and in the oil refinery processes due to atmospheric emissions and effluents (Spinelli et al., 2012).

11.2.5 Jatropha

Jatropha curcas is a 5–7 m tall tree, belonging to the Euphorbiaceae family. It is a tropical plant that is resistant to drought, capable of growing in various climatic zones with precipitations pluviometric of 250–1200 mm. It is therefore conducive to growth in arid and semi-arid conditions and has a low demand for fertility and humidity. It can also grow in moderately saline, degraded and eroded soils. It produces seeds after 1 year and reaches its maximum productivity in 5 years, being able to live from 30 to 50 years (Atabani et al., 2013). Among the countries that have the required geographical conditions, India has most jatropha plantations (Findlater and Kandlikar, 2011).

Jatropha contains highly toxic compounds, making it unsuitable for human/animal food, but it is considered a suitable source for the production of biofuels. Jatropha produces seeds that contain about 30–40% oil. However, it is noteworthy that jatropha biodiesel does not compete directly with food production, as the plant is not edible and can be grown on land unsuitable for growing food crops (Axelsson, et al., 2012).

The extraction efficiency of jatropha oil is between 60% and 99%, depending on the extraction technology applied (Achten et al., 2013). In addition, it has unsaturated components, such as linoleic acid and oleic acid, and saturated ones, such as stearic acid and palmitic acid (Singh et al., 2020). Also, it has a satisfactory agricultural yield (see Table 11.1). However, to obtain high yields, good quality of inputs and adequate soil are necessary (Carriquiry et al., 2011).

The main qualities of jatropha that make it superior to other non-edible energy crops, include drought tolerance and low demand for water and nutrients, marginal growth lands and soil recovery, high resistance to pests and diseases, greenhouse gas emission reduction, prevention of soil erosion, and better water infiltration, among others (Babazadeh et al., 2017).

Figure 11.5 describes the main stages in the process of obtaining jatropha biodiesel, which consists of (i) mechanical and chemical extraction and (ii) transesterification. Before extracting the oil, the jatropha seeds must be dried (in an oven at 105 ° C or in the sun for 3 weeks). For the mechanical extraction of seed oil, a manual plunger press or a motorized screw press can be used. The presses are fed with whole seeds to remove the oil (Achten et al., 2008). Chemical solvent extraction has high energy consumption and high greenhouse gas emission rates. After extraction, the oil goes through the transesterification process, generating crude oil (Achten et al., 2007). After this process, the raw biodiesel is collected through a washing tank to obtain pure biodiesel.

Terrestrial acidification is mainly caused by ammonia emissions related to fertilization and the combustion of biofuels due to nitrogen oxide emissions (Gmünder et al.,

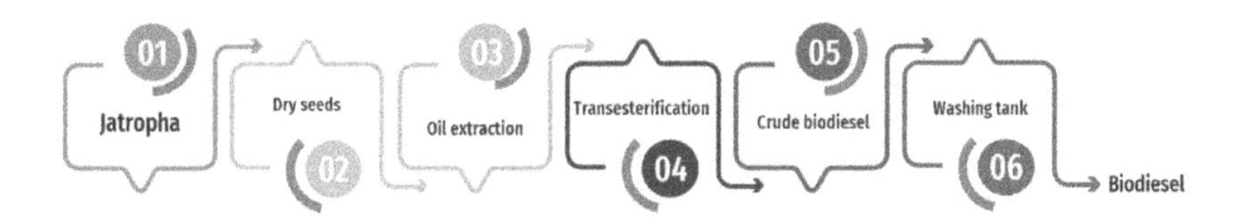

FIGURE 11.5 Flowchart of the process for obtaining second-generation biodiesel, based on the raw material of jatropha. (Adapted from Ramesh et al. (2006); Mofijur et al. 2012.)

2012). Additionally, these emissions also contribute to global warming potential. As a consequence, the global warming potential of jatropha oil is as high as that of fossil diesel. Therefore, cracking and pressing are the main contributors due to the combustion of fossil diesel and carbon dioxide emissions (Hagman et al., 2013).

The jatropha cultivation process has a relatively large eutrophication effect due to the emissions caused by phosphate, nitrate and phosphorus leaching to the surface and groundwater (Gmünder et al., 2010).

On other hand, the impact on the soil can be positive through the control of erosion and carbon sequestration. However, it can be negative if cultivated intensively with high consumption of fertilizers and use of machinery (Achten et al., 2007). Also, soil erosion contributes significantly to speeding up the flow of water, thus reducing groundwater recharge; nutrient-rich runoff can impair the quality of receiving rivers, lakes and other water reservoirs, causing eutrophication and water contamination (Pandey et al., 2012). The main advantages of jatropha are its resistance to drought and its low water requirement, thus reducing costs from irrigation and energy demand. Increasing demand for energy creates greater water needs for irrigation of plantations, resulting in conflicts, with additional environmental impact (Mofijur et al., 2012).

Finally, the success of cultivation for the production of biodiesel depends on several factors, some of which sometimes cannot be influenced or controlled, such as international demand, the price of fossil fuels, climate and agriculture. Therefore, the influential components identified must work in close coordination and take measures to ensure that cultivation is not unfavourable (Gonzáles, 2016).

11.3 Third-Generation Biodiesel

Third-generation biodiesel is derived from microorganisms such as microalgae and yeasts, which have a high capacity for the synthesis and storage of lipids in their cells. Therefore, it has become a potential as an optional fuel (Deprá et al., 2018; Mofijur et al., 2019).

Compared to the first- and second-generation biodiesels, third-generation biodiesel can alleviate food problems, as in the current form of biodiesel production, the dispute over land use and food supply becomes a problem, as it implies the loss of diversity and generates conflicts over land (Mahlia et al., 2020; Yin et al., 2020).

In this sense, this section focuses on the main third-generation raw materials, elucidating the characteristics for biodiesel production, the content of lipids and fatty acids, productivity, cultivation time and the environmental impacts of these sources.

11.3.1 Microalgae

Microalgae appear as a promising feedstock for the generation of biodiesel. They are composed of a diverse group of eukaryotic or prokaryotic microorganisms. Because of their versatile metabolism, these microorganisms can consume wasted resources, mitigate CO_2 release and the biomass generated can be used for a wide range of applications (Patidar and Mishra., 2017).

In comparison with other cultures, microalgae have significant benefits, among which they can be produced throughout the year, do not compete with arable land for food production and result in high oil productivity (Chen et al., 2018). In fact, the reaction of oil produced from microalgae is much greater than from other sources, reaching a yield that is about 25% higher (Ahmad et al., 2011). Thus, the oil yield from selected species of microalgae corresponds to approximately 200 times the yield of vegetable oils. Therefore, microalgae grown under suitable growing conditions have the capacity to yield approximately 47,000–141,000 L of seaweed oil per hectare per year (Demirbas and Demirbas, 2011).

It is known that the lipid content of microalgae is highly dependent on the microalgae species and growing conditions (Table 11.2) (Razzak et al., 2013). Therefore, in order to obtain biodiesel, it is extremely important to determine the most suitable type of microalgae strain, since the lipid content becomes the key issue in its production. However, it is important to highlight that strains with a faster growth rate generally have a relatively lower oil content (Siqueira et al., 2016).

Consequently, to obtain greater lipid production, heterotrophic systems are more efficient as compared to autotrophic systems (Mohan et al., 2015). As heterotrophic systems require an external carbon source, there is a rapid increase in biomass generation and, consequently, in the accumulation of lipids. Simultaneously, the heterotrophic route is executed in the absence of light and its construction is more accessible. Therefore, it becomes more efficient for large-scale production (Soto-Sierra et al., 2018).

On other hand, it has been reported that with autotrophic cultivation, despite being under controlled culture conditions and in common methods of increasing the lipid content, the maximum lipid accumulation remains only low to moderate. Thus, the biomass productivity is further decreased (Daroch et al., 2013).

Thus, to obtain oil for the generation of biodiesel from microalgae, cultivation, harvesting, drying, extraction and conversion techniques must be taken into account

TABLE 11.2

Oil Content in Selected Microalgal Species

Species	Oil content (% dw)
Ankistrodesmus TR-87	28–40
Chlorella sp.	29
Botryococcus braunii	29–75
Scenedesmus TR-84	45
Dunaliella tertiolecta	36–42
Isochrysis sp.	7–33
Nannochloris sp.	6–63
Nannochloropsis sp.	46 (31–68)
Hantzschia DI-160	66
Cyclotella DI-3	42
Tetraselmis sueica	15–32
Nitzschia sp.	45–47
Dunaliella primolecta	23

Source: Adapted from Demirbas and Demirbas (2011).

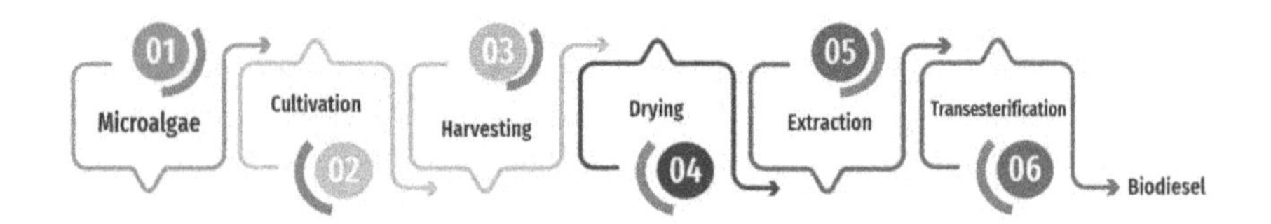

FIGURE 11.6 Flowchart of the process of the main stages for the production of biodiesel from microalgae. (Adapted from Acién Fernández 2018.)

(Menegazzo and Fonseca, 2019). In this context, Figure 11.6 shows a summary of the process flowchart.

According to Figure 11.6, it is essential to select the most appropriate type of microalgae species, and thereafter, the cultivation method is determined. Drying of biomass is carried out and later submitted to the extraction process. Oil extraction is an expensive and challenging process, in which the oil is surrounded by a rigid cell wall composed of carbohydrates and proteins. Thus, it is necessary to rupture this cell wall to obtain the oil from the microalgae. Therefore, several methods have been applied to break cell lysis, including mechanical methods, the mechanical interruption can be performed by various methods, such as ultrasound, granule counting and microwaves, where the efficiency of oil extraction through this method is related to the species of microalgae. Thus, cell disruption can also be performed by chemical methods, where commonly applied products are chloroform, hexane, benzene, acids and alkalis. Chemical extraction can be carried out in wet or dry microalgae biomass. Dry biomass has a high extraction efficiency. Finally, the desired target product is purified (Barros et al., 2015).

The main bottlenecks for the commercialization of biodiesel based on microalgae are concentrated in the great value of biomass production related to the harvest phase (Ummalyma et al., 2017).

One of the most efficient methods is centrifugation, which is responsible for almost 100% of microalgae harvesting techniques.. Unfortunately, this method can consume more energy than it produces. Therefore, it is necessary to continue studies to reduce harvesting costs, contributing to the commercialization of microalgae biodiesel becoming viable (Najjar and Abu-Shamleh, 2020).

It is worth mentioning that, related to the high photosynthetic efficiency of microalgae, it is believed that the mass cultivation of microalgae is capable of efficiently reducing the carbon dioxide emissions into the atmosphere, thus reducing the impact of global warming (Milano et al., 2016). In this sense, as mentioned earlier, microalgae-based biodiesel production systems are still far from being ecologically balanced, due to the energy and resource issues associated with lipid extraction processes. The development of microalgae requires large amounts of water, and lipid extraction involves the application of solvents, usually polar/non-polar co-solvent systems. Notoriously, these factors affect the yield of the conversion process and, therefore, it is of high importance to carry out a life cycle analysis, analysing the environmental and general economic footprint (Collotta et al., 2017).

In fact, life cycle assessment and carbon footprint appear as one of the most promising methodologies for quantifying metrics and sustainability indicators. Thus, these tools have become essential, as they assist in the recognition and discovery of solutions to mitigate greenhouse gas emissions, and the process or product over its useful life by measuring the environmental burdens imposed (Deprá et al., 2018; Patel et al., 2020). As Slade and Bauen (2013) highlight, the values found in the algae growth rates in LCA surveys are considered to be optimistic values.

Based on a study with microalgae, Clarens et al. (2010) observed that energy development has lower environmental impacts than conventional raw materials just for land use and eutrophication potential. Likewise, Yang et al. (2011) evaluated the use of water in the life cycle of microalgae-based biodiesel production, and based on this research, the water footprint during the generation of microalgae biodiesel was quantified. Thus, it was concluded that 3,726 kg of water are needed to generate 1 kg of biodiesel from microalgae if freshwater is used without recycling, indicating that the use of seawater or effluent can reduce the use of freshwater in the water life cycle by up to 90%.

In a study by Hou et al. (2011) it was shown that the generation of biodiesel based on microalgae may have a lower environmental impact for the global warming potential and ozone depletion potential categories, for example, as compared to soybeans, jatropha and fossil diesel.

11.3.2 Oleaginous Yeast

Yeasts are described as fungi that reproduce asexually by germination or fission, which results in growth composed mainly of isolated cells (multicellular growth capacity) (Kurtzman et al., 2011). Their use has been explored for years, being applied in biotechnological processes.

Currently, oleaginous yeasts have become a potential source for the generation of third-generation biodiesel because they do not need land or water. Thus, the use of yeasts has many benefits compared to other renewable sources, as they develop and store oil and can convert substrates such as CO_2, sugars and liquid acids (Chopra and Sen, 2018; Mathimani and Mallick, 2019).

To reach the accumulation of lipids (stationary phase), it takes 5–9 days for the yeast species. In addition, they have specific properties to synthesize lipids while consuming an abundance of renewable substrates, even cheap materials, such as nutritional waste from agriculture and industry (Angerbauer et al., 2008).

Due to their ability to synthesize and store a satisfactory amount of intracellular triacylglycerols, reaching up to approximately 70% of their dry weight, yeasts have shown greater promise than fungi, microalgae and bacteria (Bansal et al., 2020). In addition, the lipids accumulated by oily yeasts are chemically similar to vegetable oils and animal fats (Ageitos et al., 2011). Among the oleaginous yeasts that have the capacity to produce the largest amount of lipids in their cell compartment is *Rhodosporidium* spp. (Li et al., 2007). Unfortunately, the yeasts *Saccharomyces cerevisiae* and *Candida utilis* are unable to accumulate > 5–10% of the oil (Leong et al., 2018).

The main steps in the procedure for obtaining biodiesel from yeast oil are described in Figure 11.7. Firstly, the selection of oleaginous yeast is made, after which, the culture system for growth and accumulation of lipids is implemented. The next step is to harvest the biomass and extract the lipid from this oleaginous yeast. The yeast

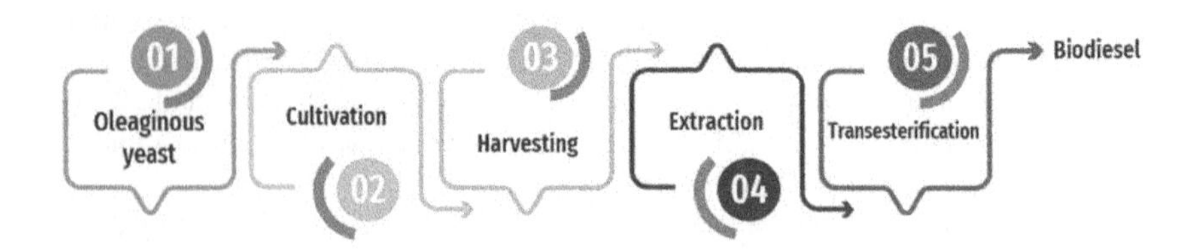

FIGURE 11.7 Schematic diagram of biodiesel production from oleaginous yeast. (Adapted from Patel 2017.)

biodiesel extraction process is similar to the microalgae process discussed earlier. Finally, the lipid extracted in fatty acid methyl esters follows (Patel et al., 2016).

However, the main bottlenecks for viable large-scale production of biodiesel from oleaginous yeast are related to the lipid extraction procedure for cell disruption (Patel et al., 2016). Thus, research is being developed mainly for biodiesel production to determine the economic and environmental viability of yeast biodiesel (Chopra et al., 2020). Therefore, according to Karlsson et al. (2016), yeast drying requires a vast amount of energy, resulting in a negative energy balance. In addition, the transesterification stage was identified as the one that causes the greatest environmental impact due to the chemicals utilized.

11.3.3 Waste-Activated Sludge

The activated sludge (AS) is formed by bacteria, archaea, viruses and eukaryotes (fungi, algae, protozoa and metazoans), forming a consortium of microorganisms, mostly formed by heterotrophic bacteria. Consequently, wastewater is used as a source of carbon and energy (Sepehri and Sarrafzadeh, 2018). Therefore, it is a technology widely used in wastewater treatment plants (WWTPs) worldwide, for the treatment of municipal wastewater, as it is a by-product developed through biological treatment (Xia et al., 2018). It is worth mentioning that this activated sludge residue is composed of several nutrients, including proteins, carbohydrates, fibres and other micronutrients (Tian et al., 2013).

In recent years, SA has been identified as a potential raw material suitable for exploration for the generation of biodiesel, as it has more sustainable energy (Raheem et al., 2018). It is known that the lipid fraction in sewage sludge originates from accessible lipid residues. Therefore, they are considered one of the most attractive candidates to overcome this problem (Sangaletti-Gerhardi et al, 2015).

Regarding the lipid content, approximately 5–12% will determine the biodiesel yield, according to the type of SA method and the extraction technique (Nazari et al., 2018). The main steps for the production of biodiesel from activated sludge from waste are shown in Figure 11.8.

Usually, the generation of biodiesel from sludge is carried out by extracting lipid transesterification. However, prior to these processes, the biomass must undergo pretreatment operations such as concentration, drying, size reduction and homogenization, which may vary according to the selected extraction process.

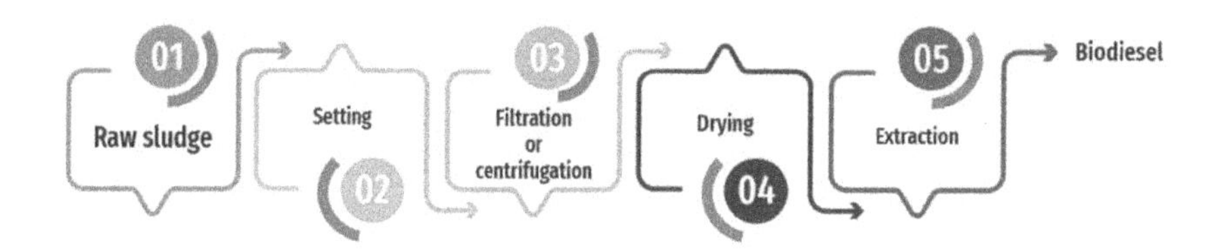

FIGURE 11.8 General scheme for the production of biodiesel from waste-activated sludge. (Adapted from Siddiquee and Rohani 2011.)

The applicability of activated sludge as a lipid raw material, being ecological, low cost and rich in inputs, makes it an essential resource for sludge treatment and biodiesel production (Zhang et al., 2013). Thus, activated sludge is capable of modifying biodiesel into a more sustainable energy carrier.

Notoriously, bottlenecks in the generation of biodiesel from activated sludge from waste include quite expensive processes such as the pre-treatment of this sludge and the freeze-drying system for obtaining dry sludge that requires a lot of energy and time. Finally, extracting lipids from sludge is an operation that requires a large number of organic solvents, which impacts on environmental issues (Siddiquee and Rohani, 2011). Thus, toxic chemicals are released into the environment, potential causing human toxicity. In the case of biodiesel production from dry sewage sludge, methanol production plays an important role in the environmental impacts associated with this energy system, especially for global warming potential and cumulative energy demand (Sharma et al., 2020).

11.4 Specification and Legal Standards for Biodiesel

It is extremely important to have evaluation resources and a base of indicators that exercise the function of guaranteeing the satisfactory quality of the biodiesel in use. In this way, several countries have established their own standards, through a rigid set of fuel specifications, highlighting legislation such as ASTM D6 751 (USA), EN 142114 (European Union) and IS (India) (Goosen et al., 2007; Hoekman et al., 2012) (Table 11.3).

Among the evaluation parameters in these laws, characteristics associated mainly with the quality of the oil used as raw material are listed. This is because the characteristics of biodiesel are influenced by several parameters, including refining processes, the composition of the raw material, the synthesis and oil extraction method. Thus, establishing quality standards involving attributes such as fatty acid methyl esters, calorific value, flash point, pour point, viscosity, iodine number, cloud point, specific gravity, plugging point of the cold filter and cetane number, it becomes indispensable (Singh et al., 2019).

On the other hand, beyond the parameters inherent to the raw material, the handling and manufacturing methodology also influences the fatty acid methyl esters profile of biodiesel (Sharma et al., 2011). Thus, the ability to cold-filter, sediment, number of

TABLE 11.3

Brazil, United States, European Union, and India Specifications for Biodiesel

Property specification	Units	Brazil	United States	European Union	India
Acid number	mg KOH/g	0.8 max.	0.5 max.	0.5 max.	0.5 max.
Ash content	% mass	–	–	–	–
BOCLE scuff	g	–	> 7000	–	–
Boiling point	°C	–	100–615	–	–
Carbon	wt%	–	77	–	–
Carbon residue	% m/m	0.05 max.	0.050 max.	0.3 max.	0.050 max.
Cetane number	–	45 min.	47 min.	51 min.	51 min.
Cloud point	°C	Report	–3 to – 12	–	–
Cold filter plugging point	°C	ANP 310	Maximum + 5	–	–
Conductivity at ambient temperature	pS/m	–	–	–	–
Copper corrosion	–	No. 1 max.	No. 3 max.	Class 1	Class 1
Density at 15 °C	kg/m^3	–	880	860–900	860–900
Diglycerides	% mass	0.25 max.	–	0.2 max.	–
Distillation temperature	°C	360 (T-95)	360	–	–
Flash point	°C	100 min.	130 min.	101 min.	120 min.
Free glycerine	% mass	0.02 max.	0.02 max.	0.02 max.	0.02 max.
Hydrogen	wt%	–	12	–	–
Iodine number	g I_2/100 g	–	–	120 max.	–
Kinematic viscosity at 40 °C	mm^2/s	ANP 310	1.9–6.0	3.5–5.0	2.5–6.0
Lubricity	m	–	520 max.	–	–
Monoglycerides	% mass	1.0 max.	–	0.8 max.	–
Oxidation stability	–	6 h min.	–	3 h min.	6 h min.
Oxygen	wt%	–	11	–	–

Pour point	°C	–	–15 to – 16	–	–
Phosphorus	%mass	10 max.	0.001 max.	0.001 max.	0.001 max.
Saponification value	mg KOH/g	–	370 max.	–	–
Sulphated ash content	% mass	0.02 max.	0.002 max.	0.02 max.	0.002 max.
Sulphur (S 50 grade)	ppm	–	–	–	50 max.
Total contamination	mg/kg	–	24	24	24
Total glycerine	% mass	0.38 max.	0.24	0.25	0.25
Triglycerides	% mass	0.25 max.	–	0.2 max.	–
Water and sediment	–	0.02 vol% max.	0.005 vol% max.	500 mg/kg	0.005 vol% max.

Sources: Adapted from Hoekman et al. (2012) and Singh et al. (2019).

acids, glycerine content, water content, methanol content and metal content are also relevant quality indicators (Cavalheiro et al., 2020).

Therefore, these standards determine the guidelines for testing biodiesel biofuels and specific values for different physical characteristics of the oil for use in an engine. Thus, all biodiesel fuels are required to comply with these specifications (Singh et al., 2019; Maroneze et al., 2014).

11.5 Beyond Sustainability

Although fossil fuels are the unquestionably backbone of socioeconomic development in the energy sector, their use results in severe impacts on the environment. As a consequence, this sector faces the double contemporary challenge in the global community whose need is to satisfy the colossal demands of energy while obtaining very significant reductions in environmental charges, mainly in greenhouse gas emissions. Thus, biodiesel has aroused the interest of researchers as a plausible alternative to conventional diesel.

However, as seen in the earlier sections, several technical bottlenecks need to be addressed before biodiesel production becomes, in fact, a mature technology. Therefore, it is necessary to consider issues beyond the environmental parameters since a series of technical limitations within the process needs to be elucidated (Zhang et al., 2020). Economic viability is one of the parameters that will generally determine the commercial effectiveness of biodiesel. In this regard, the starting point for this follow-up consists of choosing the raw material for exploration. In principle, it is known that the cost of raw material is one of the economic parameters that determine the sale price of biodiesel. Thus, it is estimated that biodiesel of vegetable origin represents a sales value of approximately 0.81 USD/L. However, studies by Zhang et al. (2020) estimate that microbial biodiesel can achieve a lower competitive price compared to vegetable biodiesel (around 0.76 USD/L). However, these scenarios represent process values calculated in the laboratory, and therefore do not reflect reality (Lee et al., 2020).

Simultaneously, another bottleneck that interferes with the strengthening of the biodiesel market consists of the need to establish robust systems for the feedstock commercialization associated with ensuring the continuous supply and a sustainable internal project. This is because dependence on domestic raw materials can harm biodiesel production due to the unpredictable nature of the climate and market-oriented agriculture. In addition, the critical factor – sustainability – is an essential element to guarantee the advancement of biofuel production in terms of international trade and the achievement of social and environmental objectives.

Associated with the previous aspect, the choice of the location of the installation of the biodiesel industry, as well as an efficient project, must be designed for material flow planning. In this regard, strategic-level decisions must encompass all requirements for quantity, location, capacity and processing of biodiesel. Likewise, tactical-level choices need to include the seasonal design (biodiesel production in different periods) together with strategic-level decisions. Thus, threats will be drastically reduced, and aspirations for the benefits of project opportunities can be achieved.

Nevertheless, evidently, the consolidation of the oil market today makes the complex and dynamic business environment of the biofuels supply chain even more precarious. As a consequence, it leads to a highly uncertain environment, and therefore the effectiveness of strengthening the biodiesel chain is compromised. Thus, whatever the manager's decision-making, it is essential to exercise caution and the most utmost care in the installations of industrial projects associated with the production of biodiesel (Habib et al., 2020).

Also, the adoption of technology adaptation techniques can serve as an insightful strategy for improving the planning of biodiesel facilities. This is because, in practice, the issue of technology transfer from developed countries to developing countries is a barrier to the sustainable application of advanced technology. Thus, to solve this bottleneck, regional innovation systems are required. Consequently, investments in the know-how of people and institutions in different activities are necessary to achieve a common goal. In terms of exemplification, the technological adaptation generated in a developed country may spread to developing countries, and not become just a model idea to be followed, but one that can be carried out in an adapted way (in economic and economic terms) by the local community (Mahlia et al., 2020).

Finally, the role of paramount importance is attributed to government incentives. Indeed, governments play an expressive role in promoting their domestic biodiesel industries. Although first-generation biodiesel technology is already established, there remain financial problems and other limitations that need to be addressed. Thus, a strategy to solve current issues consists of instituting government subsidies, promotional campaigns and biodiesel mandates in national policies that can be implemented to facilitate the development of first-generation biodiesel. Direct financial support, such as research grants for research and development, as well as an allowance for biodiesel purchases, can be included in these provisions.

11.6 Conclusion

The use of alternative first-, second- and third-generation raw materials seems to be a promising approach for the production of biodiesel. However, it is essential to ensure that the efficient production of these new fuels is environmentally friendly and economically viable in the medium to long term. Thus, circumventing the challenges mentioned above is an important factor that must be considered and extensively researched to expand the biodiesel industry throughout the globe. Finally, before deciding on the feedstock for biodiesel production, it is also imperative to see the effect of the quality of the raw material on the properties of the produced biodiesel. In addition, it is important to highlight the proposals inherent to aspects beyond sustainability, as well as a highly desirable objective with positive ramifications for the future energy needs of humanity, which need to be addressed side by side, in the search for biofuels.

REFERENCES

Achten, W. M., Trabucco, A., Maes, W., Verchot, L. V., Aerts, R., Mathijs, E., Muys, B. (2013). Global greenhouse gas implications of land conversion to biofuel crop cultivation in arid and semi-arid lands–Lessons learned from Jatropha. *Journal of Arid Environments*, 98, 135–145. DOI: 10.1016/j.jaridenv.2012.06.015.

Achten, W. M., Verchot, L., Franken, Y. J., Mathijs, E., Singh, V. P., Aerts, R., Muys, B. (2008). Jatropha bio-diesel production and use. *Biomass and Bioenergy*, 32(12), 1063–1084. DOI: 10.1016/j.biombioe.2008.03.003.

Achten, W., Muys, B., Mathijs, E., Singh, V. P., Verchot, L. (2007). Life-cycle assessment of bio-diesel from *Jatropha curcas* L. energy balance, impact on global warming, land use impact. In *Book of proceedings: The 5th international conference LCA in foods 25–26 April 2007* (pp. 96–102), Gothenburg, Sweden.

Acién Fernández, F. G., Gómez-Serrano, C., Fernández-Sevilla, J. M. (2018). Recovery of nutrients from wastewaters using microalgae. *Frontiers in Sustainable Food Systems,* 2, 59. DOI: 10.3389/fsufs.2018.00059.

Ageitos, J. M., Vallejo, J. A., Veiga-Crespo, P., Villa, T. G. (2011). Oily yeasts as oleaginous cell factories. *Applied Microbiology and Biotechnology*, 90(4), 1219–1227. DOI: 10.1007/s00253-011-3200-z.

Ahmad, A. L., Yasin, N. M., Derek, C. J. C., Lim, J. K. (2011). Microalgae as a sustainable energy source for biodiesel production: a review. *Renewable and Sustainable Energy Reviews*, 15(1), 584–593. DOI: 10.1016/j.rser.2010.09.018.

Alaboudi, K. A., Ahmed, B., Brodie, G. (2018). Phytoremediation of Pb and Cd contaminated soils by using sunflower (*Helianthus annuus*) plant. *Annals of Agricultural Sciences*, 63(1), 123–127. DOI: 10.1016/j.aoas.2018.05.007.

Aldhaidhawi, M., Chiriac, R., Badescu, V. (2017). Ignition delay, combustion and emission characteristics of diesel engine fueled with rapeseed biodiesel–A literature review. *Renewable and Sustainable Energy Reviews,* 73, 178–186. DOI: 10.1016/j.rser.2017.01.129.

Angerbauer, C., Siebenhofer, M., Mittelbach, M., Guebitz, G. M. (2008). Conversion of sewage sludge into lipids by *Lipomyces starkeyi* for biodiesel production. *Bioresource Technology*, 99(8), 3051–3056. DOI: 10.1016/j.biortech.2007.06.045.

Atabani, A. E., Silitonga, A. S., Ong, H. C., Mahlia, T. M. I., Masjuki, H. H., Badruddin, I. A., Fayaz, H. (2013). Non-edible vegetable oils: a critical evaluation of oil extraction, fatty acid compositions, biodiesel production, characteristics, engine performance and emissions production. *Renewable and Sustainable Energy Reviews*, 18, 211–245. DOI: 10.1016/j.rser.2012.10.013.

Axelsson, L., Franzén, M., Ostwald, M., Berndes, G., Lakshmi, G., Ravindranath, N. H. (2012). Jatropha cultivation in southern India: assessing farmers' experiences. *Biofuels, Bioproducts and Biorefining*, 6(3), 246–256. DOI: 10.1002/bbb.1324.

Babazadeh, R., Razmi, J., Pishvaee, M. S., Rabbani, M. (2017). A sustainable second-generation biodiesel supply chain network design problem under risk. *Omega,* 66, 258–277. Doi: 10.1016/j.omega.2015.12.010.

Balalić, I., Zorić, M., Branković, G., Terzić, S., Crnobarac, J. (2012). Interpretation of hybrid × sowing date interaction for oil content and oil yield in sunflower. *Field Crops Research,* 137, 70–77. DOI: 10.1016/j.fcr.2012.08.005.

Bansal, N., Dasgupta, D., Hazra, S., Bhaskar, T., Ray, A., Ghosh, D. (2020). Effect of utilization of crude glycerol as substrate on fatty acid composition of an oleaginous yeast *Rhodotorula mucilagenosa* IIPL32: assessment of nutritional indices. *Bioresource Technology*, 123330. DOI: 10.1016/j.biortech.2020.123330.

Barros, A. I., Gonçalves, A. L., Simões, M., Pires, J. C. (2015). Harvesting techniques applied to microalgae: a review. *Renewable and Sustainable Energy Reviews*, 41, 1489–1500. DOI: 10.1016/j.rser.2014.09.037.

Baryeh, E. A. (2001). Effects of palm oil processing parameters on yield. *Journal of Food Engineering*, 48(1), 1–6. DOI: 10.1016/S0260-8774(00)00137-0.

Bastianoni, S., Coppola, F., Tiezzi, E., Colacevich, A., Borghini, F., Focardi, S. (2008). Biofuel potential production from the *Orbetello lagoon* macroalgae: a comparison with sunflower feedstock. *Biomass and Bioenergy*, 32(7), 619–628. DOI: 10.1016/j.biombioe.2007.12.010.

Bergmann, J. C., Tupinambá, D. D., Costa, O. Y. A., Almeida, J. R. M., Barreto, C. C., Quirino, B. F. (2013). Biodiesel production in Brazil and alternative biomass feedstocks. *Renewable and Sustainable Energy Reviews*, 21, 411–420. DOI: 10.1016/j.rser.2012.12.058.

Branco-Vieira, M., Costa, D., Mata, T. M., Martins, A. A., Freitas, M. A. V., Caetano, N. S. (2020). A life cycle inventory of microalgae-based biofuels production in an industrial plant concept. *Energy Reports*, *6*, 397–402. DOI: 10.1016/j.egyr.2019.08.079

Carriquiry, M. A., Du, X., Timilsina, G. R. (2011). Second generation biofuels: economics and policies. *Energy Policy*, 39(7), 4222–4234. DOI: 10.1016/j.enpol.2011.04.036.

Cavalheiro, L. F., Misutsu, M. Y., Rial, R. C., Viana, L. H., Oliveira, L. C. S. (2020). Characterization of residues and evaluation of the physico chemical properties of soybean biodiesel and biodiesel: diesel blends in different storage conditions. *Renewable Energy*, 151, 454–462. DOI: 10.1016/j.renene.2019.11.039.

Chen, J., Li, J., Dong, W., Zhang, X., Tyagi, R. D., Drogui, P., Surampalli, R. Y. (2018). The potential of microalgae in biodiesel production. *Renewable and Sustainable Energy Reviews,* 90, 336–346. DOI: 10.1016/j.rser.2018.03.073.

Chopra, J., Sen, R. (2018). Process optimization involving critical evaluation of oxygen transfer, oxygen uptake and nitrogen limitation for enhanced biomass and lipid production by oleaginous yeast for biofuel application. *Bioprocess and Biosystems Engineering*, 41(8), 1103–1113. DOI: 10.1007/s00449-018-1939-7.

Chopra, J., Tiwari, B. R., Dubey, B. K., Sen, R. (2020). Environmental impact analysis of oleaginous yeast-based biodiesel and bio-crude production by life cycle assessment. *Journal of Cleaner Production,* 271, 122349. DOI: 10.1016/j.jclepro.2020.122349

Clarens, A. F., Resurreccion, E. P., White, M. A., Colosi, L. M. (2010). Environmental life cycle comparison of algae to other bioenergy feedstocks. *Environmental Science & Technology*, 44(5), 1813–1819. DOI: 10.1021/es1034848.

Collotta, M., Champagne, P., Mabee, W., Tomasoni, G., Leite, G. B., Busi, L., Alberti, M. (2017). Comparative LCA of flocculation for the harvesting of microalgae for biofuel production. *Procedia CIRP,* 61, 756–760. DOI: 10.1016/j.procir.2016.11.146.

da Silva César, A., Conejero, M. A., Ribeiro, E. C. B., Batalha, M. O. (2019). Competitiveness analysis of "social soybeans" in biodiesel production in Brazil. *Renewable Energy*, 133, 1147–1157. DOI: 10.1016/j.renene.2018.08.108.

Da Silva, V. P., van der Werf, H. M., Spies, A., Soares, S. R. (2010). Variability in environmental impacts of Brazilian soybean according to crop production and transport scenarios. *Journal of Environmental Management*, 91(9), 1831–1839. DOI: 10.1016/j.jenvman.2010.04.001.

Daroch, M., Geng, S., Wang, G. (2013). Recent advances in liquid biofuel production from algal feedstocks. *Applied Energy,* 102, 1371–1381. DOI: 10.1016/j.apenergy.2012.07.031.

de Almeida, V. F., García-Moreno, P. J., Guadix, A., Guadix, E. M. (2015). Biodiesel production from mixtures of waste fish oil, palm oil and waste frying oil: optimization of fuel properties. *Fuel Processing Technology,* 133, 152–160. DOI: 10.1016/j.fuproc.2015.01.041.

Demirbas, A., Demirbas, M. F. (2011). Importance of algae oil as a source of biodiesel. *Energy Conversion and Management*, 52(1), 163–170. DOI: 10.1016/j.enconman.2010.06.055.

Deprá, M. C., Dias, R. R., Severo, I. A., de Menezes, C. R., Zepka, L. Q., Jacob-Lopes, E. (2020). Carbon dioxide capture and use in photobioreactors: the role of the carbon dioxide loads in the carbon footprint. *Bioresource Technology,* 314, 123745. DOI: 10.1016/j.biortech.2020.123745

Deprá, M. C., dos Santos, A. M., Severo, I. A., Santos, A. B., Zepka, L. Q., Jacob-Lopes, E. (2018). Microalgal biorefineries for bioenergy production: can we move from concept to industrial reality? *BioEnergy Research*, 11(4), 727–747. DOI: 10.1007/s12155-018-9934-z.

El-Beltagi, H. E. D. S., Mohamed, A. A. (2010). Variations in fatty acid composition, glucosinolate profile and some phytochemical contents in selected oil seed rape (*Brassica napus L.*) cultivars. *Grasas Y Aceites*, 61(2), 143–150. DOI: 10.3989/gya.087009.

Figueiredo, F., Castanheira, É. G., Freire, F. (2017). Life-cycle assessment of irrigated and rainfed sunflower addressing uncertainty and land use change scenarios. *Journal of Cleaner Production,* 140, 436–444. DOI: 10.1016/j.jclepro.2016.06.151.

Findlater, K. M., Kandlikar, M. (2011). Land use and second-generation biofuel feedstocks: the unconsidered impacts of Jatropha biodiesel in Rajasthan, India. *Energy Policy*, 39(6), 3404–3413. DOI: 10.1016/j.enpol.2011.03.037.

Georgogianni, K. G., Kontominas, M. G., Pomonis, P. J., Avlonitis, D., Gergis, V. (2008). Conventional and in situ transesterification of sunflower seed oil for the production of biodiesel. *Fuel Processing Technology*, 89(5), 503–509. DOI: 10.1016/j.fuproc.2007.10.004.

Gmünder, S. M., Zah, R., Bhatacharjee, S., Classen, M., Mukherjee, P., Widmer, R. (2010). Life cycle assessment of village electrification based on straight jatropha oil in Chhattisgarh, India. *Biomass and Bioenergy*, 34(3), 347–355. DOI: 10.1016/j.biombioe.2009.11.006.

Gmünder, S., Singh, R., Pfister, S., Adheloya, A., Zah, R. (2012). Environmental impacts of Jatropha curcas biodiesel in India. *Journal of Biomedicine and Biotechnology*. DOI: 10.1155/2012/623070.

Gonzáles, N. F. C. (2016). International experiences with the cultivation of *Jatropha curcas* for biodiesel production. *Energy*, 112, 1245–1258. DOI: 10.1016/j.energy.2016.06.073.

González-García, S., García-Rey, D., Hospido, A. (2013). Environmental life cycle assessment for rapeseed-derived biodiesel. *International Journal of Life Cycle Assessment*, 18(1), 61–76. DOI: 10.1007/s11367-012-0444-5.

Goosen, R., Vora, K., Vona, C. (2007). Establishment of the guidelines for the development of biodiesel standards in the APEC region. *APEC Biodiesel Standard EWG*, 74, 1–136.

Habib, M. S., Asghar, O., Hussain, A., Imran, M., Mughal, M. P., Sarkar, B. (2020). A robust possibilistic programming approach toward animal fat-based biodiesel supply chain network design under uncertain environment. *Journal of Cleaner Production*, 122403. DOI: 10.1016/j.jclepro.2020.122403

Hagman, J., Nerentorp, M., Arvidsson, R., Molander, S. (2013). Do biofuels require more water than do fossil fuels? Life cycle-based assessment of jatropha oil production in rural Mozambique. *Journal of Cleaner Production*, 53, 176–185. DOI: 10.1016/j.jclepro.2013.03.039.

Havlík, P., Schneider, U. A., Schmid, E., Böttcher, H., Fritz, S., Skalský, R., Leduc, S. (2011). Global land-use implications of first- and second-generation biofuel targets. *Energy Policy*, 39(10), 5690–5702. DOI: 10.1016/j.enpol.2010.03.030.

Hoekman, S. K., Broch, A., Robbins, C., Ceniceros, E., Natarajan, M. (2012). Review of biodiesel composition, properties, and specifications. *Renewable and Sustainable Energy Reviews*, 16(1), 143–169. DOI: 10.1016/j.rser.2011.07.143.

Hosseinzadeh-Bandbafha, H., Tabatabaei, M., Aghbashlo, M., Khanali, M., Khalife, E., Shojaei, T. R., Mohammadi, P. (2020). Consolidating emission indices of a diesel engine powered by carbon nanoparticle-doped diesel/biodiesel emulsion fuels using life cycle assessment framework. *Fuel*, 267, 117296. DOI: 10.1016/j.fuel.2020.117296

Hou, J., Zhang, P., Yuan, X., Zheng, Y. (2011). Life cycle assessment of biodiesel from soybean, jatropha and microalgae in China conditions. *Renewable and Sustainable Energy Reviews*, 15(9), 5081–5091.-961. DOI: 10.1016/j.rser.2011.07.048.

Iriarte, A., Rieradevall, J., Gabarrell, X. (2011). Environmental impacts and energy demand of rapeseed as an energy crop in Chile under different fertilization and tillage practices. *Biomass and Bioenergy*, 35(10), 4305–4315. DOI: 10.1016/j.biombioe.2011.07.022.

Jaime, R., Alcantara, J. M., Manzaneda, A. J., Rey, P. J. (2018). Climate change decreases suitable areas for rapeseed cultivation in Europe but provides new opportunities for white mustard as an alternative oilseed for biofuel production. *PloS One*, 13(11), e0207124. DOI: 10.1371/journal.pone.0207124.

Karlsson, H., Ahlgren, S., Sandgren, M., Passoth, V., Wallberg, O., Hansson, P. A. (2016). A systems analysis of biodiesel production from wheat straw using oleaginous yeast: process design, mass and energy balances. *Biotechnology for Biofuels*, 9(1), 229. DOI: 10.1186/s13068-016-0640-9.

Kartika, I. A., Pontalier, P. Y., Rigal, L. (2010). Twin-screw extruder for oil processing of sunflower seeds: thermo-mechanical pressing and solvent extraction in a single step. *Industrial Crops and Products*, 32(3), 297–304. DOI: 10.1016/j.indcrop.2010.05.005.

Khanum, F., Giwa, A., Nour, M., Al-Zuhair, S., Taher, H. (2020). Improving the economic feasibility of biodiesel production from microalgal biomass via high-value products coproduction. *International Journal of Energy Research*. DOI: 10.1002/er.5768

Kumar, R., Ghosh, A. K., Pal, P. (2020). Synergy of biofuel production with waste remediation along with value-added co-products recovery through microalgae cultivation: a review of membrane-integrated green approach. *Science of the Total Environment*, 698, 134169. DOI: 10.1016/j.scitotenv.2019.134169.

Kurtzman, C., Fell, J. W., Boekhout, T. (Eds.). (2011). *The yeasts: a taxonomic study*. Elsevier.

Lee, J. C., Lee, B., Ok, Y. S., Lim, H. (2020). Preliminary techno-economic analysis of biodiesel production over solid-biochar. *Bioresource Technology*, 123086. DOI: 10.1016/j.biortech.2020.123086

Leong, W. H., Lim, J. W., Lam, M. K., Uemura, Y., Ho, Y. C. (2018). Third generation biofuels: a nutritional perspective in enhancing microbial lipid production. *Renewable and Sustainable Energy Reviews*, 91, 950. DOI: 10.1016/j.rser.2018.04.066.

Li, Y., Horsman, M., Wu, N., Lan, C. Q., Dubois-Calero, N. (2008). Biofuels from microalgae. *Biotechnology Progress*, 24(4), 815–820. DOI: 10.1021/bp070371k.

Liu, X., He, H., Wang, Y., Zhu, S., Piao, X. (2008). Transesterification of soybean oil to biodiesel using CaO as a solid base catalyst. *Fuel*, 87(2), 216–221. DOI: 10.1016/j.fuel.2007.04.013.

Mahlia, T. M. I., Syazmi, Z. A. H. S., Mofijur, M., Abas, A. P., Bilad, M. R., Ong, H. C., Silitonga, A. S. (2020). Patent landscape review on biodiesel production: technology updates. *Renewable and Sustainable Energy Reviews*, 118, 109526. DOI: 10.1016/j.rser.2019.109526.

Malça, J., Coelho, A., Freire, F. (2014). Environmental life-cycle assessment of rapeseed-based biodiesel: alternative cultivation systems and locations. *Applied Energy*, 114, 837–844. DOI: 10.1016/j.apenergy.2013.06.048.

Malca, J., Freire, F. (2009). Energy and environmental benefits of rapeseed oil replacing diesel. *International Journal of Green Energy*, 6(3), 287–301. DOI: 10.1080/15435070902886551.

Malça, J., Freire, F. (2010). Uncertainty analysis in biofuel systems: an application to the life cycle of rapeseed oil. *Journal of Industrial Ecology,* 14(2), 322–334. DOI: 10.1111/j.1530-9290.2010.00227.x.

Manik, Y., Leahy, J., Halog, A. (2013). Social life cycle assessment of palm oil biodiesel: a case study in Jambi Province of Indonesia. *The International Journal of Life Cycle Assessment*, 18(7), 1386–1392. DOI: 10.1007/s11367-013-0581-5.

Markets and Markets (2020) Biodiesel market: global industry trends, share, size, growth, opportunity and forecast 2019–2024. ID: 4894111

Maroneze, M. M., Barin, J. S., Menezes, C. R. D., Queiroz, M. I., Zepka, L. Q., Jacob Lopes, E. (2014). Treatment of cattle-slaughterhouse wastewater and the reuse of sludge for biodiesel production by microalgal heterotrophic bioreactors. *Scientia Agricola*, 71(6), 521–524. DOI: 10.1590/0103-9016-2014-0092.

Marvey, B. B. (2008). Sunflower-based feedstocks in nonfood applications: perspectives from olefin metathesis. *International Journal of Molecular Sciences*, 9(8), 1393–1406. DOI: 10.3390/ijms9081393.

Mathimani, T., Mallick, N. (2019). A review on the hydrothermal processing of microalgal biomass to bio-oil-Knowledge gaps and recent advances. *Journal of Cleaner Production*, 217, 69–84. DOI: 10.1016/j.jclepro.2019.01.129.

Menegazzo, M. L., Fonseca, G. G. (2019). Biomass recovery and lipid extraction processes for microalgae biofuels production: a review. *Renewable and Sustainable Energy Reviews*, 107, 87–107. DOI: 10.1016/j.rser.2019.01.064.

Milano, J., Ong, H. C., Masjuki, H. H., Chong, W. T., Lam, M. K., Loh, P. K., Vellayan, V. (2016). Microalgae biofuels as an alternative to fossil fuel for power generation. *Renewable and Sustainable Energy Reviews*, 58, 180–197. DOI: 10.1016/j.rser.2015.12.150.

Milazzo, M. F., Spina, F., Cavallaro, S., Bart, J. C. J. (2013). Sustainable soy biodiesel. *Renewable and Sustainable Energy Reviews*, 27, 806–852. DOI: 10.1016/j.rser.2013.07.031.

Mofijur, M., Masjuki, H. H., Kalam, M. A., Hazrat, M. A., Liaquat, A. M., Shahabuddin, M., Varman, M. (2012). Prospects of biodiesel from Jatropha in Malaysia. *Renewable and Sustainable Energy Reviews*, 16(7), 5007–5020. DOI: 10.1016/j.rser.2012.05.010.

Mofijur, M., Rasul, M. G., Hassan, N. M. S., Nabi, M. N. (2019). Recent development in the production of third generation biodiesel from microalgae. *Energy Procedia*, 156, 53–58. DOI: 10.1016/j.egypro.2018.11.088.

Moghaddam, M. J., Pourdad, S. S. (2011). Genotype× environment interactions and simultaneous selection for high oil yield and stability in rainfed warm areas rapeseed (*Brassica napus L.*) from Iran. *Euphytica*, 180(3), 321–335. DOI: 10.1007/s10681-011-0371-8.

Mohan, S. V., Rohit, M. V., Chiranjeevi, P., Chandra, R., Navaneeth, B. (2015). Heterotrophic microalgae cultivation to synergize biodiesel production with waste remediation: progress and perspectives. *Bioresource Technology*, 184, 169–178. DOI: 10.1016/j.biortech.2014.10.056.

Montoya, C., Lopes, R., Flori, A., Cros, D., Cuellar, T., Summo, M., Zambrano, J. R. (2013). Quantitative trait loci (QTLs) analysis of palm oil fatty acid composition in an interspecific pseudo-backcross from *Elaeis oleifera* (HBK) Cortés and oil palm (Elaeis guineensis Jacq.). *Tree Genetics & Genomes*, 9(5), 1207–1225. DOI: 10.1007%252Fs11295-013-0629-5.

Naik, S. N., Goud, V. V., Rout, P. K., Dalai, A. K. (2010). Production of first and second generation biofuels: a comprehensive review. *Renewable and Sustainable Energy Reviews*, 14(2), 578–597. DOI: 10.1016/j.rser.2009.10.003.

Najjar, Y. S., Abu-Shamleh, A. (2020). Harvesting of microalgae by centrifugation for biodiesel production: a review. *Algal Research*, 51, 102046. DOI: 10.1016/j.algal.2020.102046.

Nazari, L., Sarathy, S., Santoro, D., Ho, D., Ray, M. B., Xu, C. C. (2018). Recent advances in energy recovery from wastewater sludge. In *Direct thermochemical liquefaction for energy applications* (pp. 67–100). Woodhead Publishing. DOI: 10.1016/B978-0-08-101029-7.00011-4.

Pandey, V. C., Singh, K., Singh, J. S., Kumar, A., Singh, B., Singh, R. P. (2012). Jatropha curcas: a potential biofuel plant for sustainable environmental development. *Renewable and Sustainable Energy Reviews*, 16(5), 2870–2883. DOI: 10.1016/j.rser.2012.02.004.

Patel, A., Matsakas, L., Sartaj, K., & Chandra, R. (2016). Extraction of lipids from algae using supercritical carbon dioxide. In *Green sustainable process for chemical and environmental engineering and science* (pp. 17–39). Elsevier.

Patel, A. K., Joun, J., Sim, S. J. (2020). A sustainable mixotrophic microalgae cultivation from dairy wastes for carbon credit, bioremediation and lucrative biofuels. *Bioresource Technology*, 313, 123681. DOI: 10.1016/j.biortech.2020.123681.

Patel, A., Arora, N., Mehtani, J., Pruthi, V., Pruthi, P. A. (2017). Assessment of fuel properties on the basis of fatty acid profiles of oleaginous yeast for potential biodiesel production. *Renewable and Sustainable Energy Reviews*, 77, 604–616. DOI: 10.1016/j.rser.2017.04.016.

Patidar, S. K., Mishra, S. (2017). Carbon sequestration by microalgae: a green approach for climate change mitigation. In *Reference Module in Earth Systems and Environmental Sciences*, edited by Martin A. Abraham. Elsevier. DOI: 10.1016/B978-0-12-409548-9.10125-3

Prasad, S., Singh, A., Korres, N. E., Rathore, D., Sevda, S., Pant, D. (2020). Sustainable utilization of crop residues for energy generation: a life cycle assessment (LCA) perspective. *Bioresource Technology*, 303, 122964. DOI: 10.1016/j.biortech.2020.122964

Raheem, A., Sikarwar, V. S., He, J., Dastyar, W., Dionysiou, D. D., Wang, W., Zhao, M. (2018). Opportunities and challenges in sustainable treatment and resource reuse of sewage sludge: a review. *Chemical Engineering Journal*, 337, 616–641. DOI: 10.1016/j.cej.2017.12.149.

Ramesh, D., Samapathrajan, A., Venkatachalam, P. (2006). Production of biodiesel from *Jatropha curcas* oil by using pilot biodiesel plant. *Jatropha Journal*, 18(9), 1–6.

Raß, M., Schein, C., Matthäus, B. (2008). Virgin sunflower oil. *European Journal of Lipid Science and Technology*, 110(7), 618–624. DOI: 10.1002/ejlt.200800049.

Razzak, S. A., Hossain, M. M., Lucky, R. A., Bassi, A. S., de Lasa, H. (2013). Integrated CO_2 capture, wastewater treatment and biofuel production by microalgae culturing—a review. *Renewable and Sustainable Energy Reviews*, 27, 622–653. DOI: 10.1016/j.rser.2013.05.063.

Rebello, S., Anoopkumar, A. N., Aneesh, E. M., Sindhu, R., Binod, P., Pandey, A. (2020). Sustainability and life cycle assessments of lignocellulosic and algal pretreatments. *Bioresource Technology*, 301, 122678. DOI: 10.1016/j.biortech.2019.122678

Requena, J. S., Guimaraes, A. C., Alpera, S. Q., Gangas, E. R., Hernandez-Navarro, S., Gracia, L. N., Cuesta, H. F. (2011). Life cycle assessment (LCA) of the biofuel production process from sunflower oil, rapeseed oil and soybean oil. *Fuel Processing Technology*, 92(2), 190–199. DOI: 10.1016/j.fuproc.2010.03.004.

Rocha, M. H., Capaz, R. S., Lora, E. E. S., Nogueira, L. A. H., Leme, M. M. V., Renó, M. L. G., del Olmo, O. A. (2014). Life cycle assessment (LCA) for biofuels in Brazilian conditions: a meta-analysis. *Renewable and Sustainable Energy Reviews*, 37, 435–459. DOI: 10.1016/j.rser.2014.05.036.

Sangaletti-Gerhard, N., Cea, M., Risco, V., Navia, R. (2015). In situ biodiesel production from greasy sewage sludge using acid and enzymatic catalysts. *Bioresource Technology*, 179, 63–70. DOI: 10.1016/j.biortech.2014.12.003.

Saswattecha, K., Kroeze, C., Jawjit, W., Hein, L. (2015). Assessing the environmental impact of palm oil produced in Thailand. *Journal of Cleaner Production,* 100, 150–169. DOI: 10.1016/j.jclepro.2015.03.037.

Saswattecha, K., Kroeze, C., Jawjit, W., Hein, L. (2016). Options to reduce environmental impacts of palm oil production in Thailand. *Journal of Cleaner Production*, 137, 370–393. DOI: 10.1016/j.jclepro.2016.07.002.

Saydut, A., Erdogan, S., Kafadar, A. B., Kaya, C., Aydin, F., Hamamci, C. (2016). Process optimization for production of biodiesel from hazelnut oil, sunflower oil and their hybrid feedstock. *Fuel*, 183, 512–517. DOI: 10.1016/j.fuel.2016.06.114.

Schmidt, J. H. (2015). Life cycle assessment of five vegetable oils. *Journal of Cleaner Production*, 87, 130–138. DOI: 10.1016/j.jclepro.2014.10.011.

Sepehri A, Sarrafzadeh M-H. Effect of nitrifiers community on fouling mitigation and nitrification efficiency in a membrane bioreactor. (2018). *Chemical Engineering and Processing – Process Intensification,* 128, 10. DOI: 10.1016/j.cep.2018.04.006.

Severo, I. A., Siqueira, S. F., Deprá, M. C., Maroneze, M. M., Zepka, L. Q., Jacob-Lopes, E. (2019). Biodiesel facilities: what can we address to make biorefineries commercially competitive? *Renewable and Sustainable Energy Reviews*, 112, 686–705. DOI: 10.1016/j.rser.2019.06.020.

Sharma, V., Das, L., Pradhan, R. C., Naik, S. N., Bhatnagar, N., Kureel, R. S. (2011). Physical properties of tung seed: an industrial oil yielding crop. *Industrial Crops and Products*, 33(2), 440–444. DOI: 10.1016/j.indcrop.2010.10.031.

Sharma, S., Basu, S., Shetti, N. P., Kamali, M., Walvekar, P., & Aminabhavi, T. M. (2020). Waste-to-energy nexus: a sustainable development. *Environmental Pollution*, 267, 115501.

Siddiquee, M. N., Rohani, S. (2011). Lipid extraction and biodiesel production from municipal sewage sludges: a review. *Renewable and Sustainable Energy Reviews*, 15(2), 1067–1072. DOI: https://doi.org/10.1016/j.rser.2010.11.029.

Siqueira, S. F., Deprá, M. C., Zepka, L. Q., & Jacob-Lopes, E. (2018). Life cycle assessment (LCA) of third-generation biodiesel produced heterotrophically. *Open Biotechnology Journal*, 12(1).

Silalertruksa, T., Gheewala, S. H. (2012). Environmental sustainability assessment of palm biodiesel production in Thailand. *Energy*, 43(1), 306–314. DOI: 10.1016/j.energy.2012.04.025.

Singh, D., Sharma, D., Soni, S. L., Sharma, S., Kumari, D. (2019). Chemical compositions, properties, and standards for different generation biodiesels: a review. *Fuel*, 253, 60–71. DOI: 10.1016/j.fuel.2019.04.174.

Singh, D., Sharma, D., Soni, S. L., Sharma, S., Sharma, P. K., Jhalani, A. (2020). A review on feedstocks, production processes, and yield for different generations of biodiesel. *Fuel*, 262, 116553. DOI: 10.1016/j.fuel.2019.116553.

Slade, R., Bauen, A. (2013). Micro-algae cultivation for biofuels: cost, energy balance, environmental impacts and future prospects. *Biomass and Bioenergy,* 53, 29–38. DOI: 10.1016/j.biombioe.2012.12.019

Soto-Sierra, L., Stoykova, P., Nikolov, Z. L. (2018). Extraction and fractionation of microalgae-based protein products. *Algal Research*, 36, 175–192. DOI: 10.1016/j.algal.2018.10.023.

Spinelli, D., Jez, S., Basosi, R. (2012). Integrated environmental assessment of sunflower oil production. *Process Biochemistry*, 47(11), 1595–1602. DOI: 10.1016/j.procbio.2011.08.007.

Spugnoli, P., Dainelli, R., D'Avino, L., Mazzoncini, M., Lazzeri, L. (2012). Sustainability of sunflower cultivation for biodiesel production in Tuscany within the EU Renewable Energy Directive. *Biosystems Engineering*, 112(1), 49–55. DOI: 10.1016/j.biosystemseng.2012.02.004.

Sumiani, Y., Sune, B. H. (2007). Feasibility study of performing an life cycle assessment on crude palm oil production in Malaysia. *International Journal of Life Cycle Assessment*, 12(1), 50–58. DOI: 10.1065/lca2005.08.226.

Tian, Y., Zhang, J., Wu, D., Li, Z., Cui, Y. (2013). Distribution variation of a metabolic uncoupler, 2, 6-dichlorophenol (2, 6-DCP) in long-term sludge culture and their effects on sludge reduction and biological inhibition. *Water Research*, 47(1), 279–288. DOI: 10.1016/j.watres.2012.10.008.

Tomei, J., Upham, P. (2009). Argentinean soy-based biodiesel: an introduction to production and impacts. *Energy Policy*, 37(10), 3890–3898. DOI: 10.1016/j.enpol.2009.05.031.

Ummalyma, S. B., Gnansounou, E., Sukumaran, R. K., Sindhu, R., Pandey, A., Sahoo, D. (2017). Bioflocculation: an alternative strategy for harvesting of microalgae–an overview. *Bioresource Technology*, 242, 227–235. DOI: 10.1016/j.biortech.2017.02.097.

Woyann, L. G., Meira, D., Zdziarski, A. D., Matei, G., Milioli, A. S., Rosa, A. C., Benin, G. (2019). Multiple-trait selection of soybean for biodiesel production in Brazil. *Industrial Crops and Products*, 140, 111721. DOI: 10.1016/j.indcrop.2019.111721.

Xia, Y., Wen, X., Zhang, B., Yang, Y. (2018). Diversity and assembly patterns of activated sludge microbial communities: a review. *Biotechnology Advances*, 36(4), 1038–1047. DOI: 10.1016/j.biotechadv.2018.03.005

Xue, X., Collinge, W. O., Shrake, S. O., Bilec, M. M., Landis, A. E. (2012). Regional life cycle assessment of soybean derived biodiesel for transportation fleets. *Energy Policy*, 48, 295–303. DOI: 10.1016/j.enpol.2012.05.025.

Yang, J., Xu, M., Zhang, X., Hu, Q., Sommerfeld, M., Chen, Y. (2011). Life-cycle analysis on biodiesel production from microalgae: water footprint and nutrients balance. *Bioresource Technology*, 102(1), 159–165. DOI: 10.1016/j.biortech.2010.07.017.

Yin, Z., Zhu, L., Li, S., Hu, T., Chu, R., Mo, F., … Li, B. (2020). A comprehensive review on cultivation and harvesting of microalgae for biodiesel production: environmental pollution control and future directions. *Bioresource Technology*, 301, 122804. DOI: 10.1016/j.biortech.2020.122804.

Zhang, L., Loh, K. C., Kuroki, A., Dai, Y., Tong, Y. W. (2020). Microbial biodiesel production from industrial organic wastes by oleaginous microorganisms: current status and prospects. *Journal of Hazardous Materials*, 123543. DOI: 10.1016/j.jhazmat.2020.123543

Zhang, X., Yan, S., Tyagi, R. D., Surampalli, R. Y. (2013). Energy balance and greenhouse gas emissions of biodiesel production from oil derived from wastewater and wastewater sludge. *Renewable Energy*, 55, 392–403. DOI: 10.1016/j.renene.2012.12.046.

12

Role of Nanocatalysts in Biofuel Production and Comparison with Traditional Catalysts

Kamlesh Kumari, Ritu Yadav, Durgesh Kumar, Vijay Kumar Vishvakarma, Prashant Singh*, Vinod Kumar, and Indra Bahadur*

CONTENTS

* *Corresponding Authors* psingh@arsd.du.ac.in *and* bahadur.indra@gmail.com

DOI: 10.1201/9781003120858-12

12.1 Introduction

Due to the global efforts to prevent the adverse effects of climate change and due to rising oil prices, researchers around the world have attempted to reduce greenhouse gas emissions and to find clean, renewable as well as efficient fuels. Thus, biofuels have begun to replace the traditional fuels, or fossil fuels. A biofuel is a good renewable energy sources as emitted carbon dioxide (CO_2) on burning can be absorbed by biofuel crops for their growth. Biofuels are described as good sources of energy that originate from biomass, that is, the living substance present in plants or organic matter. Substances with carbohydrate compounds as the main components are biofuel resources, such as wood, oilseed plants, carbohydrate plants, fibre plants, protein plants and herbal waste. However, growing biofuel crops in farmland instead of edible food could create a problem and also, these biofuel crops and crop residues or waste material cannot produce a sufficient amount of biofuel. The objective target of the international energy agencies is to triple the global biofuel output by 2030. Biofuel may contain a mixture of chemicals, that is, 5–8 esters of fatty acids and in the case of fossils fuels, they are made up of various hydrocarbons (Caspeta et al., 2014; Chen & Dou, 2015; de Lanes, de Almeida Costa, & Motoike, 2014; Johnson, 2017; Kazamia & Smith, 2014; Ramos, Valdivia, Garcia-Lorente, & Segura, 2016; Zhang et al., 2016; Zhou & Dong, 2011).

12.2 Fuels (Alcohol/Biodiesel) Can Replace Fossil Fuels or Petroleum-Based Fuel

Biofuels like alcohol can be distinguished from petroleum-based biodiesel by the availability of oxygen. Thus, the presence of oxygen in biofuels like ethanol helps the fuel to burn more completely. During combustion, oxygen from biofuels and air binds to the hydrogen in the fuel to create water and carbon dioxide. Hence, harmful tailpipe emissions such as unburned hydrocarbons and carbon monoxide are significantly reduced. Therefore, biofuels like ethanol can be blended with gasoline to create cleaner-burning mixes and to reduce greenhouse gas emissions (Bai, Geng, Wang, & Zhang, 2019; Berhe & Sahu, 2017; Caspeta et al., 2014; Chen & Dou, 2015; Falk, Narvaez Villarrubia, Babanova, Atanassov, & Shleev, 2013; Lam, Ghaderi, Fink, & Stephanopoulos, 2014).

12.2.1 Biofuels

Biofuel is a fuel generated by slow geological processes. Basically, it is produced from biomass. The necessity for biofuels has arisen due to factors such as increased energy security requirements and to reduce the emissions of greenhouse gases from burning fossil fuels. Coal and petroleum are used primarily as energy sources because of their low price, high energy value and easy availability (Vaishali, Banty, & S., 2017) Biofuels are much more suitable than fossil fuels because they are renewable, whereas

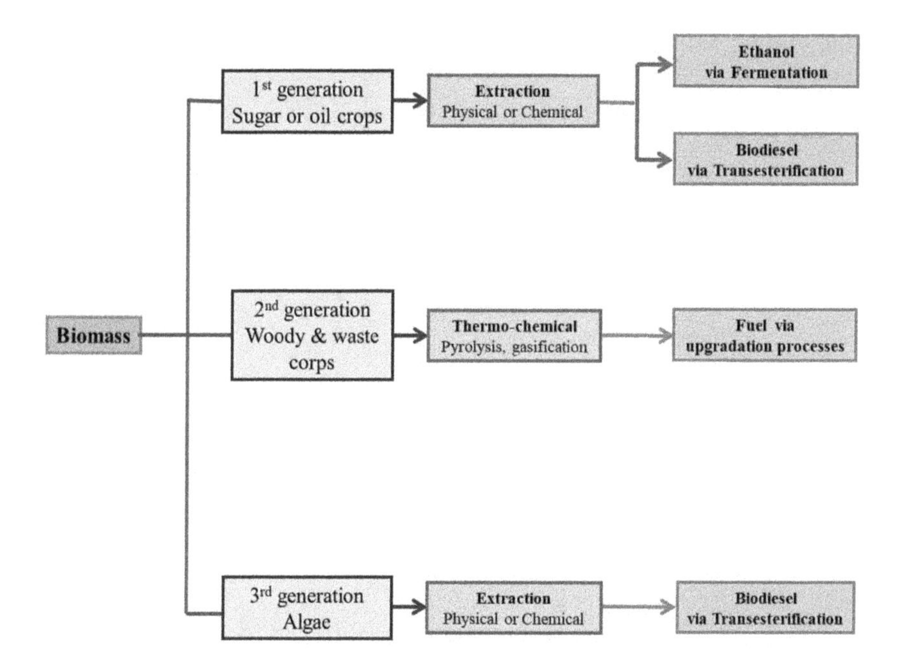

FIGURE 12.1 General classification of biofuels on the basis of sources.

fossil fuels take millions of years to form. Ethanol can be synthesized from sugarcane as well as from soybean oil but is less popular than dimethyl ether prepared from the lignocellulosic biomass; and these are the common examples of biofuels. Biofuels are categorized into three generations based on their sources of biomass (Figure 12.1). The first-generation biofuels are derived from biomass food sources such as seeds, grains and sugar, etc. On a planet where there is insufficient food for everyone, deriving biofuels from food sources is a significant disadvantage. Second-generation biofuels are obtained from non-food biomass, and are also known as advanced biofuels. These are produced from the waste obtained from food crops and agricultural residues. Then, the third-generation biofuels are considered to be the best alternative fuels as they are not derived from food sources. The source of third-generation biofuels is algae, from which a variety of fuels can be derived. There is a need for serious efforts to make these more economically viable (Adrio, 2017; Albers, Berklund, & Graff, 2016; Aro, 2015; Bai et al., 2019; Behera et al., 2015; Beller, Lee, & Katz, 2015; Bryant et al., 2020; Cai & Zhang, 2017; Caspeta et al., 2014; Chang, Yang, Lee, & Hallenbeck, 2015; Chaudhary, Gupta, Gupta, & Sharma, 2017; Chen, Lin, Huang, & Chang, 2014).

12.2.2 Classification of Biofuels

Biofuels are classified into the following types based on their method of production and the raw materials used, as illustrated in Table 12.1.

TABLE 12.1

Different Generations of Biofuels

Criteria	First generation	Second generation	Third generation	Fourth generation
Sources	Derived from food crops like wheat, sugarcane, corn, soya bean, etc.	Produced from non-food crops, i.e., lignocellulose biomass such as wood residue, organic waste, etc.	Derived from algae and other microbes. It requires biorefinery besides farmland.	An extension of third-generation biofuels, i.e., algae.
Method	Biochemical methods, like fermentation and hydrolysis	Biochemical and thermochemical method is used.	Biochemical or thermochemical and dewatering process.	Modified via genetic engineering to alter the properties and cellular metabolism and then biochemical and thermochemical process.
Examples	Ethanol, Biodiesel	Bioethanol, Fischer, Tropsch liquids (FTL); Dimethyl ether (DME), Biogas	Fuel derived from cellulosic sources, bioethanol and biodiesel	Bioethanol from genetically modified carbon-negative crops

12.2.2.1 First-Generation Biofuels

Ethanol is known for its use as a biofuel and can be obtained by the fermentation of carbohydrate (sugar/starch). The presence of oxygen in ethanol helps in its burning, making it a better and more efficient fuel with less air pollution. Ethanol can be used itself as a fuel or blended with gasoline fuel at 5–10% for use in normal cars to improve combustion and reduce pollution because ethanol is a high-octane fuel (with octane 113), with high octane meaning there is a higher resistance to knocking, and thus more power and speed. In the past, chemicals like tetraethyl lead and MTBE (methyl tert–butyl ether) were added to increase the octane number of motor fuels, but they were banned due to their highly toxic nature, and the carcinogenic nature of MTBE which increased concerns about groundwater contamination and water quality, whereas ethanol is clean, renewable, abundant and inexpensive. The United States is the largest producer of ethanol as a fuel using corn as a feedstock, followed by Brazil which uses sugarcane as the feedstock. India aims to triple its ethanol production to produce 450 crore litres of ethanol which would enable it to reduce its oil import bill by ₹12,000 crore (Chen et al., 2014; Chew & Bhatia, 2009; Costa, Calijuri, Avelar, Carneiro, & de Assis, 2016; De Bhowmick, Sarmah, & Sen, 2017; Dubini & Antal, 2015; Flores-Gomez et al., 2018; Hernandez, Solana, Riano, Garcia-Gonzalez, & Bertucco, 2014; Kang & Lee, 2015; Mehmood, 2018; Ramos et al., 2016).

12.2.2.1.1 Method of Production for Ethanol

Ethanol is produced by anaerobic fermentation of sugar by yeasts like *Saccharomyces cerevisiae*. Feedstock (sugarcane, beet or wheat) provides sugar easily for fermentation, whereas polysaccharides (starch-based feedstock from corn or wheat, barley, potato) require enzymatic hydrolysis to release the sugar for fermentation. Maltose and sucrose are rich sugars that can be hydrolysed or cleaved by enzymes to produce monosaccharides. The obtained monosaccharides are further converted into ethanol by anaerobic fermentation, as illustrated in Scheme 12.1. These enzymes are selective in nature, mainly to hexoses and pentose (Ramos et al., 2016; Sesmero, 2014; Zhang et al., 2016; Zhou, Buijs, Siewers, & Nielsen, 2014).

Monosaccharides cannot be hydrolysed to obtain a simple monosaccharide, for example glucose, fructose and riboses. The production of ethanol from starch is a two-step procedure. Sacrification is the first step, with starch undergoing amylolytic hydrolysis to release small sugars. Then, the obtained sugars result in ethanol on fermentation using *Saccharomyces cerevisiae* in the second step (Scheme 12.2).

12.2.2.1.2 Biodiesel

The production of biodiesel or bioesters can be performed based on the chemical reaction of vegetable oil with alcohol in the presence of a catalyst which may be heterogeneous or homogeneous to yield mono-alkyl esters and glycerine by using the raw

$$\underset{\text{Disaccharide}}{C_{12}H_{22}O_{11}} \xrightarrow[\text{Water}]{\text{Yeast}} \underset{\text{Monosaccharide}}{2\,C_6H_{12}O_6} \xrightarrow{\text{Yeast}} 4\,C_2H_5OH + 4CO_2$$

SCHEME 12.1 Preparation of ethanol in anaerobic fermentation.

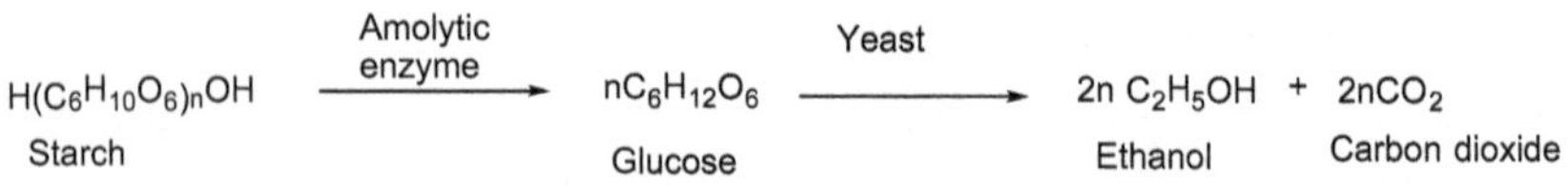

$$H(C_6H_{10}O_6)_nOH \xrightarrow[\text{}]{\text{Amolytic enzyme}} nC_6H_{12}O_6 \xrightarrow{\text{Yeast}} 2n\,C_2H_5OH + 2nCO_2$$

Starch, Glucose, Ethanol, Carbon dioxide

SCHEME 12.2 Fermentation of starch using enzymes.

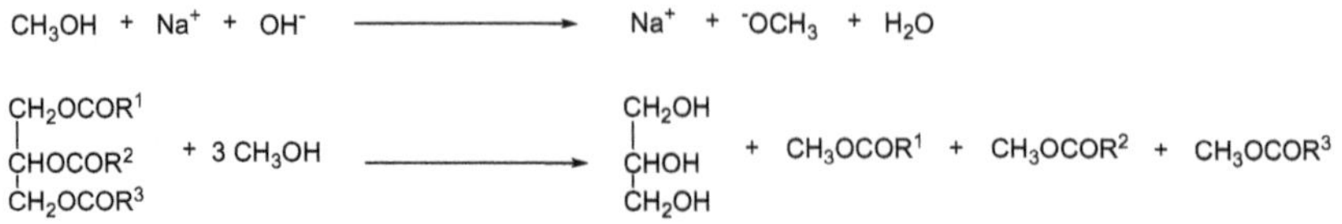

$$CH_3OH + Na^+ + OH^- \longrightarrow Na^+ + {}^-OCH_3 + H_2O$$

$$CH_2OCOR^1\text{–}CHOCOR^2\text{–}CH_2OCOR^3 + 3\,CH_3OH \longrightarrow CH_2OH\text{–}CHOH\text{–}CH_2OH + CH_3OCOR^1 + CH_3OCOR^2 + CH_3OCOR^3$$

SCHEME 12.3 Mechanism for hydrolysis in basic medium.

material from oily seeds of rapeseed, soya bean, jatropha, palm, etc. This is blended with petroleum diesel to create biodiesel blended fuel (Kang & Lee, 2015; Karimi, Othman, Uzunoglu, Stanciu, & Andreescu, 2015; Kazamia & Smith, 2014; Kim, Jia, & Wang, 2006; Kim et al., 2020; Lam et al., 2014; Puri, Barrow, & Verma, 2013). In homogeneous catalysis, the catalyst is dispersed in a gaseous or liquid phase as the reactant. The hydrolysis of esters can be done by acid solutions, acid catalysis, organo-metallic catalysis and enzymatic catalysis. In heterogeneous catalysis, the reactants as well as the catalyst are not in the same phase and have a boundary. The catalysts can be solids, while the reactants can be in a gaseous or liquid state.

12.2.2.1.3 Production of Biodiesel

12.2.2.1.3.1 Use of Homogeneous Catalyst Transesterification can be performed by esterification of triglyceride (RCOOR') using alcohol. It is reported to be a reversible reaction and performed between the reactant and the catalyst (acid–base). This gives a mixture of methyl-ester of fatty acid along with glycerine. Glycerine can be removed easily by centrifugation. Biodiesel must be purified and freed from a high glycerine content. Otherwise, the fuel tank can become clogged, with injector fouling and valve deposits in the vehicle engine.

12.2.2.1.4 Base-Catalysed Transesterification

The mechanism for the transesterification in basic medium is very simple, efficient and interesting. The reaction is carried out in a polar medium, therefore, sodium hydroxide or potassium hydroxide is used as the catalyst. Herein, methanol is treated with sodium hydroxide and results in sodium methoxide. Then, it reacts with the triglycerides, and undergoes hydrolysis resulting in esters of fatty acids along with glycerine as a by-product (Scheme 12.3).

12.2.2.1.5 Acid-Catalysed Transesterification

The catalyst used in this method is a mineral acid such as sulphuric acid, and H^+ of the mineral acid protonated the carbonyl group of the ester group as the ketonic group has a lone pair of electrons. Further, resonance takes place to obtain a more stable intermediate: a carbocation. also, a molecule of alcohol reacts to this carbocation and

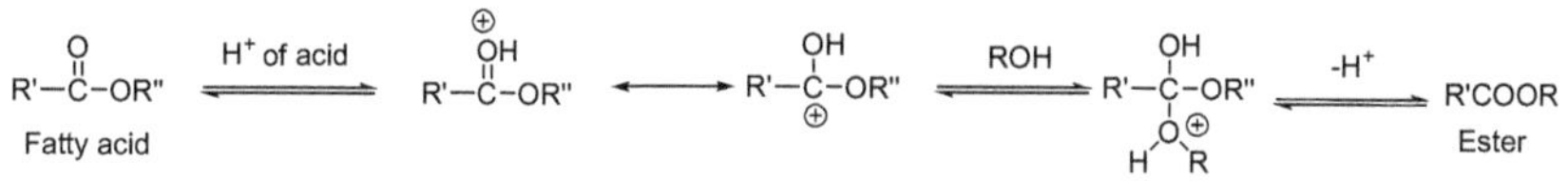

R'–C(=O)–OR" (Fatty acid) ⇌ (H⁺ of acid) R'–C(=$\overset{\oplus}{O}$H)–OR" ⟷ R'–$\overset{\oplus}{C}$(OH)–OR" ⇌ (ROH) R'–C(OH)(–$\overset{\oplus}{O}$(H)R)–OR" ⇌ (-H⁺) R'COOR (Ester)

SCHEME 12.4 Mechanism for hydrolysis in acidic medium.

gives another reaction intermediate, finally producing the ester of interest with some by-products (Scheme 12.4).

12.2.2.1.6 Role of Heterogeneous Catalyst

Solid catalysts are for oils with a high free fatty acid (FFA) content that can catalyse the transesterification as well as esterification of the available fatty acids to give the methyl ester of fatty acids. The mechanism for the formation of methyl ester is explained next. Initially, there is carbocation formed by the interaction of the acidic site of catalyst and carbonyl oxygen of (FFA) or monoglyceride as the oxygen of the carbonyl group has a lone pair of electrons. Then, the double bond of the carbonyl group breaks and an electron cloud approaches the oxygen as it is more electronegative than carbon, and carbocation is formed. Then, oxygen in the alcohol has a lone pair of electrons that attack the carbocation, which is considered to be a nucleophilic attack, producing the tetrahedral intermediate. During esterification, water molecules are eliminated from the tetrahedral geometry to form a mole of ester.

First-generation biofuels are produced by contemporary processes using biomass and are also called conventional biofuels. The sources of these fuels are sugar, grains, starch and seeds and they are prepared by fermentation, distillation and transesterification processes. Fermentation of sugar is done to obtain ethanol mainly, along with butanol and propanol in small amounts. In the United States, ethanol is used as an additive to gasoline as it releases less greenhouse gases on burning. During transesterification, alcohol, such as methanol, in the presence of catalyst is exposed to animal fat or plant oil. The main product is separated from the by-products of the reaction in the distillation process, as shown in Figure 12.2. Despite all these advantages, first-generation fossil fuels also have certain drawbacks from the sustainability perspective. The first and foremost disadvantage is to the competition with food crops, such as corn and sugar beet. In recent years, this has contributed to higher prices for food and animal feeds worldwide. Not only this, there is also a risk of greenhouse gas emission, air pollution, cost of production and implementation, etc. (Jiang et al., 2019; Kang & Lee, 2015; Kazamia & Smith, 2014; Khan, Shin, & Kim, 2018; Khanna, Wang, Hudiburg, & DeLucia, 2017; Kim et al., 2006; Lam et al., 2014; Liu, Marks, & Li, 2019; Mehmood, 2018).

12.2.2.2 Second-Generation Biofuels

The other term used for second-generation fuels is advanced biofuels, which can be synthesized from non-food biomass, that is, non-edible residues from food crop/whole plants. At present, no countries are producing second-generation fuels, and the preparation of these fuels involves several steps including hydrolysis, fermentation, water-treatment, etc. Further, algae also fall under this category and are an important

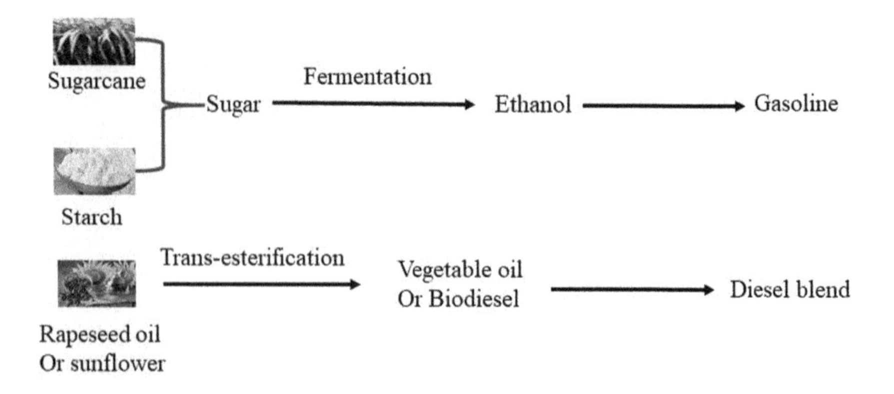

FIGURE 12.2 Various processes involved in the production of first-generation biofuels.

fuel. These fuels could easily replace fossil-derived fuels in comparison to biodiesel because hydrotreated biodiesel is produced from vegetable oil (HVO) and does not have issues such as increased emissions of NO_2, deposit formation and compatibility (De Bhowmick et al., 2017; de Lanes et al., 2014; Falk et al., 2013; Hernandez et al., 2014; Hood, 2016; Jiang et al., 2019; Kang & Lee, 2015; Kazamia & Smith, 2014; Khan et al., 2018; Kim et al., 2006; Lam et al., 2014; Mehmood, 2018).

12.2.2.2.1 Biochemical Fuels Like Bioethanol

This is blended with gasoline to produce an oxygenated fuel, with lower hydrocarbon and greenhouse gas emissions. It is prepared by biochemical conversion. Polysaccharides can be converted into sugars via pretreatment. It involves the separation of xylose and lignin from cellulose. Lignin and hemicelluloses cannot be fermented easily, but only with the help of enzymes and this also requires water extraction. Fermentation is used to produce ethanol using yeast or bacteria and simple distillation is carried out to obtain pure ethanol (Figure 12.3).

12.2.2.2.2 Biofuel Obtained through Thermochemical Conversion

The production of biofuels like Fischer-Tropsch liquid (FTL), DME and various alcohols through this process begins with gasification or pyrolysis. Figure 12.4 illustrates the replacement of various biofuels derived from petroleum. Alcohol-based fuels have the ability to replace gasoline in spark ignition engines, and green diesel and DME are appropriate for compression ignition engines. Second-generation biofuels address problems related to first-generation fuel as second-generation fuels come from distinct biomass. Higher energy yield per acre is generated by second-generation biofuels in comparison to first-generation biofuels. With increases and changes in scientific advances used for second-generation biofuels, the opportunities for cost reductions may increase. A poorer quality of land is also used, where edible food crops may not be grown. Cellulosic sources that grow along with food crops which are used as a biomass can remove nutrients from the soil. Without a doubt, the process for the production of second-generation biofuels is more complicated than that for first-generation biofuels.

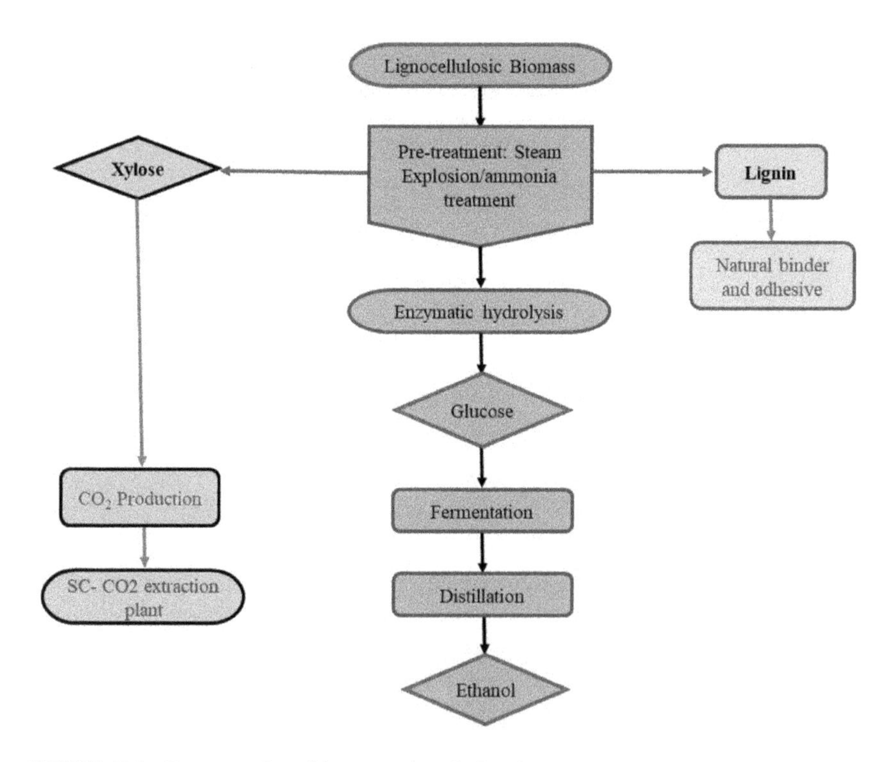

FIGURE 12.3 Representation of the generation of ethanol.

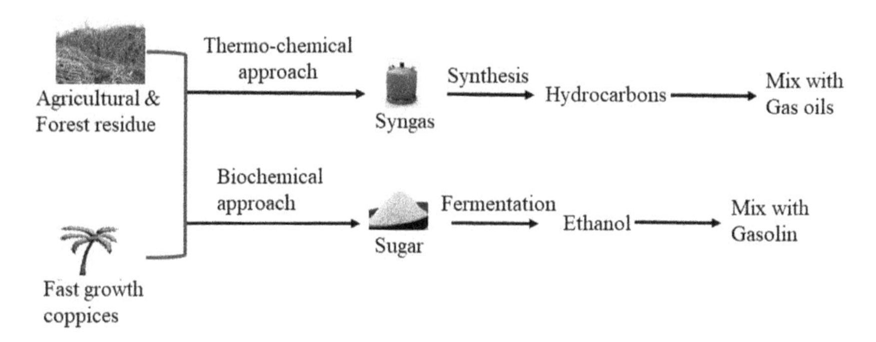

FIGURE 12.4 Different biological and chemical processes involved in the production of second-generation biofuels.

12.2.2.3 Third-Generation Biofuels

The third-generation biofuels have proved to be the best alternative to fossil fuels. The greatest advantage of third-generation biofuels is that they are not derived from food sources. In third-generation biofuels, the main energy source is algae. Oil is extracted from algae and then converted into biofuels in the same way as the first-generation biofuels are converted. The best advantage of the use of algae is that they

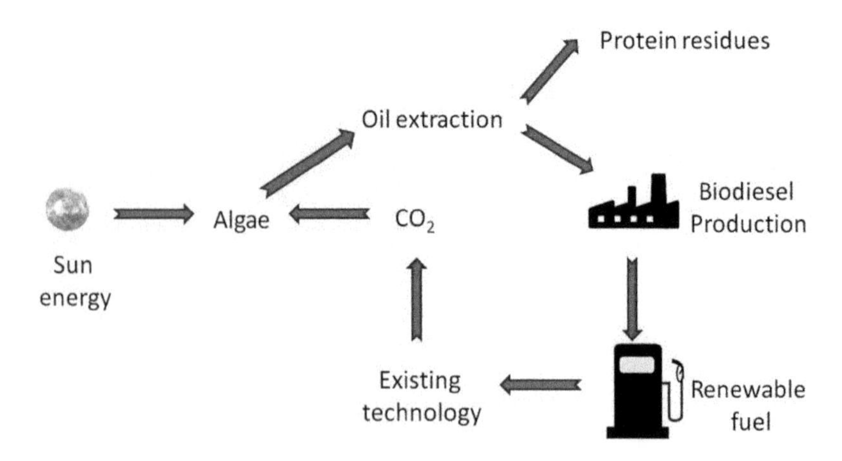

FIGURE 12.5 Techniques involved in the production of third-generation biofuels.

are cultured in such a way that they have the potential to produce more energy per acre. Undoubtedly, third-generation biofuels are the best alternatives to petroleum fuels. Algae can be grown in wastewater, salt water and even using sewage. However, attempts are being made to improve the extraction process so that it can be economically feasible. Figure 12.5 shows the general steps for the production of third-generation biofuels (Chang et al., 2015; Chaudhary et al., 2017; Chen & Dou, 2015; Chen et al., 2014; Hood, 2016).

The global energy requirement is increasing daily and the scarcity of fossil fuels has resulted in a focus on renewable energy sources. Thus, the scale of transportation fuel needs much greater than that for cooking. The first and foremost concern has arisen due to the expansion of biofuels production, causing the diversion of land away from uses for food, preservation of biodiversity and other important purposes (Abo, Odey, Bakayoko, & Kalakodio, 2019; Albers et al., 2016; Aro, 2015; Asveld, 2016; Bagnoud-Velasquez, Refardt, Vuille, & Ludwig, 2015).

12.3 Types of Promising Catalysts Used for the Production of Biofuels

A catalyst is an agent or substance that causes a chemical reaction without being changed. Herein, the chemical reaction proceeds at a faster rate in the manufacturing of biofuels or rather in the process of transesterification by homogeneous, heterogeneous and enzymatic catalysts. Nanocatalysts have overtaken these three in the production of biofuels, with advances in nanocatalysts showing improvements in the effective surfaces of reactants as shown in Figure 12.6 (Adrio, 2017; Babadi, Bagheri, & Hamid, 2016; Berhe & Sahu, 2017; Byun & Han, 2016; Inamuddin, Shakeel, Imran Ahamed, Kanchi, & Abbas Kashmery, 2020; Jiang, Zhou, Liao, Zhang, & Jin, 2017; Khan et al., 2018; Khanna et al., 2017; Liu et al., 2019; Pakapongpan, Tuantranont, & Poo-Arporn, 2017; Pramanik & Bhaumik, 2013; Prashanth et al., 2017; Yang et al., 2015; Zhang et al., 2017).

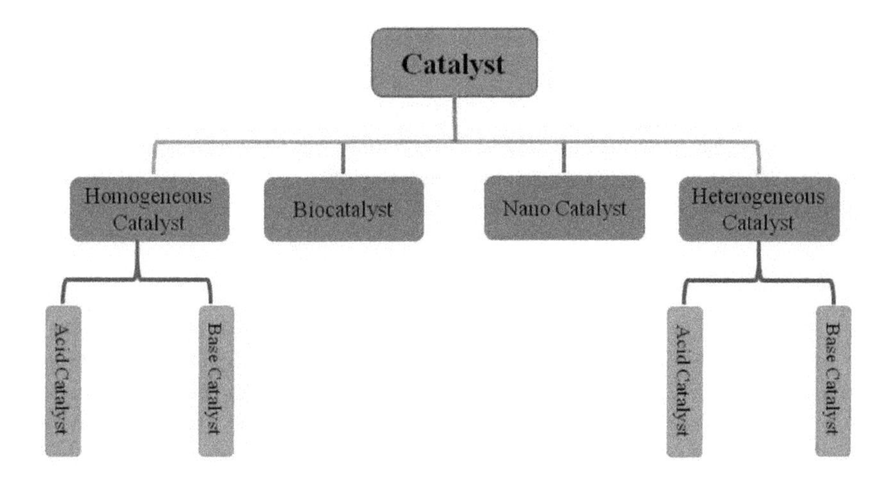

FIGURE 12.6 Types of catalysts used in the processing of biofuels.

12.3.1 Homogeneous Catalysts

The homogeneous catalysts used in the manufacture of biofuels are divided into two categories: homogeneous alkaline catalysts and homogeneous acidic catalysts. Homogeneous catalysts require moderate temperature and atmospheric pressure during the process of transesterification of oil and fats. However, this method is not a complete success because the catalyst used cannot be recycled.

12.3.2 Homogeneous Alkaline Catalysts

Some of the homogeneous alkaline catalysts used in the synthesis of biofuels are NaOH, KOH, $NaOCH_3$, $KOCH_3$, $NaOC_2H_5$, Na_2O_2 and C_4H_9Ona, etc.

12.3.3 Homogeneous Acidic Catalysts

Basically, homogeneous acid-based catalysts helps to lower the amount of FFAs used in cooking oil and animal fat with an alkali catalyst. Special equipment is needed because of their corrosive nature. The rate of the reaction is nearly 4000 times lower with homogeneous acidic catalyst than a homogeneous alkaline catalyst. This requires high temperature as well as a high molar ratio of alcohol to oil.

12.3.4 Heterogeneous Catalysts

Generally, heterogeneous catalysts act different than homogeneous catalysts as they appear in solid form. The workings of heterogeneous catalysts are completely different to those of homogeneous catalysts. Homogeneous catalysts are further divided into alkaline and acidic types. These catalysts can be easily recycled for reuse as they are easily separated from the final product. This is the only reason to use heterogeneous catalysts, which are claimed to be eco-friendly.

12.3.5 Heterogeneous Alkaline Catalysts

Examples of heterogeneous alkaline catalysts are CaO, MgO, (CaO/ZnO), (Li/CaO) and other mixed metal oxides (MgZnAlO, K_2CO_3/Al–Ca, Mg-Al hydrotalcites). The source of heterogeneous alkaline catalyst is waste material. They are used to overcome the main shortcoming of homogeneous alkaline catalysts. Alkaline catalysts are less soluble, therefore, they can be easily separated and reused.

12.3.6 Heterogeneous Acidic Catalysts

Heterogeneous acidic catalysts are comprised of zeolite, heteropolyacids, pure oxides or modified transition metals like zeolite-X 4.2, SO_4^{2-}/TiO_2-SiO_2, carbon-based from starch, SO_4^{2-}/ZrO_2-SiO_2, $Zr_{0.7}H_{0.2}PW_{12}O_4$, ferric-manganese-doped tungstated/ molybdena, sulphated PAHs, KSF clay amberlyst, Fe-Zn cyanide complexes, carbon-based solid acid and WO_3/ZrO_2. The stability of heterogeneous acidic catalysts is high in comparison to homogeneous catalysts, and the price is low. Acidic catalysts cauee minimal diffusion. Acidic heterogeneous catalysts have a huge number of active sites and modest acidity. the elimination of various processes of the separation of hazardous wastewater can be done by using homogeneous catalysts in preference to heterogeneous catalysts. Good conversion yields can be attained by heterogeneous acidic catalysts at high temperatures.

12.4 Biocatalysts

The use of biocatalyst is a method that has solved problems for both humankind and nature. Biocatalysts are used in the production of chemical products, medicines, biofuels, detergents, food additives, biosurfactants, functionalized biological polymers, etc. The first and foremost factor to highlight is the ability to produce high-purity biofuels from low-value materials such as cooking oil, with no soap formation. This factor distinguishes biocatalysts from others. The problems of FFAs and water in raw materials are eradicated by using enzymes as a catalyst. Biofuels obtained from waste oil, beef and pork suet with enzymatic transesterification are considered to be most efficient and this is the reason for the combination of FFAs and enzymatic catalysts, which turn it into alkaline esters. Generally, based on their activity there are two types of enzymatic catalyst: extracellular and intracellular lipases. The factors that contribute in making biocatalysts effective are thermal stability, selectivity, high level of efficiency, etc. Despite such positive aspects, biocatalysts cannot be described as the best option for producing biofuels as the biocatalyst production costs are very much higher and their reaction time is also longer.

12.5 Nanocatalysts

Nanocatalysts are those that link homogeneous and heterogeneous catalysts. Over the past few years, researchers have been working on their usage, including carbon nanotubes, nanoclays, nanofibres, nanowires and nanoparticles. Nanomaterials have a size of less than 100 nm. Nanoparticles are selective, highly durable and reliable, due to which they play an integral part in the catalytic process. In other words, nanoparticles have high selectivity, huge reactivity and the ability for ultra-low energy consumption. On decreasing the particle size, an increase in the rate of the reaction is found. This method is not a complete success however, because of certain obstacles such as the removal of products from the waste and recycling them for their reuse. Furthermore, optimization of catalysts is important and is required to enhance the number of active sites and the surface area. Advanced or novel catalysts have several active phases and include an appropriate base to enable an increase in their extraordinary abilities (Adrio, 2017; Babadi et al., 2016; Berhe & Sahu, 2017; Byun & Han, 2016; Inamuddin et al., 2020; Jiang et al., 2017; Khan et al., 2018; Khanna et al., 2017; Liu et al., 2019; Pakapongpan et al., 2017; Pramanik & Bhaumik, 2013; Prashanth et al., 2017; Yang et al., 2015; Zhang et al., 2017)

12.6 Nanomaterials Used in Biofuel Production

The dominance of conventional fossil fuels such as oil, gas, coal, etc. in daily life has created a sustainability problem. Due to the use of fossil fuels, the amount of these resources is decreasing daily. Inflation in the price of these fossil fuels is also more common, along with the adverse impacts on the environment. Some adverse effects of these fossil fuels such as global climate change, greenhouse effect, etc. create demands for other energy resources to replace these conventional resources. Hence, researchers are trying to develop alternative energy resources such as biomass, biofuels, etc. These resources can be stored in the desired manner and utilized to generate fuels, including electricity, as needed (Baloch, Upaichit, & Cheirsilp, 2020; Booramurthy, Kasimani, Pandian, & Ragunathan, 2020; Chen et al., 2018; Gardy, Rehan, Hassanpour, Lai, & Nizami, 2019; Mamo & Mekonnen, 2019). These resources are collectively described as bioenergy resources, and they can be found in different forms including solid, liquid and gas, of biological origin. Some examples of bioenergy are bioalcohol, biodiesel, etc. Bioalcohol is obtained from corn, wheat, sugar beet and sugarcane, while biodiesel is obtained by transesterification of oils extracted from rapeseed, palm, soybean and sunflower. Techniques like gasification, combustion and pyrolysis covert biomass to bio-oil, biochar and biogas. These can be used to produce electricity very easily. However, anaerobic conditions are required to be maintained to produce biogas commercially. To provide gas for cooking and lighting in rural areas it is more suitable. Microbial fuel cells also produce bioenergy using organisms, and have the ability to generate electricity. There are numerous problems in producing bioenergy from microbial sources, including production, cultivation, etc., although have been some developments in this field. There are also some difficulties in existing processes such as infrastructure, cost and the need for the development of new methodologies to produce higher bioenergy yields. These developments include pre-treatment processing, fermentation, etc., which enable bioenergy to be produced more efficiently and effectively.

Currently, new developments also lead to the use of the principles and fundamentals of nanotechnology. Nanotechnology deals with numerous branches of science in an interdisciplinary manner. Nanoscale material is defined as matter which has at least one dimension at the level of 100 nm. This tiny material possesses a unique size which produces a large surface area to volume ratio which can increase the active sites. Nanoparticles have excellent abilities due to their different size, shape and morphology; and have been explored in different scientific disciplines. (Genwa, Kumari, & Singh, 2019; Kumar et al., 2008; Kumar, Singh, Kumari, Mozumdar, & Chandra, 2011; Kumari, Singh, Bauddh, Mallick, & Chandra, 2019; Kumari, Singh, Mehrotra, & Chandra, 2012; Singh, Katyal, Kalra, & Chandra, 2008; Singh, Kumari, Katyal, Kalra, & Chandra, 2009). Apart from this, nanomaterials have a higher reaction rate in comparison to other molecules. The unique scalable optical properties of nanomaterials have a great impact on fields such as electronic, medicines, ionic liquids, bioenergies, etc. Bioenergy production has reported a number of methods in which nanotechnology is directly involved in increasing the production of bioenergy. High catalytic activity, adsorption capacity, durability and crystallinity make it much more suitable for bioenergy production. Nanosystems like metallic nanoparticles, nanotubes, nanofibres, etc. enhance the production of bioprocesses to obtain an excellent amount of biofuels. However, nanocrystals, nanodroplets, nanomagnets, etc. also have been used in the production of biofuels. Herein, this chapter describes the recent developments in the use of nanomaterials in the production of biofuels.

Ajala et al. reported the use of waste-iron-filling (WIF) to prepare a pro-catalyst via the co-precipitation approach. It was converted to solid acid catalysts of RBC500, RBC700 and RBC900 by calcination at very high temperatures of 500, 700 and 900°C respectively. Furthermore, it was sulphonated. They studied the efficacy of the catalysts in the production of biodiesel production using waste cooking oil (WCO). It is important to mention that the WCO has a low amount of fatty acids. The catalyst used is efficient and promising, even after a few cycles. The work reported is novel as they report the synthesis of nanocatalysts in solid state, which is a cost-effective and promising alternative (Ajala, Ajala, Ayinla, Sonusi, & Fanodun, 2020). Ananthi et al. reported the isolation of microalgal isolates and yeast isolates as they are the sources for the production of biodiesel. The production of these products is characterized by the FTIR spectrum used to investigate lipid property. They established the capability of these isolates for the production of biodiesel (Ananthi et al., 2018). Anchupogu et al. studied the synergistic potential of an Al_2O_3 NPs and *Calophyllum inophyllum* biodiesel blend. Further, it helps the gas recirculation on the combustion, performance and emission characteristics of a diesel engine. The results obtained concluded that the brake thermal efficiency of CIB20ANP40 fuel is enhanced by 7.71% (Anchupogu, Rao, & Banavathu, 2018). Baloch et al. optimized the production via the cost-effect approach of *Magnusiomyces capitatus* A4C extracellular lipase (ECL). They developed a simple, easy, cost-effective and green methodology for biomimetic mineralization to synthesize multilayered nano-entrapped ECL, which was used as a biocatalyst for biodiesel production. Synthesized nanocatalysts were found to be cheap and promising catalysts for biodiesel production as an alternative to petrol or diesel (Baloch et al., 2020). Blair et al. worked on the secondary organic aerosols obtained after the oxidation of diesel and biofuel in the presence of sunlight. They observed the formation of sulphuric acid and organosulphur (Blair et al., 2016). Booramurthy et al. reported the use of waste from tannery sheep fat industries to prepare the biodiesel using nano-sulphated zirconia. They studied the synthesized nanocatalyst using several advanced techniques such as SEM, pXRD, FTIR, etc. They optimized the conditions and methanol to fat (15:1) with 8% of catalyst at 65^{0}C to give the maximum yield of ~99%. Further, the recyclability and reuse of the catalyst was studied (Booramurthy et al., 2020). Chen et al. developed a heterogeneous catalyst for transesterification for the preparation of biodiesel from the *Jatropha curas* L. seed oil. The synthesized catalyst was based on sulphated TiO_2 and well characterized by XRD, SEM, FTIR, TG, etc. The catalytic activity was characterized after use. The catalyst was recycled and reused (Chen et al., 2018). Damanik et al. wrote a review on a cleaner fuel, that is, they discussed the popularity and acceptability of biodiesel. They explained how the amount of carbon dioxide and unburnt hydrocarbons can be reduced drastically by using biodiesel. Further, other oxides of nitrogen, carbon, etc. can be reduced and the efficiency of the engines can be improved (Damanik, Ong, Tong, Mahlia, & Silitonga, 2018). Degfie et al. reported the synthesis of biodiesel from waste cooking oil using catalysts based on nanomaterials and their recycling, as well as their reuse. The catalyst was based on CaO and synthesized by the thermal decomposition method of methanol. It may be cheap and with a low level of environmental pollution. It gives biodiesel in a high amount of 96% at optimized conditions as evaluated by the American fuel standards (Degfie, Mamo, & Mekonnen, 2019). Dehghani et al. reported on the preparation of zirconia supported on MCM-41-based nanocatalyst using an ultrasonic approach. They varied the strength of irradiation and sonication to optimize the efficiency of the catalyst, which was studied for the preparation of biodiesel. Further, the recycling and reusability of the catalyst were studied for up to five cycles with the biodiesel having a good yield (~75%) (Dehghani & Haghighi, 2016). Dehghani et al. used MCM-41 and Ce-based bifunctional nanocatalysts for the preparation of biodiesel. The catalyst is highly stable and well characterized. It yields biodiesel in a high yield of 94% and the efficacy of the catalyst is maintained until the seventh cycle; yielded to ~89%. It showed high stability (Dehghani & Haghighi, 2019b). Further, Dehghani et al. prepared Al-MCM-412- and Ce-based bifunctional nanocatalysts for the production of biodiesel in high yield. The best performance was obtained with 5% weight at 70°C for 6 hours. Of the different samples, Mg/ACM-U (Si/Ce=25) was found to have the best conversion and reusability for five cycles (Dehghani & Haghighi, 2019a). Devarajan et al. investigated the use of silver and silver oxide nanoparticles at different concentrations to treat the methyl ester of neem oil. An improvement in engine performance has been observed after the addition of 5/10 ppm of silver oxide nanoparticles (Devarajan, Munuswamy, & Mahalingam, 2018). Douki et al. reported that diesel exhaust increases pollution and is a concern of human health as it causes various diseases including cancer. They used rats in their study and optimized the parameters like filtering of gasoline. The toxic effects were studied. They suggested few significant effects after the filtration of particles; indicating that the toxicity was due to gases and nanoparticulates (Douki et al., 2018). Gardy et al. investigated the effects of several catalysts in the synthesis of biodiesel. It is known that the use of homogeneous catalysts in preparation of biodiesel has the disadvantage that they cannot be recovered and pollute the environment. Furthermore, enzymes can also be used as they do not cause pollution, however they are costly at an industrial scale. Heterogeneous catalysts can be simply recycled, and reused in production. Nanomaterials have been reported as suitable catalysts because of the lower cost, easy recyclability and reuse (Gardy et al., 2019). Liu et al. reported on nano-adsorbents prepared by a derivative of boronic acid and silica. These can easily eliminate the remaining glycerol from the activity of lipase in the production of biodiesel with good recyclability (Liu, Ma, Zeng, Du, & Yuan, 2018). Madhuvilakku et al. reported a simple transesterification methodology using promising heterogeneous mixed metal oxides, based on oxides of titanium and zinc. Prepared catalysts were well characterized by their surface characterization. The catalysts showed excellent performance after optimization of concentration, temperature and time. The maximum obtained yield of biodiesel using the nanocatalysts was 98% (Madhuvilakku & Piraman, 2013). Mamo et al. used nanoparticles of calcium oxide from waste in transesterification. For this, the catalyst was heated to a high temperature of 900 °C and well characterized. The size of the nanocatalyst was about 44 nm, with no regular shape or structure in the production of biodiesel. The maximum yield of biodiesel obtained was 92% at 60 °C with 2% catalysts, a methanol:oil ratio of 11:1 and time of 3 hours (Mamo & Mekonnen, 2019). Marella et al. used silica-based nanocatalysts and their results illustrated their promising ability in the production of biodiesel (Marella, Parine, & Tiwari, 2018). Mofijur et al. investigated various nanocatalysts in the production of biodiesel via transesterification. The methodology showed good activity and selectivity, and they developed cheap and efficient catalysts for biodiesel production (Mofijur et al., 2020). Moghzi et al. reported the synthesis of nanocatalysts based on barium which were thoroughly characterized and used in biodiesel production. Soybean is converted to biodiesel using the barium-based catalyst through transesterification (Moghzi & Soleimannejad, 2018). Nematian et al. synthesized hybrid nanocatalysts based on magnetite nanoparticles and graphene oxide, which were derivatized using small molecules. Activity, stability and kinetic studies were performed. The efficiency of the catalysts was good enough for up to five cycles of transesterification (Nematian, Shakeri, Salehi, & Saboury, 2020). Pandian et al. reported on titanium nanooxide nanoparticles in their study of their potential in the production of biodiesel and their effects on the emission of gases. The catalysts used had a very high surface areas with a particle size of 60 nm. A significant reduction in emissions (oxides of carbon and nitrogen, along with smoke) was observed using nanoparticles of oxide of titanium (Pandian, Ramakrishnan, & Devarajan, 2017). Pandit et al. reported the synthesis of nanocatalysts of calcium oxide (~6 nm) in transesterification of dry biomass to biodiesel. The optimization of conditions is carried out to obtain the maximum production of biodiesel by considering the catalyst amount and time. The maximum biodiesel yield obtained was ~86% with the use of this catalyst (Pandit & Fulekar, 2017). Pourkhesalian et al. investigated the chemical content of biodiesel and the activity of the nanocatalyst (Pourkhesalian et al., 2014). Rasouli et al. reported on the production of biodiesel from goat fat using nanoparticles of oxide of magnesium (~5.5 nm) via transesterification. The maximum biodiesel yield obtained was ~93% using this catalyst (Rasouli & Esmaeili, 2019). Shahraki et al. described a new and efficient nanocatalyst based on alumina. It was used for the production of biodiesel via the transesterification of soyabean oil with methanol at a low ultrasonic frequency and stirring. Various parameters such as concentration, catalyst amount and time were optimized to get the maximum conversion into biodiesel (95%) (Shahraki, Entezari, & Goharshadi, 2014). Soudagar et al. synthesized nanoparticles of zinc oxide and explored their use in biodiesel production to reduce the emissions of oxides of carbon and nitrogen, and of hydrocarbons (Soudagar et al., 2020). Tahvildari et al. used nanoparticles of calcium and magnesium as catalyst for biodiesel production through the reaction of soybean oil and methanol, i.e. transesterification. This catalyst can be recycled. The methodology was optimized by considering the concentration, amount of catalyst, time, etc. Oxides of calcium were found to be promising in transesterification and gave a good yield, and they can also be recycled and reused (Tahvildari, Anaraki, Fazaeli, Mirpanji, & Delrish, 2015). Vedagiri et al. described the use of oxides of cerium and zinc in transesterification for biodiesel production (Vedagiri, Martin, Varuvel, & Subramanian, 2019).

12.6.1 Nanomaterials

The engineered materials at a molecular level have novel properties and are completely different from the bulk material. These materials have distinct properties at the nanoscale due to their large surface area to volume ratio and it also triggers the quantum effect. This enhanced surface area compared to volume provides enhanced chemical reactivity and makes it special for the production of biofuels. Nanomaterials are also classified as zero-, one-, two- and three-dimensional (3-D) nanostructures. Apart from this broad classification of nanomaterials, they also include dendrimers, fullerenes, quantum dots and nanotubes, which are also used in the production of biofuels. Nanoparticles with different morphology and size show remarkable applications in the production of biofuels. The fabrication of nanoparticles offers superior quality of nanoparticles with excellent applicability but they require development to improve their quality. Organic nanoparticles have been mainly focused in fields of polymer, drug delivery, micelles, gene delivery, etc. Metal nanoparticles such as gold, silver and copper, metal oxide nanoparticles such as nickel oxide, iron oxide, tin oxide and magnetite are the most useful nanostructures in the field of biofuel production. Some enzymes that are used in biofuel production are immobilized on the surface of these nanosystems to achieve excellent catalytic property. The recyclability of the catalyst plays a significant role in any catalytic process. Immobilized nanoparticles possess excellent activities over a number of cycles. Moreover, it has been found that enzymes act as carriers when immobilized over the surface of nanoparticles. Nanoparticles possess promising paramagnetic ability and coercivity in biofuel production. However, some metal nanoparticles (Ni/Co) also exert a toxic effect in oxidation reactions. Hence, more research in this field is required to overcome these problems.

12.6.2 Nanoparticles Used in Biofuel Production

12.6.2.1 Magnetic Nanoparticles

Magnetic nanoparticles have been used in various fields of science such as environmental, biomedical, pharma, biotechnology and industrial chemistry. They also frequently used in the field of biofuel production as their unique properties include high stability and recyclability. There is a broad range of magnetic nanoparticles (e.g., iron, platinum, cobalt) that can be used as catalysts and also have the ability to produce biofuel. Immobilization of magnetic nanoparticles is easy in comparison to other nanoparticles and they also enable the removal of the immobilization by the application of some external magnetic fields without leaving any toxic effect. Their potential can be understood by bioethanol production using lignocellulose. The reusability of these immobilized enzymes is greater than in non-immobilized enzymes. Apart from the immobilization of the catalyst, magnetic nanoparticles can also be used for the coating of other nanoparticles to produce a new nanomaterial with better catalytic ability. Immobilization improves the thermo-stability of the enzyme compared to the bare enzyme.

12.6.2.2 Carbon Nanotubes (CNTs)

Carbon nanotubes (CNTs) possesses unique applicability in various branches of science, such as optics, biotechnology, genomics, pharmacy, catalysis, etc. There are numerous methods including chemical vapor deposition, laser ablation and arc discharge to synthesize carbon nanotubes. The structures of carbon nanotubes can be understood as rolled sheets of graphite in a cylindrical shape which possesses a diameter in the range of 1–100 nanometres. Carbon nanotubes have excellent biocompatibility to immobilize enzymes on their surface to make them more efficient than the bare enzyme. CNTs have unique applicability in the processing of biofuels due to their unique structure, thermal and mechanical stability and excellent biocompatibility. In the field of sensor development and biofuel production, CNTs have attracted a great deal of attention in recent years. CNTs possesses a high surface area and have the capability to bind with a greater number of enzymes in terms of enzyme loading. Recently, it has been found that these enzymes enhance the stability and activity of CNTs upon binding. It has been found by a number of research groups that functionalized CNTs alter the catalytic activity of the enzyme immobilized on the surface more significantly. Carbon nanotubes are of two types: single-walled nanotubes and multiwalled nanotubes. Multiwalled CNTs have greater efficiency as compared to single-walled nanotubes in the immobilization of enzymes, resulting in greater applicability in the processing of biofuels. Functionalized multiwalled CNTs have attracted more attention in the field of biofuel production as they enhance the catalytic efficiency of the immobilized enzymes many-fold. An electron transfer reaction can be directed by the enzyme, with the rooted active site attached to the CNTs. The main interest in CNTs in biofuel production and processing is due to their high electron active area. This area is useful to increase the enzyme concentration, as well as other compounds, on the surface of CNTs. However, the porosity and conductivity are responsible for better immobilization of the enzymes for biosensor and biofuel processing (Koo, Das, & Yoon, 2016; Zhang et al., 2017).

12.6.2.3 Other Nanoparticles

In different cases of biofuel production, heterogeneous catalysts are used as they have some special abilities over other catalysts. These advances include easy separation, less contamination, environment-friendly, non-corrosive, long life and high selectivity. Biofuel production, such as biodiesel and bioethanol, from the lignocellulose can be achieved with the use of heterogeneous nanocatalysts. When nanoparticles introduced into porous material they have potential to deconstruct the cellulose to esters. Inorganic nanoparticles can be used in heterogeneous catalysis due to their high acidic strength, specific surface area, etc., which make them useful for the fundamental application. Mesoporous nanoparticles can be synthesized by various techniques to control the size, uniformity, morphology, etc. and also to tune the properties of nanoparticles as is wished.

12.7 Conclusion

Biofuels have the capability to decrease greenhouse gas emissions, lower pollution, decrease poverty and create more employment, and thus could bring sustainable development to the world. In addition to these beneficial factors, they also have some negative effects, such as the conflict between energy crops and edible crops. They has also a negative impact on tropical forests, savannahs and biodiversity as these can be replaced by energy crops. On the social-economical side, there are important concerns for the importance of biofuel production for labour practices and food security. Therefore, there is a need to focus on enzymes that can convert or ferment hemicellulose and lignin easily. This results in the direct production of fuel in one step. Furthermore, there is a need to focus on the generation of biofuels to meet energy requirements by replacing fossil fuels. One thing must be kept in mind is that the source of energy changing from fossil fuels to biofuels must decrease the burden on the economy. Third- and fourth0generation fuels require biotechnology and genetic engineering, thus researchers should focus on the development of new methods of synthesizing cheap biofuel to fulfil the needs. Nanocatalysts are preferred over tradition catalysts as they can greatly increase the point of mass transfer loss; reduce the price of biodiesel; have many active sites; can be synthesized easily; and have huge surface area. Nanomaterials can be used as effective and highly active catalysts for transesterification of different oils and fats to obtain a green energy source via the formation of biodiesel. This is a way to enhance the catalytic activity with a huge surface area. High selectivity, reactivity, controlled rate of reaction, easy retrieval and recyclability make nanomaterials unique catalysts for the production of biofuels. The main focus of this chapter has been to evaluate and design biofuels, i.e. biochemical fuels like bioethanol. Biofuels can be obtained through various thermochemical methods.

REFERENCES

Abo, B. O., Odey, E. A., Bakayoko, M., & Kalakodio, L. (2019). Microalgae to biofuels production: A review on cultivation, application and renewable energy. *Rev Environ Health, 34*(1), 91–99.

Adrio, J. L. (2017). Oleaginous yeasts: Promising platforms for the production of oleochemicals and biofuels. *Biotechnol Bioeng, 114*(9), 1915–1920.

Ajala, E. O., Ajala, M. A., Ayinla, I. K., Sonusi, A. D., & Fanodun, S. E. (2020). Nano-synthesis of solid acid catalysts from waste-iron-filling for biodiesel production using high free fatty acid waste cooking oil. *Sci Rep, 10*(1), 13256.

Albers, S. C., Berklund, A. M., & Graff, G. D. (2016). The rise and fall of innovation in biofuels. *Nat Biotechnol, 34*(8), 814–821.

Ananthi, V., Siva Prakash, G., Mohan Rasu, K., Gangadevi, K., Boobalan, T., Raja, R., et al. (2018). Comparison of integrated sustainable biodiesel and antibacterial nano silver production by microalgal and yeast isolates. *J Photochem Photobiol B, 186*, 232–242.

Anchupogu, P., Rao, L. N., & Banavathu, B. (2018). Effect of alumina nano additives into biodiesel-diesel blends on the combustion performance and emission characteristics of a diesel engine with exhaust gas recirculation. *Environ Sci Pollut Res Int, 25*(23), 23294–23306.

Aro, E. M. (2015). From first generation biofuels to advanced solar biofuels. *Ambio, 45 Suppl 1*, S24–31.

Asveld, L. (2016). The need for governance by experimentation: The case of biofuels. *Sci Eng Ethics, 22*(3), 815–830.

Babadi, A. A., Bagheri, S., & Hamid, S. B. (2016). Progress on implantable biofuel cell: Nano-carbon functionalization for enzyme immobilization enhancement. *Biosens Bioelectron, 79*, 850–860.

Bagnoud-Velasquez, M., Refardt, D., Vuille, F., & Ludwig, C. (2015). Opportunities for Switzerland to contribute to the production of algal biofuels: The hydrothermal pathway to bio-methane. *Chimia (Aarau), 69*(10), 614–621.

Bai, W., Geng, W., Wang, S., & Zhang, F. (2019). Biosynthesis, regulation, and engineering of microbially produced branched biofuels. *Biotechnol Biofuels, 12*, 84.

Baloch, K. A., Upaichit, A., & Cheirsilp, B. (2020). Multilayered nano-entrapment of lipase through organic-inorganic hybrid formation and the application in cost-effective biodiesel production. *Appl Biochem Biotechnol, 193*(1), 165–187.

Behera, S., Singh, R., Arora, R., Sharma, N. K., Shukla, M., & Kumar, S. (2015). Scope of algae as third generation biofuels. *Front Bioeng Biotechnol, 2*, 90.

Beller, H. R., Lee, T. S., & Katz, L. (2015). Natural products as biofuels and bio-based chemicals: Fatty acids and isoprenoids. *Nat Prod Rep, 32*(10), 1508–1526.

Berhe, T., & Sahu, O. (2017). Chemically synthesized biofuels from agricultural waste: Optimization operating parameters with surface response methodology (CCD). *MethodsX, 4*, 391–403.

Blair, S. L., MacMillan, A. C., Drozd, G. T., Goldstein, A. H., Chu, R. K., Pasa-Tolic, L., et al. (2016). Molecular characterization of organosulfur compounds in biodiesel and diesel fuel secondary organic aerosol. *Environ Sci Technol, 51*(1), 119–127.

Booramurthy, V. K., Kasimani, R., Pandian, S., & Ragunathan, B. (2020). Nano-sulfated zirconia catalyzed biodiesel production from tannery waste sheep fat. *Environ Sci Pollut Res Int, 27*(17), 20598–20605.

Bryant, N. D., Pu, Y., Tschaplinski, T. J., Tuskan, G. A., Muchero, W., Kalluri, U. C., et al. (2020). Transgenic poplar designed for biofuels. *Trends Plant Sci, 25*(9), 881–896.

Byun, J., & Han, J. (2016). Catalytic production of biofuels (butene oligomers) and biochemicals (tetrahydrofurfuryl alcohol) from corn stover. *Bioresour Technol, 211*, 360–366.

Cai, W., & Zhang, W. (2017). Engineering modular polyketide synthases for production of biofuels and industrial chemicals. *Curr Opin Biotechnol, 50*, 32–38.

Caspeta, L., Chen, Y., Ghiaci, P., Feizi, A., Buskov, S., Hallstrom, B. M., et al. (2014). Biofuels. Altered sterol composition renders yeast thermotolerant. *Science, 346*(6205), 75–78.

Chang, J. S., Yang, J. W., Lee, D. J., & Hallenbeck, P. C. (2015). Editorial. Advances in biofuels and chemicals from algae. *Bioresour Technol, 184*, 1.

Chaudhary, N., Gupta, A., Gupta, S., & Sharma, V. K. (2017). BioFuelDB: A database and prediction server of enzymes involved in biofuels production. *PeerJ, 5*, e3497.

Chen, C., Cai, L., Shangguan, X., Li, L., Hong, Y., & Wu, G. (2018). Heterogeneous and efficient transesterification of *Jatropha curcas* L. seed oil to produce biodiesel catalysed by nano-sized SO4 (2-)/TiO2. *R Soc Open Sci, 5*(11), 181331.

Chen, R., & Dou, J. (2015). Biofuels and bio-based chemicals from lignocellulose: Metabolic engineering strategies in strain development. *Biotechnol Lett, 38*(2), 213–221.

Chen, W. H., Lin, B. J., Huang, M. Y., & Chang, J. S. (2014). Thermochemical conversion of microalgal biomass into biofuels: A review. *Bioresour Technol, 184*, 314–327.

Chew, T. L., & Bhatia, S. (2009). Effect of catalyst additives on the production of biofuels from palm oil cracking in a transport riser reactor. *Bioresour Technol, 100*(9), 2540–2545.

Costa, T. O., Calijuri, M. L., Avelar, N. V., Carneiro, A. C. O., & de Assis, L. R. (2016). Energetic potential of algal biomass from high-rate algal ponds for the production of solid biofuels. *Environ Technol, 38*(15), 1926–1936.

Damanik, N., Ong, H. C., Tong, C. W., Mahlia, T. M. I., & Silitonga, A. S. (2018). A review on the engine performance and exhaust emission characteristics of diesel engines fueled with biodiesel blends. *Environ Sci Pollut Res Int, 25*(16), 15307–15325.

De Bhowmick, G., Sarmah, A. K., & Sen, R. (2017). Lignocellulosic biorefinery as a model for sustainable development of biofuels and value added products. *Bioresour Technol, 247*, 1144–1154.

Degfie, T. A., Mamo, T. T., & Mekonnen, Y. S. (2019). Optimized biodiesel production from waste cooking oil (WCO) using calcium oxide (CaO) nano-catalyst. *Sci Rep, 9*(1), 18982.

Dehghani, S., & Haghighi, M. (2016). Sono-sulfated zirconia nanocatalyst supported on MCM-41 for biodiesel production from sunflower oil: Influence of ultrasound irradiation power on catalytic properties and performance. *Ultrason Sonochem, 35*(Pt A), 142–151.

Dehghani, S., & Haghighi, M. (2019a). Sono-dispersed MgO over cerium-doped MCM-41 nanocatalyst for biodiesel production from acidic sunflower oil: Surface evolution by altering Si/Ce molar ratios. *Waste Manag, 95*, 584–592.

Dehghani, S., & Haghighi, M. (2019b). Sono-dispersion of MgO over Al-Ce-doped MCM-41 bifunctional nanocatalyst for one-step biodiesel production from acidic oil: Influence of ultrasound irradiation and Si/Ce molar ratio. *Ultrason Sonochem, 54*, 142–152.

de Lanes, E. C., de Almeida Costa, P. M., & Motoike, S. Y. (2014). Alternative fuels: Brazil promotes aviation biofuels. *Nature, 511*(7507), 31.

Devarajan, Y., Munuswamy, D. B., & Mahalingam, A. (2018). Influence of nano-additive on performance and emission characteristics of a diesel engine running on neat neem oil biodiesel. *Environ Sci Pollut Res Int, 25*(26), 26167–26172.

Douki, T., Corbiere, C., Preterre, D., Martin, P. J., Lecureur, V., Andre, V., et al. (2018). Comparative study of diesel and biodiesel exhausts on lung oxidative stress and genotoxicity in rats. *Environ Pollut, 235*, 514–524.

Dubini, A., & Antal, T. K. (2015). Generation of high-value products by photosynthetic microorganisms: From sunlight to biofuels. Introduction. *Photosynth Res, 125*(3), 355–356.

Falk, M., Narvaez Villarrubia, C. W., Babanova, S., Atanassov, P., & Shleev, S. (2013). Biofuel cells for biomedical applications: Colonizing the animal kingdom. *Chemphyschem, 14*(10), 2045–2058.

Flores-Gomez, C. A., Escamilla Silva, E. M., Zhong, C., Dale, B. E., da Costa Sousa, L., & Balan, V. (2018). Conversion of lignocellulosic agave residues into liquid biofuels using an AFEX-based biorefinery. *Biotechnol Biofuels, 11*, 7.

Gardy, J., Rehan, M., Hassanpour, A., Lai, X., & Nizami, A. S. (2019). Advances in nano-catalysts based biodiesel production from non-food feedstocks. *J Environ Manage, 249*, 109316.

Genwa, M., Kumari, K., & Singh, P. (2019). Nanotechnology: Beneficial or harmful. *Nanosci Nanotech—Asia 9*(1), 1–3.

Hernandez, D., Solana, M., Riano, B., Garcia-Gonzalez, M. C., & Bertucco, A. (2014). Biofuels from microalgae: Lipid extraction and methane production from the residual biomass in a biorefinery approach. *Bioresour Technol, 170*, 370–378.

Hood, E. E. (2016). Plant-based biofuels. *F1000Res, 5*.

Inamuddin, Shakeel, N., Imran Ahamed, M., Kanchi, S., & Abbas Kashmery, H. (2020). Green synthesis of ZnO nanoparticles decorated on polyindole functionalized-MCNTs and used as anode material for enzymatic biofuel cell applications. *Sci Rep, 10*(1), 5052.

Jiang, L., Zhou, P., Liao, C., Zhang, Z., & Jin, S. (2017). Cobalt nanoparticles supported on nitrogen-doped carbon: An effective non-noble metal catalyst for the upgrade of biofuels. *ChemSusChem, 11*(5), 959–964.

Jiang, Y., Wu, R., Zhou, J., He, A., Xu, J., Xin, F., et al. (2019). Recent advances of biofuels and biochemicals production from sustainable resources using co-cultivation systems. *Biotechnol Biofuels, 12*, 155.

Johnson, F. X. (2017). Biofuels, bioenergy and the bioeconomy in North and South. *Ind Biotechnol (New Rochelle N Y), 13*(6), 289–291.

Kang, A., & Lee, T. S. (2015). Converting sugars to biofuels: Ethanol and beyond. *Bioengineering (Basel), 2*(4), 184–203.

Karimi, A., Othman, A., Uzunoglu, A., Stanciu, L., & Andreescu, S. (2015). Graphene based enzymatic bioelectrodes and biofuel cells. *Nanoscale, 7*(16), 6909–6923.

Kazamia, E., & Smith, A. G. (2014). Assessing the environmental sustainability of biofuels. *Trends Plant Sci, 19*(10), 615–618.

Khan, M. I., Shin, J. H., & Kim, J. D. (2018). The promising future of microalgae: Current status, challenges, and optimization of a sustainable and renewable industry for biofuels, feed, and other products. *Microb Cell Fact, 17*(1), 36.

Khanna, M., Wang, W., Hudiburg, T. W., & DeLucia, E. H. (2017). The social inefficiency of regulating indirect land use change due to biofuels. *Nat Commun, 8*, 15513.

Kim, J., Jia, H., & Wang, P. (2006). Challenges in biocatalysis for enzyme-based biofuel cells. *Biotechnol Adv, 24*(3), 296–308.

Kim, Y., Thomas, A. E., Robichaud, D. J., Iisa, K., St John, P. C., Etz, B. D., et al. (2020). A perspective on biomass-derived biofuels: From catalyst design principles to fuel properties. *J Hazard Mater, 400*, 123198.

Koo, M. H., Das, G., & Yoon, H. H. (2016). Electrochemical performance of glucose/oxygen biofuel cells based on carbon nanostructures. *J Nanosci Nanotechnol, 16*(3), 3054–3057.

Kumar, A., Singh, p., Saxena, A., De, A., Chandra, R., & Mozumdar, S. (2008). Nano-sized copper as an efficient catalyst for one pot three component synthesis of thiazolidine-2, 4-dione derivatives. *Catal Comm, 10*, 17–22.

Kumar, P., Singh, P., Kumari, K., Mozumdar, S., & Chandra, R. (2011). A green approach for the synthesis of gold nanotriangles using aqueous leaf extract of *Callistemon viminalis*. *Mat Lett, 65*, 595–597.

Kumari, K., Singh, P., Bauddh, K., Mallick, S., & Chandra, R. (2019). Implications of metal nanoparticles on aquatic fauna: A review. *Nanosci Nanotech – Asia, 1*(3), 30–43.

Kumari, K., Singh, P., & Mehrotra, G. K. (2012). A facile one pot synthesis of collagen protected gold nanoparticles using Na-malanodialdehyde. *Mat Lett, 79*, 199–201.

Lam, F. H., Ghaderi, A., Fink, G. R., & Stephanopoulos, G. (2014). Biofuels. Engineering alcohol tolerance in yeast. *Science, 346*(6205), 71–75.

Liu, D. H., Marks, T. J., & Li, Z. (2019). Catalytic one-pot conversion of renewable platform chemicals to hydrocarbon and ether biofuels through tandem $Hf(OTf)_4$ + Pd/C catalysis. *ChemSusChem,* 12(24), 5217–5223.

Liu, L., Ma, G., Zeng, M., Du, W., & Yuan, J. (2018). Renewable boronic acid affiliated glycerol nano-adsorbents for recycling enzymatic catalyst in biodiesel fuel production. *Chem Commun (Camb), 54*(88), 12475–12478.

Madhuvilakku, R., & Piraman, S. (2013). Biodiesel synthesis by TiO_2-ZnO mixed oxide nanocatalyst catalyzed palm oil transesterification process. *Bioresour Technol, 150*, 55–59.

Mamo, T. T., & Mekonnen, Y. S. (2019). Microwave-assisted biodiesel production from microalgae, *Scenedesmus* species, using goat bone-made nano-catalyst. *Appl Biochem Biotechnol, 190*(4), 1147–1162.

Marella, T. K., Parine, N. R., & Tiwari, A. (2018). Potential of diatom consortium developed by nutrient enrichment for biodiesel production and simultaneous nutrient removal from waste water. *Saudi J Biol Sci, 25*(4), 704–709.

Mehmood, M. A. (2018). Editorial: Recent trends in biofuels and bioindustry. *Protein Pept Lett, 25*(2), 98.

Mofijur, M., Siddiki, S. Y. A., Shuvho, M. B. A., Djavanroodi, F., Fattah, I. M. R., Ong, H. C., et al. (2020). Effect of nanocatalysts on the transesterification reaction of first, second and third generation biodiesel sources: A mini-review. *Chemosphere*, 270, 128642.

Moghzi, F., & Soleimannejad, J. (2018). Sonochemical synthesis of a new nano-sized barium coordination polymer and its application as a heterogeneous catalyst towards sono-synthesis of biodiesel. *Ultrason Sonochem, 42*, 193–200.

Nematian, T., Shakeri, A., Salehi, Z., & Saboury, A. A. (2020). Lipase immobilized on functionalized superparamagnetic few-layer graphene oxide as an efficient nanobiocatalyst for biodiesel production from *Chlorella vulgaris* bio-oil. *Biotechnol Biofuels, 13*, 57.

Pakapongpan, S., Tuantranont, A., & Poo-Arporn, R. P. (2017). Magnetic nanoparticle-reduced graphene oxide nanocomposite as a novel bioelectrode for mediatorless-membraneless glucose enzymatic biofuel cells. *Sci Rep, 7*(1), 12882.

Pandian, A. K., Ramakrishnan, R. B. B., & Devarajan, Y. (2017). Emission analysis on the effect of nanoparticles on neat biodiesel in unmodified diesel engine. *Environ Sci Pollut Res Int, 24*(29), 23273–23278.

Pandit, P. R., & Fulekar, M. H. (2017). Egg shell waste as heterogeneous nanocatalyst for biodiesel production: Optimized by response surface methodology. *J Environ Manage, 198*(Pt 1), 319–329.

Pourkhesalian, A. M., Stevanovic, S., Salimi, F., Rahman, M. M., Wang, H., Pham, P. X., et al. (2014). Influence of fuel molecular structure on the volatility and oxidative potential of biodiesel particulate matter. *Environ Sci Technol, 48*(21), 12577–12585.

Pramanik, M., & Bhaumik, A. (2013). Organic-inorganic hybrid supermicroporous iron(III) phosphonate nanoparticles as an efficient catalyst for the synthesis of biofuels. *Chemistry, 19*(26), 8507–8514.

Prashanth, G. K., Prashanth, P. A., Nagabhushana, B. M., Ananda, S., Krishnaiah, G. M., Nagendra, H. G., et al. (2017). Comparison of anticancer activity of biocompatible

ZnO nanoparticles prepared by solution combustion synthesis using aqueous leaf extracts of *Abutilon indicum, Melia azedarach* and *Indigofera tinctoria* as biofuels. *Artif Cells Nanomed Biotechnol, 46*(5), 968–979.
Puri, M., Barrow, C. J., & Verma, M. L. (2013). Enzyme immobilization on nanomaterials for biofuel production. *Trends Biotechnol, 31*(4), 215–216.
Ramos, J. L., Valdivia, M., Garcia-Lorente, F., & Segura, A. (2016). Benefits and perspectives on the use of biofuels. *Microb Biotechnol, 9*(4), 436–440.
Rasouli, H., & Esmaeili, H. (2019). Characterization of MgO nanocatalyst to produce biodiesel from goat fat using transesterification process. *3 Biotech, 9*(11), 429.
Sesmero, J. P. (2014). Cellulosic biofuels from crop residue and groundwater extraction in the US Plains: The case of Nebraska. *J Environ Manage, 144*, 218–225.
Shahraki, H., Entezari, M. H., & Goharshadi, E. K. (2014). Sono-synthesis of biodiesel from soybean oil by KF/gamma-Al(2)O(3) as a nano-solid-base catalyst. *Ultrason Sonochem, 23*, 266–274.
Singh, P., Katyal, A., Kalra, R., & Chandra, R. (2008). Copper nanoparticles in ionic liquid: An easy and efficient catalyst for the coupling of thiazolidine-2,4-dione, aromatic aldehyde and ammonium acetate. *Cat Comm 9*, 1618–1623.
Singh, P., Kumari, K., Katyal, A., Kalra, R., & Chandra, R. (2009). Synthesis and characterization of silver and gold nanoparticles in ionic liquid. *Spectrochim Acta A Mol Biomol Spectrosc, 73*(1), 218–220.
Soudagar, M. E. M., Banapurmath, N. R., Afzal, A., Hossain, N., Abbas, M. M., Haniffa, M., et al. (2020). Study of diesel engine characteristics by adding nanosized zinc oxide and diethyl ether additives in Mahua biodiesel-diesel fuel blend. *Sci Rep, 10*(1), 15326.
Tahvildari, K., Anaraki, Y. N., Fazaeli, R., Mirpanji, S., & Delrish, E. (2015). The study of CaO and MgO heterogenic nano-catalyst coupling on transesterification reaction efficacy in the production of biodiesel from recycled cooking oil. *J Environ Health Sci Eng, 13*, 73.
Vaishali, S., Banty, K., & Singh, P. (2017). Biofuels for Sustainable Development: A Global Perspective. In R. Singh &. S. Kumar (Eds.), *Green Technologies and Environmental Sustainability*. Springer.
Vedagiri, P., Martin, L. J., Varuvel, E. G., & Subramanian, T. (2019). Experimental study on NOx reduction in a grapeseed oil biodiesel-fueled CI engine using nanoemulsions and SCR retrofitment. *Environ Sci Pollut Res Int, 27*(24), 29703–29716.
Yang, Y., Ochoa-Hernandez, C., de la Pena O'Shea, V. A., Pizarro, P., Coronado, J. M., & Serrano, D. P. (2015). Transition metal phosphide nanoparticles supported on SBA-15 as highly selective hydrodeoxygenation catalysts for the production of advanced biofuels. *J Nanosci Nanotechnol, 15*(9), 6642–6650.
Zhang, H., Zhang, L., Han, Y., Yu, Y., Xu, M., Zhang, X., et al. (2017). RGO/Au NPs/N-doped CNTs supported on nickel foam as an anode for enzymatic biofuel cells. *Biosens Bioelectron, 97*, 34–40.
Zhang, Z., O'Hara, I. M., Mundree, S., Gao, B., Ball, A. S., Zhu, N., et al. (2016). Biofuels from food processing wastes. *Curr Opin Biotechnol, 38*, 97–105.
Zhou, M., & Dong, S. (2011). Bioelectrochemical interface engineering: Toward the fabrication of electrochemical biosensors, biofuel cells, and self-powered logic biosensors. *Acc Chem Res, 44*(11), 1232–1243.
Zhou, Y. J., Buijs, N. A., Siewers, V., & Nielsen, J. (2014). Fatty acid-derived biofuels and chemicals production in *Saccharomyces cerevisiae. Front Bioeng Biotechnol, 2*, 32.

Index

C

G

H

I

J

K

L

M

N

O

T

Ingram Content Group UK Ltd.
Milton Keynes UK
UKHW020624180523
421935UK00001B/2